Computer Science, Technology and Applications

nova
science publishers
www.novapublishers.com

Computer Science, Technology and Applications

Situational Modeling: Definitions, Awareness, Simulation
Alexander Fridman, PhD (Editor)
2023. ISBN: 979-8-88697-590-1 (Hardcover)
2023. ISBN: 979-8-88697-725-7 (eBook)

Applications of Artificial Intelligence in the Healthcare Sector
Jyoti Prakash Patra, PhD (Editor)
Yogesh Kumar Rathore (Editor)
2023. ISBN: 979-8-88697-502-4 (Hardcover)
2023. ISBN: 979-8-88697-541-3 (eBook)

Speech Recognition Technology and Applications
Vasile-Florian Păiş (Editor)
2022. ISBN: 978-1-68507-929-1 (Hardcover)
2022. ISBN: 979-8-88697-179-8 (eBook)

Internet of Everything: Smart Sensing Technologies
T. Kavitha, PhD (Editor)
V. Ajantha Devi, PhD (Editor)
S. Neelavathy Pari, PhD (Editor)
Sakkaravarthi Ramanathan, PhD (Editor)
2022. ISBN: 978-1-68507-865-2 (Hardcover)
2022. ISBN: 978-1-68507-943-7 (eBook)

A Beginner's Guide to Virtual Reality (VR) Modeling in Healthcare Applications with Blender
Ho Lun Ho, Ka Yin Chau, PhD, Yan Wan, Yuk Ming Tang, PhD (Editors)
2022. ISBN: 978-1-68507-811-9 (Softcover)
2022. ISBN: 978-1-68507-945-1 (eBook)

More information about this series can be found at
https://novapublishers.com/product-category/series/computer-science-technology-and-applications/

L. Ashok Kumar
D. Karthika Renuka
V. Chandrasekar
and Sonali Agarwal
Editors

Blockchain Enabled Secure Big Data Computing for Smart Cities Using Internet of Things

nova
science publishers

www.novapublishers.com

Library of Congress Cataloging-in-Publication Data

ISBN: 979-8-89113-243-6

Published by Nova Science Publishers, Inc. † New York

Dedication

To my wife Ms. Y. Uma Maheswari and daughter A. K. Sangamithra, for their constant support and love.

Dr. L. Ashok Kumar

To my parents Mr. N. Dhanaraj and Ms. D. Anuradha, who laid the foundation for all my success, to my husband Mr.R.Sathish Kumar and my daughter P.S.Preethi for their unconditional love and support for competition of this book and to my friends and brothers, who has been highly credible and a greatest source of motivation and as a pillar of inspiration.

Dr. D. Karthika Renuka

I would like to convey her sincere thanks to her family and friends for encouraging her to start the work, preserve it and publish it.

Dr .Sonali Agarwal

I acknowledge the teachers, students and research staff who have helped make all the scientific accomplishments.

Dr. V. Chandrasekar

Contents

Foreword

The rapid progress of technological innovations has greatly expedited the growth and expansion of smart cities. The advancements in technology have successfully tackled numerous infrastructure challenges that were previously associated with aging systems and the growing expectations of citizens for improved services. As a result, the advantages of smart technology have become increasingly apparent. In today's rapidly urbanizing world, the concept of a sustainable smart city has emerged as a promising solution to enhance the quality of life of urban dwellers. By leveraging cutting-edge technologies, such as improved data security, privacy, efficient information sharing, effective decision-making and high-quality services, a sustainable smart city aims to create a more livable and sustainable urban environment. This book explores the key principles and practices that underpin the development of a sustainable smart city, providing valuable insights for urban planners, policymakers, and researchers alike. In this modern era, the concept of a smart city has emerged as a promising solution to enhance urban living standards. By leveraging cutting-edge technologies such as Big Data, Internet of Things (IoT), Artificial Intelligence, and advanced data security measures, a smart city aims to optimize its operations and resources for maximum efficiency. This book delves into the intricacies of smart city technology and its potential to revolutionize the way we live, work, and interact with our urban environments. In the modern era of technology, it is commonplace for intelligent applications to produce vast quantities of data from a variety of disparate sources. In today's digital age, the management and secure storage of large amounts of data is of utmost importance. This publication delves into the fundamental domains of smart cities, including banking, insurance, healthcare, and supply chain management. This book delves into the exciting realm of blockchain ecosystems and their potential to revolutionize smart and sustainable city living.

Best Wishes

Dean, College of Innovative Design and Management, National Taipei University of Business,
Creative Technologies and Product Design, Taiwan

Preface

Smart cities are becoming increasingly popular worldwide as cities grow; technology evolves and improves daily "Smart Cities" envision to provide a healthy, happy and sustainable urban ecosystem by integrating the information and communication technology (ICT), Internet of things (IoT) and citizen participation to effectively manage and utilize city infrastructure and services. A smart city uses the Internet of Things (IoT), artificial intelligence, and other data-gathering technology to help it run more efficiently. All smart cities have multiple layers working together. For example, a technology base consisting of sensors and smartphones connected to high-speed networks can produce raw data, which computers then process to provide insights and give alerts. The internet of things devices are used in smart cities that are all interconnected to run critical systems that a city requires to function correctly. The Internet of Things controls traffic lights, security cameras, weather, infrastructure, meters, and other data collection devices to run the city. Because technological devices do not sleep, eat, or take breaks like humans, the systems can be relied on to operate continuously. The combination of blockchain and IoT has a lot of potential in creating a marketplace of connected devices and for companies to create value from the data collected. By using smart technologies, cities could ultimately connect and integrate their various services and sectors—such as utilities, energy, healthcare, transportation, governance, and security—onto digital platforms.

Over the last few years, the growing big data and block chain technology has made it feasible to associate complex thing to the internet. It has led to a digital disturbance by alternating the way the technology is being utilized. Several industries, from smart cities to transportation and smart grid have been metamorphosed and are now smarter than ever via cloud of things. Blockchain is a system of decentralized and immutable electronic ledgers or databases in which anyone can securely store and access public records while maintaining information integrity. Blockchain offers a method of communication, transactions, security, and governance that is transparent and available to all

Acknowledgement

We bow our head before "The God Almighty" who blessed us with health and confidence to undertake and complete the book successfully. We express our sincere thanks to the Principal and Management of PSG College of Technology, Indian Institute of Information Technology, Allahabad, and Colorado State University, USA for their constant encouragement and support.

We thank our family and friends who always stood beside us and encouraged us to complete the book.

Dr. L. Ashok Kumar is thankful to his wife, Y. Uma Maheswari, for her constant support during writing. He is also grateful to his daughter, A. K. Sangamithra, for her support; it helped him a lot in completing this work.

Dr. D. Karthika Renuka would like to express gratitude to her parents, for their constant support. Her heartfelt thanks to her husband, Mr. R. Sathish Kumar, and her dear daughter, Ms. P. S. Preethi, for their unconditional love which made her capable of achieving all her goals.

Dr. Sonali Agarwal would like to convey her sincere thanks to her family and friends for encouraging her to start the work, preserve it and publish it. This book is based on the research conducted on the big data approaches used for smart cities and smart buildings.

Dr. V. Chandrasekar acknowledges the teachers, students and research staff who have helped make all the scientific accomplishments.

We would like to acknowledge the help of all the people involved in this project. First, we would like to thank each one of the authors for their contributions. Our sincere gratitude goes to the chapter's authors who contributed their time and expertise to this book. We thank all the authors of the chapters for their commitment to this endeavor and their timely response to our incessant requests for revisions.

Second, the editors wish to acknowledge the valuable contributions of the reviewers regarding the improvement of quality, coherence, and content presentation of chapters. Most of the authors also served as referees; we highly appreciate their double task.

Next, the editors would like to recognize the contributions of editorial board in shaping the nature of the chapters in this book. In addition, we wish to thank the editorial staff at Nova Publication for their professional assistance and patience. Sincere thanks to each one of them.

Dr. L. Ashok Kumar, Professor, PSG College of Technology, India
Dr. D. Karthika Renuka, Professor, PSG College of Technology, India
Dr. Sonali Agarwal, Associate Professor, IIIT - Allahabad, Allahabad, India
Dr. V. Chandrasekar, University Distinguished Professor and Associate Dean for International Programs, Colorado State University, USA

Chapter 1

Internet of Things for Smart Cities: Overview and Challenges

A. Suganya*, ME
and S. Manoj Kumar[2], PhD
[1]Department of Information Technology, Karpagam College of Engineering, Coimbatore, India
[2]Department of Computer Science and Engineering, KPR Institute of Engineering and Technology, Coimbatore, India

Abstract

International Smart City goals are achieved by the new Internet of Things (IoT). It enables the remote monitoring, administration, and control of equipment as well as the generation of new insights and helpful information through the use of enormous real-time data streams. Two of the fundamental characteristics of a smart city are its extensive use of information resources and high degrees of information technology integration. The cornerstones of urban growth for a smart city should include smart technology, smart industry, smart services, smart administration, and smart life. The Internet of Things requires installing sensors (RFID, IR, GPS, laser scanners, etc.) for everything and connecting them to the internet through specific protocols for information exchange and communications in order to achieve intelligent recognition, location, tracking, monitoring, and management. For smart cities to operate technically with IoT support, they must possess the three qualities of being instrumented, networked, and intelligent. Only by combining all these intelligent components can a Smart City be constructed at the most advanced stage of IOT development. In

* Corresponding Author's E-mail: suganyaap93@gmail.com.

In: Blockchain Enabled Secure Big Data Computing …
Editors: L. Ashok Kumar, D. Karthika Renuka, V. Chandrasekar et al.
ISBN: 979-8-89113-243-6

particular, novel research initiatives from both academia and business are required for the development of an efficient, scalable, and reliable Smart City based on IoT. These problems are a result of the Internet of Things and Smart City apps' rapid growth. New protocols, methods, and services are urgently required to overcome these issues.

Keywords: Internet of Things, Smart City, block chain

Introduction

Smart citieshave tremendously advanced and greatly increased their potential in recent years. In actuality, recent developments in the Internet of Things (IoT) have paved way for fresh opportunities and also serve as a set of essential enabling technologies for smart cities and enabling the creation and automation of cutting-edge services and advanced applications for the many city stakeholders (Cheng Le Zhong et al., 2015). A key enabler in the current smart cities scenario, the Internet of Things (IoT) and Internet of Everything (IoE) technologies' growing growth and adoption has elevated smart city to a big scale. One of the most valuable disruptive technologies will be the Internet of Things (IoT), which will offer up new horizons, opportunities, and difficulties for the development of smart services and applications. Smart cities are intricate sociotechnical infrastructures made up of both digital and human agents. The wide range of varied approaches, contexts, application domains, and technological solutions that have been put forth in the literature for the implementation and administration of smart cities reflect this complexity. The major objective of putting IoT technologies into place and integrating them is to enable smart cities to progress further, offering new capabilities and features while drastically minimising human interaction. The demands of supporting numerous different data providers, dealing with numerous different protocols and data formats, assuring interoperability and scalability, and supporting the sharing of components represent some of the major technical challenges for contemporary IoT-enabled smart cities. This is essential to prevent the adoption of redundant data ingestion, storage, and analysis solutions, which would increase operational expenses and decrease the sustainability of the city. Smart urbanism uses the Internet of Things (IoT), information and communication tools (ICT), and other technology to address urban difficulties. Making the best possible use of the tools and technology at hand to create smart cities is the main objective.

An IoT-enabled application is a crucial part of this, but it has a single point of failure and several security, privacy, latency, and reliability issues. The emerging technology known as blockchain is able to handle the aforementioned security and privacy concerns while still providing high-quality services because of its many attributes, including transparency, trustlessness, decentralisation, immutability, and others.

Everything and every device in today's digital world is interconnected. The internet transforms these devices into smart ones when they are connected to each another through internet. The Internet of Things (IoTs) is a wide network of interconnected things that can store and collect data about their immediate surroundings; as a result, the IoT is sometimes referred to as an ecosystem of interconnected things. The IoT is essential within the current framework for remotely accessing and managing digital equipment. This creates the opportunity to advance technology, community, and trade for both tangible and intangible goods. The devices, which could be anything from a nano chip to a router, communicate with one another via sensors, actuators, and software. The Internet of Things (IoT) is expanding quickly and has a wide range of uses. IoT system is given in Figure 1.1.

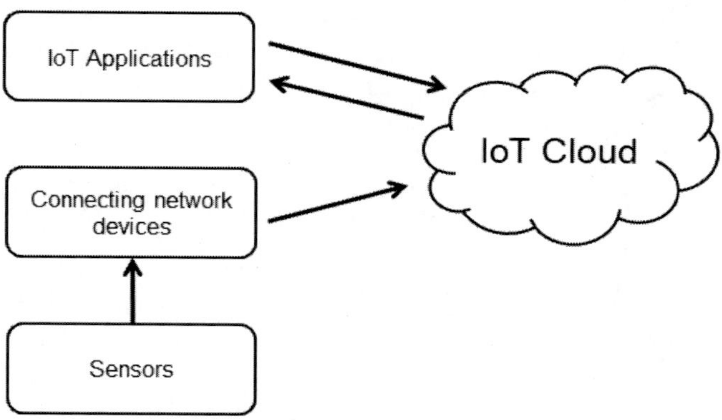

Figure 1.1. Internet of Things.

Almost all applications in the digital age are entirely dependent on wireless sensor networks (WSNs) due to their undeniable advantages, such as lower cost, less infrastructure, various network topologies, less maintenance, etc. Wireless sensors and WSN are monitoring almost every element of life, including weather, natural disasters, traffic, forests, healthcare, location, etc. Control, monitoring, regulation, warning, and tracking the patient's actions are the main objectives of the sensors used in healthcare services. By immediately evaluating and analysing these actions, sensors enable the healthcare system to be autonomous and without physical interference.

At the end, real-time healthcare industry will advance greatly. The goal of numerous IoT-related healthcare applications (AbderahmanRejeb et al., 2022, A. Gharaibeh et al., 2017, Mohamed MedhatGaber et al., 2019, Hind Bangui et al. 2018,Mark Deakin et al., 2018) is to make life simpler for patients, medical staff, and administrators. These applications contribute to the improvement of the current healthcare system by providing real-time intensive care regarding the patient's sickness, handling medical crises, etc. Patients can track and submit information about their current health problems, such as blood pressure, insulin levels, heart issues, physical fitness, etc., to a specific medical facility or doctor with the help of IoT applications. Healthcare midpoints and clinics frequently employ IoT apps to determine the location and health status of patients and to deliver quick services.

Using cloud technology, the IoT can more affordably process, analyse, and manage large volumes of data with increased fault tolerance. Traditional data processing is not advantageous for modern technology, hence the Internet of Things (IoT) relies on big data technologies for processing and analysis. Context-aware technologies are integrating with IoT and processing information in a manner similar to this to assist with searching, filtering, and comprehending enormous volumes of sensor data.

IoT Overview

IoT, a recent technology, includes smart objects that contain physical components like sensors and actuators for sensing the internal status of the object or the external environment and acting based on the received data. Decisions can be made as a result of further analysis of the data generated by the smart object's sensors. They also contain software that regulates numerous elements and activities produced by these intelligent gadgets.

Basic Parameters

It is crucial to comprehend the fundamentals of IoT before learning about its applications and hardware. Researchers claim that identification, sensing inputs through sensors, communication, computing, services, and decision-making are common components. For IoT to work effectively, these criteria are necessary. Figure 1.2 displays the fundamental IoT parameters.

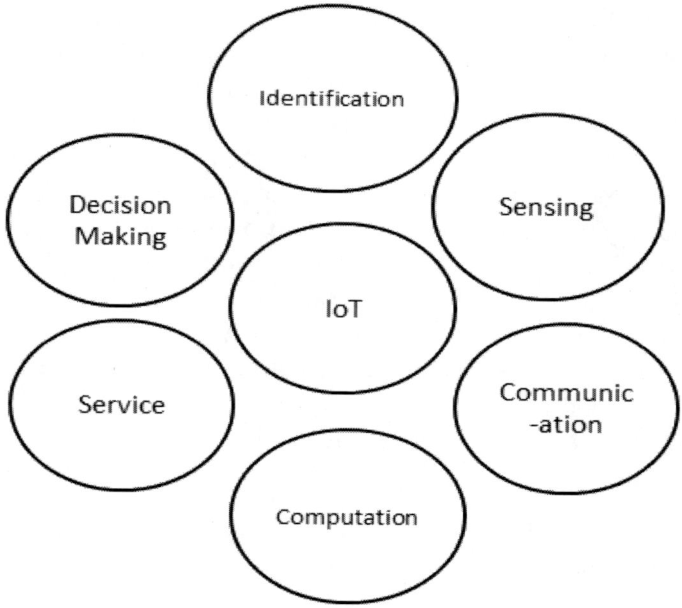

Figure 1.2. Basic Components of IoT.

Identification

When a user asks it, the IoT must be able to identify a customer using their name and address and must accurately detect every object in the network. Numerous methods can be used to identify the Internet of Things. In IoT, addressing is essential for identification since it allows for the separation of an object ID from an address. IPV6 is extremely beneficial for the Internet of Things with the introduction of low power wireless networks. Any physical IoT object, including sensors, actuators, etc., may have an object ID, whereas object address designates the location of the object inside the network's communication range. Additionally, communication network objects can employ public IP addresses as opposed to private ones.

Sensors

A sensor is a part used in the Internet of Things (IoT) that detects and measures environmental or physical conditions, such as temperature, humidity, light, pressure, motion, and sound. As they gather information about their surroundings and transfer it to a central processing unit for further analysis, sensors are an essential part of IoT devices (Martin Bauer, et al., 2021).

In the Internet of Things, there are numerous distinct kinds of sensors, each with special qualities and capacities. As an illustration, temperature sensors monitor temperature, whereas humidity sensors measure the amount of moisture in the air. Light sensors assess the amount of light in a certain area, whereas motion sensors detect movement. Sensors can be used to control Internet of Things (IoT) devices in addition to gathering data. For instance, a motion sensor may be used to turn on and off lights in a room while a temperature sensor can be used to manage the heating and cooling systems in a building. Overall, sensors are a crucial part of IoT devices because they allow them to gather and send data that is necessary for making wise decisions and increasing productivity across a range of applications.

Communication

One among the essential elements of the Internet is the Intelligent Healthcare Network, or "IoThNet". It is mostly used for sending and receiving information.A scenario in which the sensors and the medical device are placed in or near to the patient's body to collect medical data is where different IoT Network building blocks are placed to intelligently inspect on the person while adhering to its own set of protocols. Despite the existence of the IoThNet, a variety of communication protocols, including, ultra-wide bandwidth (UWB), Wi-Fi, radio frequency identification (RFID), Bluetooth, IEEE 802.15.4, long term evolution (LTE), Z-wave, near field communication (NFC), etc., are used in the Internet of Things to allow communication between heterogeneous objects.

In order to establish connectivity between IoT devices and the portable access medium, LTE is the most widely utilised communication protocol. A wireless standard for high-speed data transport on mobile phones is called LTE. The advanced version of LTE bandwidth support employs broadcasting and multicasting and extends uplink and downlink spatial multiplexing up to 100 MHz, which might be helpful in a highly dynamic environment with frequent topology changes.A short-wavelength communication method called Bluetooth is used to transfer data between nearby devices. It is very advantageous for the Internet of Things because of the decreased energy consumption and quick transmission over small distances. WiFi can be used to effectively exchange data across heterogeneous items at distances of up to 100 meters in a mobile ad-hoc environment without the requirement for a router.

Computation

Processing units, hardware platforms, and software all show that IoT can communicate. The most important software platform is the operating system, which runs while the system is being activated. Real-time operating systems are commonly used in the development of Internet of Things applications. The purpose of the cloud is to make it simpler for smart devices to send data to cloud for processing. The data that was obtained from Big Data is then made available to the user.

Services

Identity-related, pervasive, collaborative-aware, and information aggregation services are some of the several subcategories of IoT services. The foundational services required for the operation of additional services are identity-related services. If real-world things are to be presented in a virtual environment, every programme must be able to distinguish them. Through information aggregation services, the sensor data is gathered, analysed, and reported to IoT applications. Collaborative-aware services make decisions and take appropriate action based on the information offered by services linked to aggregate. Ubiquitous services are accessible to everyone and everywhere since they serve as a service provider for cooperatively aware services.

Decision Making

This ability enables the consumer to receive the required services after data from various IoT devices has been gathered. Extraction encompasses information finding, information use, and information modelling. Before fulfilling the correct demand for the proper user or client, it contains decision-making analysis and utilises the proper data extraction service.

Architecture of IoT

A new paradigm for digital device communication and interaction with the real world and human users has been established with the help of IoT (A.H. Alavi et al., 2018). IoT architectures combine a numerous technologies to provide cloud, fog, and edge computation for data sensing, actuating, sending, processing, analysis, and final exploitation.

Although various functional architectures for IoT frameworks have been presented in the literature, they all generally base their designs on a condensed version of the OSI model. The perception/sensing layer, the network layer, and the application layer are the three layers that some studies suggest should make up architecture. The perception/sensing layer, the transportation/network layer, the middleware/processing layer, the application layer, and the business layer are some extensions of this strategy that result in five-layer architecture.

Both the three-layer and five-layer systems (Muhammad Burhan et al., 2018, Pallavi Sethi et al., 2017) lowest layer, perception/sensing, is where sensors are utilised to gather information about the outside world. The data gathered by sensors must be transferred from the network/transportation layer to the processing and analysis layer, which may be cloud-based or edge-based. While the application layer is in charge of offering services and applications to the end users, the middleware/processing layer is in charge of processing and analysing the data. The management and commercialization of IoT services and applications are dealt with by the business layer.

In general, the IoT architecture is a crucial part of IoT systems, enabling the smooth integration of various technologies and devices to ensure effective and efficient data gathering, processing, analysis, and exploitation. Figure 1.3 and Figure 1.4 depict three and five layer architecture respectively.

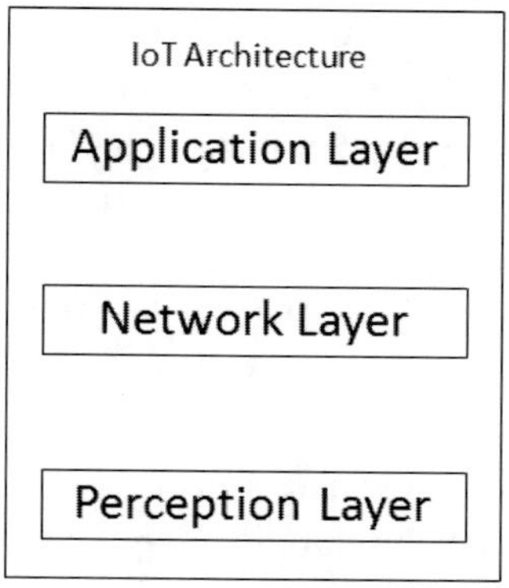

Figure 1.3. Three layer architecture of Internet of Things.

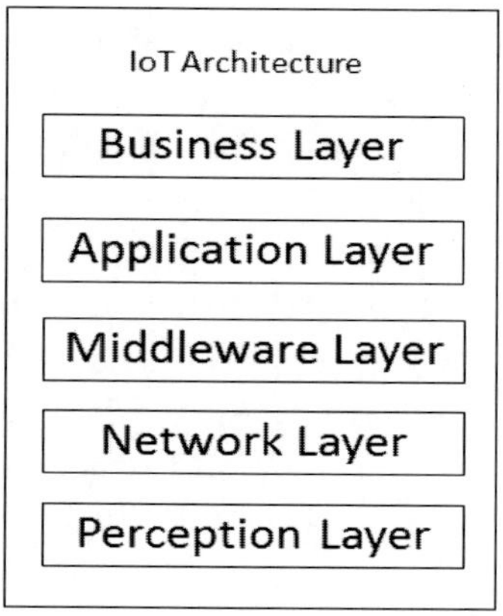

Figure 1.4. Five layer architecture of Internet of Things.

Perception layer/Sensing Layer

Underneath the other levels is this layer, sometimes known as the physical or hardware layer. Before preparing data and transmitting it to the network layer, this layer manages data collection and signalling.

Network layer

IoT's network layer is in charge of enabling communication between various sensors, devices, and other parts. This layer is necessary for data transmission between devices and data routing to the correct location.

In order to facilitate communication between devices, the network layer often makes use of a variety of networking technologies, including Wi-Fi, Bluetooth, Zigbee, LoRaWAN, and cellular networks. The choice of networking technology is influenced by a number of variables, including price, data rate, range, and power consumption.

The network layer manages data transmission between devices as well as security, dependability, and scalability concerns. Data is protected during transmission between devices by security techniques like encryption and authentication, while data delivery dependability is ensured by mechanisms like error repair and resend. The use of scalability technologies, such as routing protocols and network topologies, allows the network to expand and support more devices.

Data management is a crucial component of the network layer. In order to do this, data must be stored and processed as it is transferred between devices. In some circumstances, data may be temporarily held in the network layer to lower latency and boost productivity. IoT systems rely on the network layer, which makes it possible for devices to connect with one another and makes data transmission between devices easier. Additionally, it is essential to the security, dependability, and scalability of IoT networks.

Middleware

The software layer between the application and the network layer is referred to as middleware in the Internet of Things. The middleware layer is in charge of controlling the data flow between devices and offering a standardised interface for communication between them.

Providing for interoperability between various systems and devices is one of middleware's primary responsibilities in the Internet of Things. IoT systems frequently contain a diverse range of hardware and software, which can make it challenging for them to communicate with one another. Regardless of the technology used, middleware offers a common interface and accepted protocols for devices to communicate with one another (S. Bansal et al., 2020).

Middleware offers additional services including data management, data processing, and device management in addition to facilitating interoperability. While data processing entails analysing and changing data in real-time, data management entails storing, retrieving, and transporting data between devices. Monitoring and regulating the actions of devices, such as updating firmware or altering configuration settings, is known as device management.

Depending on the requirements of the system, middleware can be installed in many places in an IoT system. It may be set up, for instance, on a single device, at the edge, or in the cloud.

As a result, middleware is a crucial part of IoT systems since it enables seamless communication and collaboration between various devices and technologies. It offers data management, data processing, and device management features in addition to providing a standardised interface for devices to communicate with one another.

Application Layer/Business Layer

The IoT application layer is in charge of providing end users with services and apps. It is built on top of the middleware layer and is the top layer in the IoT architecture. End users can connect with IoT systems and devices and carry out various tasks like monitoring, controlling, and automating processes thanks to the application layer.

The individual use cases and business models of an IoT system are implemented at the application layer. It is the location where the IoT system's value proposition is realised. For instance, the application layer of a smart home system might have programmes for regulating temperature, lighting, and security. The application layer of a smart manufacturing system might have programmes for observing and improving production procedures.

Depending on the needs of the system and the end users, the apps in the application layer may be web-based, mobile, or desktop-based. To offer cutting-edge functionality and insights, the applications can also make use of diverse technologies like machine learning, artificial intelligence, and big data analytics.

The application layer is also in charge of overseeing the system's user interface and user experience. This entails creating user-friendly user interfaces that make it simple for end users to interact with the system as well as providing feedback on how the system and its devices are doing.

Because it enables end users to engage with and gain value from the system, the application layer is an essential part of IoT systems. Additional technologies like machine learning and big data analytics are used to deliver additional functionality and insights in this setting, along with the system's specialised use cases and business models.

Impact of IoT in Smart City

The rise of smart cities is a prime example of how the Internet of Things (IoT) has the potential to transform how we live and work. Smart cities use IoT technologies to raise the sustainability, livability, and efficiency of metropolitan areas. We shall examine how IoT is altering smart cities in this post.

The ability to gather and analyse data from a range of sources is one of the primary advantages of IoT in smart cities. IoT sensors can be installed all around a city to track energy use, traffic patterns, and air quality. Then, using real-time analysis, this data may be used to spot trends, patterns, and potential improvement areas.

IoT sensors, for instance, can be utilised to track traffic patterns and modify traffic lights in real-time to ease congestion and enhance traffic flow. In order to cut costs and increase efficiency, smart waste management systems can employ IoT sensors to monitor trash levels in bins and optimise collection routes. IoT-enabled streetlights can change their brightness levels in response to the volume of foot traffic, which saves energy and enhances safety.

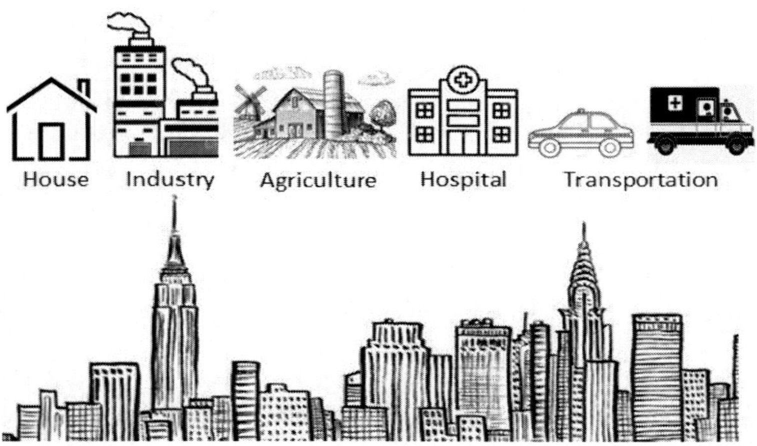

Figure 1.5. IoT in Smart city.

Enhancing sustainability is one of IoT's major advantages for smart cities. Smart cities can implement energy-saving measures that minimise greenhouse gas emissions and energy costs by monitoring energy use and identifying places where energy is being wasted. IoT sensors, for instance, can be used in smart buildings to track energy consumption and modify temperature and lighting levels to maximise energy efficiency.

Smart cities safety and security can also be enhanced using IoT. Smart traffic management systems can warn emergency services to hazards and accidents, speeding up response times and lessening the effects of accidents. Intelligent surveillance systems can keep an eye on public areas and spot unusual behaviour, alerting law authorities to potential threats. Figure 1.5 describe the impact of IoT in making cities smart.

Finally, by making smart cities more livable and accessible, IoT can raise the quality of life there. In order to help citizens navigate the city more easily, smart transportation systems can offer real-time information regarding public transportation schedules and availability. By directing motorists to available parking places, intelligent parking systems can ease traffic congestion and annoyance. Smart public places can be created with accessibility in mind, making sure that persons with disabilities can use and enjoy them.

In conclusion, there are many ways that smart cities are changing as a result of the Internet of Things. Cities are becoming more livable, workable, and sustainable as a result of IoT technologies, which also improve efficiency and sustainability as well as safety and accessibility. We can anticipate much more innovation and advancement in the creation of smart cities as IoT continues to grow and mature.

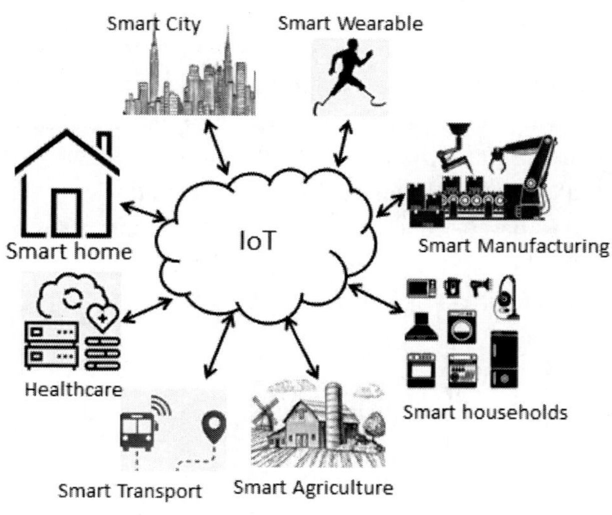

Figure 1.6. Applications of IoT in Smart city.

Applications of IoT in Smart Cities

There are several uses for the Internet of Things (IoT) in smart cities. Smart cities are metropolitan areas that use cutting-edge technologies to improve efficiency, sustainability, and quality of life. We will now examine some of the major IoT applications in smart cities (Pallavi Sethi et al., 2017). Figure 1.6 suggests few applications of IoT in smart cities.

Transportation

Smart transportation is one of the most significant IoT uses in smart cities. IoT sensors can be installed all around a city to keep tabs on parking availability, public transportation timetables, and traffic flow. This information can then be utilised to improve access to public transportation, lessen traffic congestion, and lessen the impact on the environment. Additionally, autonomous vehicle technology can be incorporated into smart transportation networks, lowering accident risks and enhancing traffic flow.

Energy Management

Smart energy management is one of the IoT's most important applications in smart cities. IoT sensors can be used to track energy use in infrastructure and buildings, allowing for more effective resource management. For instance, IoT-enabled streetlights may change their brightness depending on foot traffic, which saves electricity (Janani et al., 2021). By adjusting temperature and lighting settings based on occupancy patterns, smart buildings can save energy. Smart cities can decrease expenses and increase sustainability by optimising their energy use.

Public Safety

Public safety in smart cities is significantly impacted by IoT. Smart surveillance systems can keep an eye on public areas and spot suspicious activity, allowing law enforcement to act swiftly and decisively in the event of an emergency. Intelligent traffic management systems are able to identify dangers and accidents, alerting emergency services to provide quick aid. Smart cities can make their citizens' lives safer and more secure by utilizing IoT technologies.

Smart Government

In smart cities, smart government is a crucial IoT application. Governments may promote citizen participation, enhance services, and streamline operations by utilisingIoTtechnologies. In this essay, we'll examine some of the most important ways that IoT can be applied to build a smart city with a smart government.

Data gathering and analysis is one of the most crucial ways that IoT may be used to build a smart government. IoT sensors can be installed all around the city to gather information on a variety of city functions, including traffic flow, air quality, and energy use. The government can then use this data to better allocate resources and provide better services by analysing it to find trends and patterns. IoT can also be utilised to improve public participation in politics. Governments can inform citizens about local services and events in real-time and collect feedback by using IoT-enabled communication channels.

Smart government also entails transparent and effective procedures. IoT can be utilised to reduce expenses and waste by optimising resource consumption. IoT sensors, for instance, can be used to track water use in public buildings and spot locations where water might be saved. Energy waste can be decreased by smart buildings adjusting temperature and lighting levels based on occupancy patterns.

Smart Economy

A smart economy is one that makes use of cutting-edge technology, such as IoT, to maximise resource utilisation, boost productivity, and promote economic growth. We'll look at some of the most important ways that IoT can be applied to build a smart economy in this essay.

Resource utilization optimisation is one of the most crucial ways that IoT can be used to build a smart economy. IoT sensors can be used to track and manage how resources like electricity, water, and materials are used in manufacturing and other sectors of the economy. In the long run, this can result in cost savings and enhanced profitability by reducing waste and improving efficiency.

The creation of smart cities is a significant IoT application in building a smart economy. IoT technologies are used in "smart cities" to develop more effective and sustainable urban environments, enhancing citizen quality of life and luring businesses and investment. IoT-enabled solutions can be used to monitor and manage traffic flow, lower building energy consumption, and enhance security and safety for the general public. By making the environment more appealing and competitive for firms and investors, this can promote economic growth.

In conclusion, a smart economy can be created using a variety of IoT applications. Utilising IoT technologies, organisations and governments may increase customer satisfaction, optimise resource use, boost supply chain management, and build more effective and sustainable urban settings. We may anticipate seeing even more cutting-edge applications in the creation of a smart economy in the future as IoT continues to evolve (Waleed Ejaz, 2019).

Supply Chain Management

Additionally, IoT can be used to enhance supply chain management, boosting effectiveness and cutting costs. Supply chain managers may optimise logistics and transportation, decreasing delays and increasing delivery times, by tracking the movement of items using IoT-enabled sensors. Costs can be cut and customer satisfaction can increase as a result.

Last but not least, IoT may be utilised to improve the customer experience, boosting loyalty and happiness. Smart homes and wearable technology are examples of IoT-enabled gadgets that can offer individualised and practical experiences, increasing customer engagement and loyalty. In order to provide a more cosy and pleasant environment, IoT-enabled smart home gadgets, for instance, can automatically alter lighting and temperature based on user preferences.

Industry 4.0

Smart manufacturing and production, commonly referred to as Industry 4.0, is the process of enhancing manufacturing and production operations through the use of cutting-edge technology, such as IoT. In this post, we'll examine some of the most important ways that IoT may be applied to develop a smart manufacturing and production environment.

The improvement of manufacturing processes is one of the most significant ways that IoT may be employed in smart industry and production. IoT sensors can be used to monitor and manage manufacturing processes to make sure they are running as efficiently as possible. In the end, this can result in cost savings and higher profitability because it can assist decrease waste and boost productivity.

Predictive maintenance is a significant IoT use in smart industry and production. Manufacturers can prevent problems from occurring by monitoring equipment with IoT sensors, allowing for proactive maintenance and repair. This can lower maintenance costs and downtime, increasing productivity and overall efficiency.

Additionally, IoT can be utilised to enhance quality control in production procedures. Manufacturing companies may quickly remedy mistakes and flaws by closely monitoring the production process in real-time. This can save waste and raise product quality, which will ultimately increase consumer happiness.

In the manufacturing sector, IoT can also be utilised to improve supply chain management. Manufacturers may optimise logistics and transportation, cutting down on delays and enhancing delivery times, by tracking the movement of products and materials using IoT-enabled sensors. Costs can be cut and customer satisfaction can increase as a result.

And finally, IoT can be applied to improve worker safety in industrial settings. IoT-enabled sensors can be used to track and identify dangerous situations, including high temperatures or toxic substances, and warn personnel of potential dangers. This can contribute to making the workplace safer and lowering the likelihood of accidents and injuries.

In summary, the Internet of Things (IoT) offers several uses for developing a smart manufacturing and industry. Manufacturers may boost worker safety, optimise production procedures, enhance quality control, lower downtime and maintenance costs, and more by utilizing IoT technologies. We may anticipate seeing even more cutting-edge applications in the creation of a smart industry and manufacturing in the future as IoT continues to evolve.

Smart Healthcare Technology

Sensors for gathering data, cloud computing for long-term data processing and wireless body area networks for providing communication between the sensors on or near the patient and the communication or data processing devices like routers and gateways are the standard and widely used technologies in the majority of IoT healthcare applications. IoT is slowly but surely entering the healthcare and wellness industries. According to a study, almost 60% of respondents used sensors to monitor people by checking their blood pressure, sugar levels, etc. in order to provide the proper health care. The emphasis is shifting to wellness and preventive medicine. People are living longer as a result of the adoption of IoT, which also permitted long-term self-reliance and shifted the focus to chronic care. IoT has a significant role to play in the transition of healthcare to smart healthcare. Due to their ability to continuously monitor numerous physiological indicators like heart rate, blood pressure, temperature, and respiration rate, wearable sensors have gained popularity in the healthcare industry. For patients and healthcare providers, wearable sensors can give useful data that enables early detection of health issues and more effective management of chronic disorders. Wearable devices include, pulse sensors, smart watches, fitness trackers, Glucose monitors, ECG monitors, Blood pressure monitors, temperature sensors, Oxygen sensors, Accelerometers, nutrition tracking, smart wheel chair, Accelerometers etc., Additional applications like remote patient monitoring, smart pill management, rehabilitation like smart prosthetic, fatal monitoring, telemedicine and implantable devices like pace makers, insulin pumps, Neurostimulators, Prosthetic limbs.

Internet of Medical Things

The Internet of Things (IoT) has transformed the healthcare sector by making it possible to create medical devices that can gather and send real-time data, improving patient outcomes and making healthcare systems more effective. Examples of IoT-related medical equipment include:

Wearable Technology

These gadgets, which include smartwatches, activity trackers, and health monitors, can gather information on a patient's vital signs, level of activity, and sleeping habits. Healthcare professionals can receive this data and utilize it to monitor a patient's health and deliver individualized care.

Remote Monitoring Systems

These platforms, like telemedicine ones, enable remote patient-provider connections. They can gather information on a patient's health, offer real-time input to medical professionals, and facilitate remote consultations and diagnosis.

Smart Hospital Equipment

These platforms, like telemedicine ones, enable remote patient-provider connections. They can gather information on a patient's health, offer real-time input to medical professionals, and facilitate remote consultations and diagnosis.

Challenges in IoT

While IoT has the potential to revolutionize smart cities, there are several challenges that need to be addressed for successful implementation (Abbas Shah Syed. et al., 2021, Mohd Abdul Ahad, 2020)

Security

IoT devices are susceptible to cyberattacks since they are connected to the internet. The safety and privacy of the populace as well as the city as a whole may be jeopardised by this. In addition, a lot of IoT devices have weak security measures, which makes them easy targets for attackers. Prioritising cyber security and putting in place robust security measures are essential for smart cities to fend off potential threats.

Data Privacy

There is a massive amount of data being produced every day as a result of the development of IoT devices. This information can be used to identify people and follow their whereabouts because it contains sensitive and private information. To protect citizen privacy and guarantee that data is gathered and utilised responsibly and transparently, it is crucial for smart cities to enact data privacy legislation.

Interoperability

It can be challenging to verify that the many IoT systems and devices being produced by various manufacturers are interoperable and can connect with one another without interruption. This may result in inefficiencies and reduce IoT's ability to enhance smart city operations. Prioritising interoperability and working towards the creation of common standards and protocols are crucial for smart cities.

Power Management

IoT devices frequently rely on batteries for power, thus it's critical to optimise power consumption to increase battery life and lessen the frequency of battery replacements. This can be accomplished by utilising low-power IoT technology and putting effective power management techniques into place.

Funding and Resourcing

IoT technology implementation necessitates a large investment in hardware and infrastructure, as well as continual upkeep and supervision. The creation and upkeep of IoT infrastructure must be supported by sufficient funds and resources in smart cities.

While IoT offers a lot of potential to improve smart cities, there are a number of obstacles that must be overcome for this to happen. These difficulties include interoperability, power management, data protection, cybersecurity, and finance and resources. Smart cities can fully realise the IoT's potential to build more effective, sustainable, and livable urban settings by tackling these issues.

Conclusion

The Internet of Things, or IoT, is a network of real-world objects like machinery, cars, and other machinery that have sensors, software, and network connectivity built into them. To enable automation and provide new efficiencies in a variety of sectors and applications, these gadgets gather data and interchange it with other devices and systems.

For the Internet of Things to be successfully implemented, however, there are a number of issues that must be resolved. These difficulties include interoperability, power management, data protection, cybersecurity, and finance and resources. To ensure the safe, effective, and long-term usage of IoT, smart cities and organisations need to prioritise these issues and work on finding solutions (Ramón Martínea et al., 2020).

References

Ahad, Mohd Abdul, Sara Paiva, Gautami Tripathi, Noushaba Feroz, *Enabling technologies and sustainable smart cities*, Sustainable Cities and Society, 2020.

Alavi, A.H., Pengcheng Jiao, William G. Buttlar, Nizar Lajnef. *Internet of Things-enabled smart cities: State-of-the-art and future trends Measurement*, 129 (2018), pp. 589-606.

Bangui, Hind, Said Rakrak, Said Raghay, Barbora Buhnova, Moving towards Smart Cities: A Selection of Middleware for Fog-to-Cloud Services, *Applied Sciences,* 8 (11) (2018), p. 2220.

Bansal, S., Dilipkumar, IoT Ecosystem: A Survey on Devices, Gateways, Operating Systems, Middleware and Communication. *International Journal of Wireless Information Networks,* 27 (3) (2020), pp. 340-364.

Bauer, Martin, Luis Sanchez, Jaeseung Song, *IoT-enabled smart cities: Evolution and outlook*. Sensors 2021.

Burhan, Muhammad, Rana Asif Rehman, Bilal khan, Byung Seokim, IoT elements, layered architectures and security issues: A comprehensive survey. *Sensors.* 2018, 18, 2796.

Deakin, Mark, Alasdair Reid., Smart cities: Under-gridding the sustainability of city-districts as energy efficient-low carbon zones, *Journal of Cleaner Production,* 173 (2018), pp. 39-48.

Ejaz, Waleed, Alagan Anpalagan, *Internet of Things for Smart Cities: Technologies, Big Data and Security*; Springer: Berlin/Heidelberg, Germany, 2019.

Gaber, Mohamed Medhat, Adel Aneiba, Shadi Basurra, Oliver Batty, Ahmed M. Elmisery, Yevgeniya Kovalchuk, Muhammad Habib Ur Rehman, Internet of Things and data mining: From applications to techniques and systems, *WIREs Data Mining and Knowledge Discovery,* 9 (3) (2019), p. e1292.

Gharaibeh, A., Ammar Gharaibeh, Mohammad A. Salahuddin, Sayed J. Hussini, Abdallah Khreishah, Issa Khalil, Member, Mohsen Guizani, *Smart Cities: A Survey on Data Management, Security, and Enabling Technologies IEEE Communications Surveys Tutorials*, 19 (4) (2017), pp. 2456-2501.

Janani, R.P,, Renuka K, Aruna A, Lakshmi Narayanan K, IoT in smart cities: A contemporary survey. *Glob. Transit. Proc.* 2021, 2, 187–193.

Le Zhong, Cheng, ZhenZhu, Ren-Gen Huang, *Study on the IoT architecture and gateway technology*. In *Proceedings of the 14th International Symposium on Distributed Computing and Applications for Business Engineering and Science,* Guiyang, China, 18–24 (2015).

Martínez, Ramón, NuriaVela, Abderrazak el Aatik, Eoin Murray, Patrick Roche andJuan M. Navarro. *On the use of an IoT integrated system for water quality monitoring and management in wastewater treatment plants.* Smart Urban Water Networks, 2020.

Rejeb, Abderahman, Karim Rejeb, Steve Simske, Horst Treiblmaie, Suhaiza Zailani, The big picture on the internet of things and the smart city: a review of what we know and what we need to know, *Internet of Things,* Volume 19, August 2022.

Sethi, Pallavi, Smruti R. Sarangi, Internet of Things: Architectures, protocols, and applications. *J. Electr. Comput. Eng.* 2017.

Shah Syed, Abbas, Daniel Seirra-sosa, Anup Kumar, Adel Elmaghraby, IoT in smart cities: A survey of technologies, practices and challenges. *Smart Cities* (2021), 4- 24.

Chapter 2

Blockchain and Tiny Machine Learning Algorithms for Smart City Infrastructure Using IoT

M. P. Karthikeyan[1,*], MCA, MPhil, PhD
K. Krishnaveni[2], PhD
N. Muthumani[3], PhD
and P. Vanithamani[4], MSc, MPhil, B.Ed., PhD
[1]School of Computer Science and IT, JAIN (Deemed-to-be University), Bengaluru
[2]Department of Computer Science, Sri S Ramasamy Naidu Memorial College, Sattur, Virudhunagar, Tamil Nadu
[3]PPG College of Arts and Science Coimbatore, Tamil Nadu
[4]Department of Computer Science, PPG College of Arts and Science, Coimbatore, Tamil Nadu

Abstract

Ever since the advent of technology human beings have made deliberate efforts to mold it out for its own convenience, comfort, and pleasure; resulting in consumption of resources that were tendered by nature upon us. Smart city systems have gained a lot of traction in recent decades because they increase people's comfort and quality of life. Smartphones and microcontrollers are used to control the majority of smart city systems. Using wireless communication technology, a smartphone application is utilized to control and monitor house, cap, and city services. We analysis the smart home with integrated IoT applications and cloud computing by embedding into creating a network of IoT using technology, sensors and actuators, increasing computation storage,

* Corresponding Author's Email: karthi.karthis@gmail.com.

In: Blockchain Enabled Secure Big Data Computing …
Editors: L. Ashok Kumar, D. Karthika Renuka, V. Chandrasekar et al.
ISBN: 979-8-89113-243-6

power, and increasing data exchange accuracy. So, this thesis guides up through few among many methods to which energy can be utilized at its best using modern day technology incorporating IoT as a fundamental working principle for carrying out the most important task. Module second illustrates the problem faced in respect to resources depletion as the result of heavy consumption by Population growth and unplanned city development. All of this resulted in further nefarious activities such as power theft (to meet electricity demands) and an unpleasant situation (e.g., overcrowded roads due to heavy population uses automobiles). As a result, in order to address the aforementioned issue, we have proposed a module that combines intelligence and current technology through Automation and Information Engineering based on the Internet of Things. IoT here also helps to mitigate the installation and running cost and covering it all within acceptable boundaries and scope of the economy.

Keywords: Internet of Things, MQTT protocol, Google Assistance, gas monitoring, temperature device monitoring

Introduction

The building components of our suggested advanced smart home integrated compound include the traditional smart green home, the internet of things, cloud computing, and rule-based event processing. To the proposed composition, each component provides its essential qualities and technology. The internet of things (IoT) contributes internet connectivity and remote management of mobile appliances equipped with a range of sensors. Home equipment, such as air conditioning, lights, and other environmental devices, can be fitted with sensors. As a result, it incorporates computer intelligence into home equipment to allow for the measurement of home conditions and the monitoring of home appliance operation. Cloud computing makes it possible to scale processing power, storage space, and apps for developing, managing, and running home services, as well as accessing home devices from anywhere at any time (Soliman et al., 2013). The advanced smart home composition is controlled and orchestrated using a rule-based event processing system. The building pieces of our suggested advanced smart home integrated compound are the classic smart home, internet of things, cloud computing, and rule-based event processing. The proposed composition is enhanced by each component's basic qualities and technologies. Traditional computing, which was formerly limited to a desktop or laptop, is no longer the case. It is currently gaining traction and includes intelligent objects such as machinery, infrastructures, the environment, devices and peripherals for consumer utility in everyday life, among other things which are linked together via the internet Previously, the Internet was limited to information exchange between individuals and groups of users, whether local or worldwide, but now the Internet is poised to revolutionize civilization (Quelch and Klein., 2007). The internet's ultimate purpose is to give timely, relevant information on real-world applications and things. With a constant developmental and analytical research-based strategy, online applications and utilities have grown in popularity, surpassing human imagination. As a result, the Internet of Things (IoT) was born (Internet of Things). The Internet of Things (IoT) is a global network of intelligent items that aspires to connect the majority of common in-hand equipment and devices to the internet for the purpose of providing comfort, luxury, and a form of fortification to crippled and elderly people (Li et al., 2015). This chapter focuses on the idea of IoT's widespread application as a boon in the building of modern infrastructure at discrete levels, such as:

At the domestic commercial level, specific consumers are covered as needed (Smart Home). It also emphasizes the importance of energy conservation and optimum use of available energy in order to conserve it for future generations. The Internet of Things (IoT) is made up of two words: "Internet" and "Things." The phrase "Internet" refers to an infrastructure that allows for scalable, flexible, and reliable connections between various intelligent things, whereas the term "Things" refers to the intelligent objects themselves. As a result, the Internet of Things (IoT) is a global network of a large number of unique networks that allows users to access real-time and reliable information about real-world applications and things (Jin Jiong et al., 2014). Figure 2.1 shows how this global relationship can be expressed.

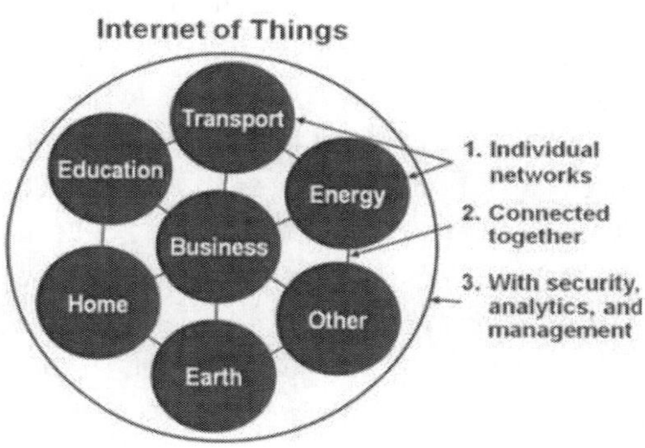

Figure 2.1. Global network of individual networks.

As a result, the paper includes outstanding techniques to help approach makers push forward and progress the beneficial aspects of the IoT while reducing challenges and ensuring more comprehensive goals, such as the following:

Encourage the use of IoT in the private sector and improve the conditions for the creation of innovative businesses and plans of action based on the IoT's new opportunities. At times, value chains could take advantage of IoT opportunities between enterprises, and toll sharing could have multiplier effects. For example, the Internet of Things enables businesses to distribute more broadly benefit-based plans of action. Small and large businesses will gradually rent their assets and compete on the whole cost of ownership rather than the fundamental purchase price. Adapt research and development methodologies across a broad range of topics. As a result, the IoT is a well-organized aspect of the overall research effort, including by providing financial assistance. This will, for example, allow measure and assess to progress, with the purpose of adjusting plans to present and future IoT advancements. While gains from improvements in IoT's foundational components, such as better M2M correspondences, information handling, sensors, and actuators, will be obvious and quantifiable, estimating profits from interest in IoT's development, application, and mix, as with many emerging exploration subjects, will be all the more difficult (Moin Sana et al., 2019).

Estimate and survey current arrangements and applications to see if they are IoT-ready and don't pose an unintentional impediment to possible IoT benefits. If existing policies and processes are based on assumptions that may limit the use of IoT, they may need to be adjusted. For example, human services decisions to reimburse medical professionals for physical visits or to demand a physical mark should be examined in light of the use of remote monitoring and treatment. Encourage the use of global specialized IoT gauges developed by benchmarking agencies or industry consortiums. The improvement of an interoperable IoT biological community relies heavily on institutionalization. It is critical for enabling the development of new frameworks, accelerating progress, and increasing aggressiveness. After some time, the most encouraging institutionalization approaches will be recognized by inventive development and end-user decision (Toseland et al., 2002).

Assess a range of assets to meet current and future IoT requirements. Diverse IoT components, from machinery to sensors, necessitate a diverse range of assets that are suited for use. Future requests for range should be surveyed, and the methods by which range could be made available for a variety of jobs, including the IoT, should be audited. With modification assistance and re-skilling programmers, promote abilities to increase job opportunities and support workers whose duties are becoming displaced by IoT-enabled and mechanical machines and frameworks. For example, new jobs in IoT-related administrations, such as data analysis, will be created, while old tasks may be improved as new apparatuses become available. In some industries, such as warehousing, the Internet of Things may improve the nature of jobs, while in more "roboticized workplaces," fewer personnel may be required (Belfo Fernando and António Trigo, 2013).

Construct trust on the Internet of Things by monitoring computerized security and protection risks in accordance with the OECD's 2015 Reference on Digital Security Risk Management for Economic and Social Prosperity, as well as the OECD Privacy Guidelines. Expanded cross-outskirt and cross-segment interoperability of strategy systems would benefit trust, particularly for IoT things in the purchaser market. The IoT's inevitability and longevity have an impact on security, security, responsibility, buyer assurance, and well-being. Governments might re-energize assistance discourse among administrative organizations, as well as with enterprises that were previously not firmly engaged with correspondences, such as transportation and utility administrations. For example, what rights or controls should a client have over data collected by a connected vehicle or a smart meter, and what is a suitable level of granularity for rights or controls? (Kalimuthu et al., 2021).

Create open information systems that allow for the reuse of government informational indexes and encourage businesses to submit non-sensitive data for public benefit. To take advantage of the IoT, open experts and the frameworks they supervise may need to update their parts and procedures. For example, transportation companies may benefit from continuing information on street conditions, but they could also report this information to street support machine drivers and those in charge of maintaining such foundations. Interfacing movement lights, for example, could increase the activity stream over a metropolis in urban planning. These activities should consider the security and protection issues that may arise (Singleton et al., 2016).

Numbering requires flexibility because different administrations or Machine to Machine customers may have different needs. Industry uses national numbers in a more regional way (e.g., extraterritorial use of national numbers) as well as universal numbers to communicate IoT-related administrations. Furthermore, controllers should examine introducing additional, and removing existing, constraints or management boundaries associated with the task and use of numbered assets, as this could act as a barrier to the take-off of a global M2M market. Encourage the implementation of IPv6 as an enabling factor for IoT. With the current state of address consumption, IPv6 configuration is unavoidable, and promoting the IPv6 transition is the greatest way to assist the IoT. Several legislatures have formally established development initiatives and alter government purchasing (Laoudias et al., 2018).

The following are the primary contributions of the work in this thesis: The first chapter discussed the introduction to IoT smart homes, the necessity for IoT smart home process analysis, and the issue statement for IoT smart homes. The available approaches for home automation work are described in Chapter 2. Various sources, inferences from literature, and smart home automation methods are discussed. The performance of the proposed system design methodologies is examined in Chapter 3. A quick examination of the various sensors and software employed in the proposed system is offered. The outcomes are then compared and discussed. In Chapter 4, you'll learn how to use results and discussion images to create a fresh approach. An overview of the anticipated work's outcome is provided in Chapter 5. The potential for future developments is also highlighted.

Literature Review

Innovation can simply provide us with new opportunities. It is up to us to make use of these in all of our development strategies. Traditional business structures must be reconsidered. Other research fields must incorporate the Internet of Things into their routine considerations. This chapter contains details on previous studies done by several researchers in this topic. A smart home's appliances have a global connectivity to the outside world to make communication easier.

Geels Frank (2004) explored the impact of energy usage on home behavior in a smart city and proposed the conceptual CLIOS model (A Complex, Large scale, Interconnected, Open and Socio-technical model). In their proposed paradigm, they used social science theories, cognitive science, and existing learning approaches. Furthermore, they believe that a smart city can drive individuals to cut energy use and adopt renewable energy systems by combining the use of renewable technologies, awareness campaigns, and comparative system analysis. Muhammad Raisul Alam et al., (2012) used a geo-statistical algorithm to monitor and adjust the temperature of a smart house. Two significant geo statistical methods are variography and kriging.

Variography can be used to calculate the spatial dependency of the observations, while kriging can be used to interpolate the predicted values.

The Internet of Things maintains its fundamental position in terms of information and communication technologies, as well as societal transformation (Miorandi et al., 2012). Only the most relevant improvements include distinguishing proof and following advancements, wired and remote sensor and actuator systems, updated correspondence conventions, and conveyed information for keen goods. Any genuine commitment to the advancement of the Internet of Things, as one can imagine, should essentially be the result of synergetic exercises led in numerous disciplines of learning, such as media communications, informatics, hardware, and sociology. In such an uncertain circumstance, this research is aimed at those who need to approach this perplexing control and contribute to its improvement.

The load shedding and power consumption in a smart house were researched by Wang et al., (2016). They used the "Frequency Response Grid Friendly Appliance System (FRGFAS)" to track, manage, and automate household appliances and outdoor lighting power consumption. The FRGFAS's parallelism and flexibility are enhanced by FPGA parallelism and LabVIEW flexibility. Their proposed load shedding framework is well suited for ZigBee applications and other related applications. Vollaro et al., (2014) employed two distinct software's, AermecMC11300 and TRNSYS, to evaluate the effect of heat capacity of the building manufacturing material and thermal inertia of the building. The aggregate thermal load is calculated by the AermecMC11300 software, whereas the hourly thermal load is calculated by the TRNSYS programme.

Paolina Ferrulli (2016) The rise of infrastructural, operational, and natural concerns, such as environmental change, noisy contaminations, and failure, has increased the demand for strong, small, operationally adaptable, and astute checking frameworks. Smart sensor systems are a growing subject of research in this context, as they combine many of the challenges of contemporary software engineering, distant communication, and flexible registering. The use of a remote embedded processing framework to check clamor contamination levels in infrastructural conditions is proposed in this paper. Because of their enormous applicability, the arrangement integrates advancements that have been increasing in the field of portable registering and also Internet of Things (IoT). The detecting frameworks are linked to the installed registering framework to monitor any deviations in clamor contamination parameters from their normal behavior. This methodology is adaptable and distributive to any infrastructure condition that necessitates continuous observation and behavioral analysis. The suggested display's performance is evaluated using a model that includes Intel Galileo and sensor sheets, as well as inserted programming. The application is tested for a few factors and associated behavioral examples, as well as client-provided information, to provide a controlled contamination checking to influence the earth to be wise.

Marcus Buckingham and Curt Coffman (2014) Water is one of the most essential elements on the planet. People nowadays are constantly looking for ways to make their lives easier. This concept is used to characterize water monitoring frameworks such as tank water level detection checking, water contamination monitoring, and water pipeline spillage monitoring. We keep a strategic distance from the enormous amount of water squandered by uncontrolled use of large lofts/workplaces by employing Wireless Sensor Technology. Water level checking using a microprocessor (PID) is used to show the level of water in the tank to specialists. Sensor-Based Water Pollution Detection will assess the water quality using these factors, such as pH, turbidity, and temperature, which will be measured continually by the sensors and monitored by a professional. We use the weight into the pipe employing on drive delicate resistors (FSR) produced from a hole is detected, it will be shown by an expansion in the LED meter and a hurried sound will be heard in the headset. Our motivation for this proposal is to conserve the water by utilizing innovation, and the checking framework makes use of everyday life gadgets such as a PC or a cell phone.

The advancement of technology has paved the way for new types of devices that can communicate with one another to deliver results over long distances (Jaiganesh, et al., 2017). The Internet of Things (IoT) connects remote transmitting implanted gadgets in a single connection (Internet of Things). If all objects and people in everyday life were given unique identities, computers would be able to track and store them. Other than RFID, innovations such as close field correspondence, scanner tags, QR codes, and computerized watermarking can be used to label items. Arduino is used as an installed controller to connect Ethernet shield with a PC/Laptop to provide IoT over Ethernet in this new technique for employing inserted innovation to provide such application. This stopping service is available in the air terminal and is provided by an air terminal professional with a client ID and password. When a client needs to verify the state of a vehicle in the parking lot, they log on to the air terminal web connection with their ID and secret word and use IoT to view the status of the vehicle. The purpose of this article is to discuss an IoT-based airport parking system in order to put Arduino into action as an IoT application.

For home energy management and load control in a network, Mahapatra, and Nayyar, 2019 suggested a Home Energy Management System (HEMS). With the following three modules, this model integrates non-conventional energy resources, a visual interface to monitor and control the system, and a visual interface to monitor and control the system.

- HEMS software.
- A centralized module to manage home energy consumption.
- A load scheduler for the network of appliances.

The home power management problem was investigated by Ransing and Rajput, 2015, who used wireless sensor networks, LabVIEW, and the ZigBee protocol. They discovered that proper administration, monitoring, and frequent notifications to home appliances result in reduced power consumption. The absence of physical cabling is a fundamental benefit of their suggested framework. An approach for modelling and evaluating home appliances has been proposed by Van Nguyen et al., 2009. In their proposed technique, home appliances can be accomplished in the following two ways.

- Thermal home appliances can be simulated by applying constant numeric values of heating and cooling.
- Rest of the home appliances can be simulated in its time duration profile.

System Design

Security has now become a major concern in both public and private organizations, with various security frameworks proposed and created for some significant procedures. Homes in the twenty-first century will become increasingly self-controlled and robotized, particularly when used in a private home, due to the convenience it provides, and security has now become a major concern in both public and private organizations, with various security frameworks proposed and created for some significant procedures. Security frameworks are required for data, property, and preventing burglary or wrongdoing in the home. A smart home is a collection of sensors and controllers that work together to allow a client to control numerous devices in their house via the Internet of Things. The sensors detect various changes, screen them, store the data, and display it all at once for analysis and control

(Tripathi et al., 2022). This motivates us to make changes to our home to suit the needs of each family. This is a clever framework that allows us to control our home's machinery using locally available components such as a PIC controller, light sensors, and PIR sensors. The Internet of Things is closely linking everyday goods to the Web in order to facilitate communication between things and people, as well as between things themselves. Current smart home systems include a variety of entertainment systems to keep homeowners happy and entertained (Strengers and Kennedy, 2020). The steps below outline the hardware and design approach for a smart city home monitoring and automation system.

Difficulties in urban settlement planning, development, and operation are allowing for new thinking in a range of disciplines. Design, urban planning, building, development, data innovation, frameworks and natural science, property development, back, and municipal government experts have a better understanding of partners and learn how to connect with them (Grant Robert M. and Charles Baden Fuller., 2004). For intelligent city efforts, framework models capable of seeing deeply into how urban areas function, how people use the city, how they feel about it, where the city has difficulties, and what types of progress can be linked could be used In the electrical and electronics sector, there is an overall expansion and technological growth as we reach the era of modernization and automation. Modern electrical equipment and fixtures are becoming more opulent and simpler to use, such as via Android or a personal computer. As a result, we identified a demand that might be met by slightly altering existing technology. Adopting this new technology would provide benefits such as the capacity to control appliances using Android, a personal computer, or autonomously, as well as the ability to calculate the amount of energy wasted by a specific load. Currently, IoT devices are used to monitor the level of comfort in real estate sites in apartments and residences. The Internet of Things sensor can detect temperature, humidity, light, and acceleration, as well as record audio and video. Despite being in the same building, an upper-floor apartment received two hours more sunlight than a lower-floor one. Furthermore, it was discovered that while looking away from the source of the noise, apartment complexes near a train or highway emit very little noise. Using this knowledge, we'd be able to prevent biased apartment ratings and generate more objective judgments (Obuchi et al., 2018).

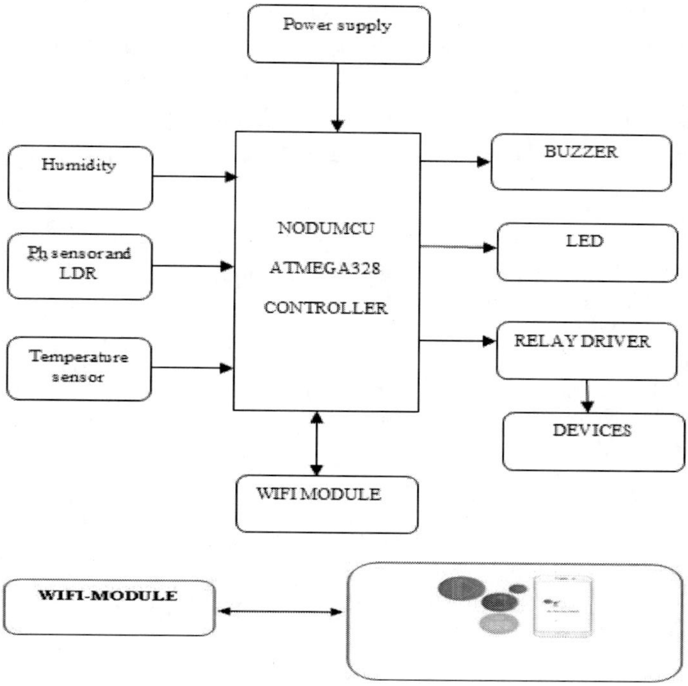

Figure 2.2. Proposed system block diagram.

The system is made up of a controller and a Nedemcu board. We needed internet connectivity to transmit the message, which was provided by the Wi-Fi module. Vibration sensors are used to find sensor readings such as temperature, gas, and PIR sensor values. We utilized an app called Open Nedemcu idea to keep track of the live data. The internet connectivity for the system is provided through a common access point, such as a router. The general block diagram of the proposed system is shown in Figure 2.2. The values of the cloud server are cleared at first, so that a previous value does not have to be stored after each restart. The VNC Viewer app is then launched, while the Wi-Fi module is powered up and a link is created between the Wi-Fi module and the access point, allowing the sensor value to be uploaded and accessed over the internet. The controller collects the data through the Wi-Fi module and conducts the required action, depending on the control signal provided by the user, using the VNC Viewer application and uploading it to the cloud server. The block diagram of impact detection by the sensors whenever temperature values occur is shown in Figure 2.2. It begins with the ADC's initialization, as the DHT22 sensor's output is an analogue signal. The

ADC converts the temperature values sensed by the sensor into digital data and sends it to the microcontroller. If the impact value is less than the threshold value, the microcontroller will disregard it; if the impact value is greater than the threshold value, the microcontroller will interpret it as an overheating condition and switch on the fan, sending a signal to the module and base station.

```
1. Is the humidity sensor temperature and PH > threshold
2. If true, wait (seconds) and check for sent MQTT protocal status.
3. If the reset button is pushed, proceed to step 1; otherwise, proceed to step 4.
4. Gather data from the sensors.
5. Collect current device details on/off condition from home
6. If (exIt temperature information)
{ Fan or FAN device on ();
}
If (exit humidity information)
{
User intimation and turn on motor();
}
If (exit PH information)
{
User intimation();
}
7. Else
{
Sensor value continuously monitoring();
}
```

Figure 2.3. Pseudocode for proposed system design.

The temperature sensor is an analogue sensor that monitors temperature and sends an analogue signal to the Nedemcu controller, which then sends the signal to the Nedemcu controller over the MQTT protocol (mosquito protocol). MQTT protocol is a kind of communication between a server and a controller that uses a Wi-Fi module to update data in real time. Google Assistant was used to turn on/off MQTT protocol devices. This sensor will be used to detect soil values and will be used to detect ph levels. Which is used to monitor continuously to the soil ph values store to the database? Humidity sensor used to detect the humidity values of both soil and air, which help to monitor the environment moisture content and needy of water content in the irrigation parts working principle humidity is converted into electrical signal, electrical signal converts both analog and digital signal. Normally analog signal threshold-based set point set and make it to digital signal. This digital signal handles the microcontroller continuously monitor the signal if the signal getting means of low values immediately respond and alert the user and turn on the motor. The pseudocode for proposed system design is presented in Figure 2.3.

Result and Discussion

Proposed system working module consist of three different types of architecture inbuilt in single modules they are, it consists of three module they are,

First module: Temperature sensor detects the temperature values and fan will be on automatically otherwise fan is ideal state this will be very useful to energy consumption without manpower.

Second module: The temperature, humidity and PH values monitoring system, LM35, dht22 sensor used to detect the values continuously any temperature, humidity and PH is detecting intimation given to the controller. Depends on this level of sensor on and off the motor pumps.

Third module: Third module is humidity monitoring system. DHT22 sensor used to monitor the humidity; DHT22 sensor monitors both temperature as well as humidity. The temperature values are continuously sent via server-based database. We check the details for users through server pages. ldr sensor used to monitor the day and night light intensity will automatically on/off the lights. Based on this sensor values irrigation motor will be turned on/off automatically. User to energy consumption.

If temperature values are exiting the particular values user will be turned on the fan using through voice signal-based assistant's android application. Google assistance-based control module is shown in Figure 2.4. Server and monitoring section applications are demands on the Wi-Fi based MQTT communication protocols. The proposed design consists of the following advantages of smart city implementation.

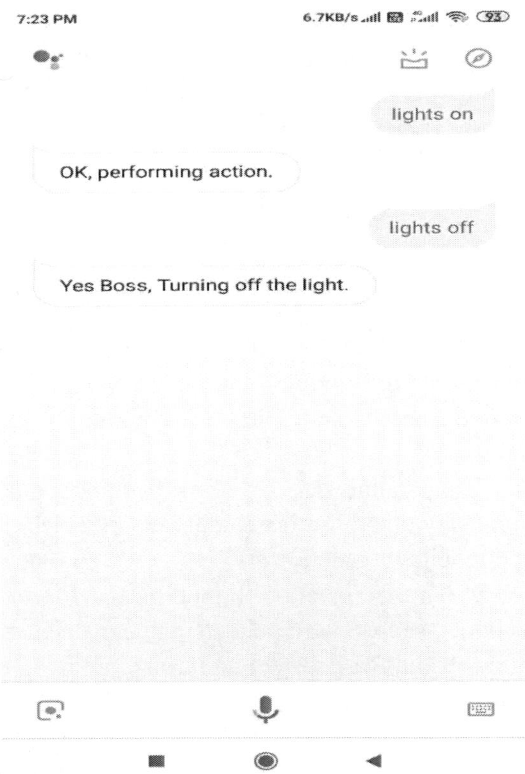

Figure 2.4. Google assistance-based control module.

First and foremost, the IoT preferences listed above will have an impact on your principal issue by lowering costs and increasing capability. In any event, IoT data analysis will soon enable you to detect new business boundaries—and thus new job opportunities. In the coming decade, the Internet of Things may be the "X factor" that offers many organizations a critical advantage over competitors (Madupu and Karthikeyan, 2018).

Information: The greater the records, the less complex it is to determine the right choice. Information about what to get from the staple at the same time as you're out, without checking by myself, reduces time and further is worthwhile as well.

Following: The computers preserve a track each on the nice and the opportunity of things at domestic. Knowing the end date of factors before one uses them upgrades prosperity and man or woman success. in addition, you may by no means pass over the mark on whatever when you require it at closing.

Time: The measure of time stored in looking and the number of journeys carried out for the most elements would be large.

Cash: The cash related point is the high-quality desired outlook. This development may want to supplant people who are answerable for looking and looking after measures.

In light of the manner that IoT engineers save time and money, IoT sort out benefits not one, but rather everyone, i.e., persons, society, associations, and so on. IoT frameworks pass on more quickly and unmistakably while using the least amount of power. This brings the status of a person's fulfilment up to date. It is used for understanding watching, in which various types of wireless sensors are implanted on the patient's body, which communicate with the IoT system and supply all of the patient's essential data. IoT applications are used as a type of home security device that can be examined and managed locally or remotely using simple software packages that can be downloaded to PDAs or phones. Security alerts, cameras, sensors, entryway bolts, and other common IoT gadgets are used as part of the home computerization situation.

IoT is used in a variety of applications, including benefit and individual tracking, stock control, energy preservation, and shipping. It's similar to M2M, but it has broader applications. M2M is only used for machine-to-machine communication. Things on the Internet of Things (IoT) communicate with their owners, displaying their location and conditions. Self-ruling structures are the genuine guarantee of IoT and make robotization a stride assist by completely tackling the capacities of cutting-edge network. This happens when different frameworks focus on delivering robotized results that all originate from one sensor yet change somewhat in light of particular situations. Security experts aren't outsiders to the conceivable outcomes of mechanization and depend on advancements like examination to recognize movement and trigger an activity. They will assume an essential part in changing spaces into profoundly associated, productive, independent situations.

Gadgets, for example, cameras and sensors, which have for some time been vital to security frameworks, will assume an indispensable part in the self-sufficient building. They can possibly interface with each other's framework in the building, making new coordination openings. As of now, the move from robotized structures to independent structures is changing the way security experts must consider IoT, particularly sensors. Sensors and associated gadgets will incredibly upgrade wellbeing and security in all spaces. For self-ruling structures to be effective, their parts ought to be incorporated physically with the spaces, lifting the space's knowledge, and worked to stand the trial of time.

Gadgets must have a considerable measure of processing power and can be redesigned or revived. This will expand the gadget's lifecycle and give a relocation way as new advances wind up noticeably accessible. As structures end up noticeably quicker witted and depend all the more vigorously on client gave information, the trustworthiness of basic foundation is basic to guaranteeing information is secure. IoT robotizes information security by investigating typical movement and raising alarms when it shows up a gadget could have been traded off. The possibility of the self-governing structures is simply being investigated now in the customer showcase, and the venture market won't be extremely a long way behind. Friendliness and office structures might be the first to profit by the propelled network and knowledge of full self-governance. By joining propelled network with the use of down to earth however dynamic keen building systems, these spaces will be changed into more secure, more secure situations for individuals and information, with improved profitability, proficiency, solace and manageability (Sivanantham, 2021).

The expansion of Internet of Things (IoT) gadgets is significantly extending the biological community in which physical security lives, opening up energizing new coordination potential outcomes and business openings inside the undertaking market. Much like the advancement from simple to IP video security, IoT will indeed change the diversion for security experts, yet on a considerably more amazing scale. Though the change from simple to IP was restricted to security and access control – the IoT upset makes the likelihood for each framework, from HVAC to lighting to AV, to interface with physical security. As these frameworks turn out to be more interwoven, the domain of the security proficient enormously grows enabling them to convey new reconciliations that have genuine incentive for the end client. IoT will bring upgraded profitability, productivity, solace, security and manageability to the spaces in which we live and work. Maybe the most substantial advantage of IoT will be the more noteworthy degree of profitability since gadgets like sensors will fill numerous needs inside an assigned space, costing less to introduce.

IoT will enable building tenants to remain associated in incalculable ways and gadgets should likewise encourage everyday existence without entangling things or acting as a burden. Effective IoT establishments will be anything but difficult to utilize and convey, while completely coordinating with each other on a protected stage. Finish combination is critical, and wise gadgets and sensors will empower physical security to be additionally overseen without human intercession. The change from simple to IP video security has, from multiple points of view, set the phase for the IoT upheaval's effect on security. The presentation of IP cameras obscured the lines amongst security and IT divisions, and the expansion of more associated gadgets will keep on eroding these storehouses further. Working intimately with IT offices will be basic to executing effective arrangements that are additionally secure. As structures wind up plainly quicker witted and depend all the more intensely on client gave information, the trustworthiness of fundamental foundation is basic to guaranteeing information is shielded.

Conclusion

This investigation begins by setting the stage for rapid urbanization, which continuously places pressure on urban areas to deliver great solutions in order to satisfy manageable development needs. The shrewd city foundation is extremely setting specific, and the article looked at several major differences between the concept of sharp framework in developed and developing country settings. Smart City innovations are particularly significant in developing nation settings, according to the investigation, as their idea for new streets for development bouncing. The Internet of Things (IoT) is a paradigm that transforms every object into an intelligent object. Intelligent objects contain properties such as identification, sensing, communication, and processing that enable them to communicate with other things, software, and internet-based services. The IoT's backbone is intelligent things, which help to improve a city's living quality. Many IoT-based applications run in different places across the city in a smart house. In this thesis, we used IoT to investigate concerns such as power usage and security in a home in order to make it a smart home. People are always inventing and discovering new advancements and products, therefore IoT will have a significant impact on our daily lives. Things that are accomplished using the internet will unquestionably make our lives less demanding than they are now. We would have the ability to work different contraptions all while the IoT will keep running far with having the ability to interface with different sorts of devices and because of the straightforwardness of task that it offers to some person who uses it. We can keep an eye out for things that were past our degree like when we are at office, what is new with our home and youngsters, notwithstanding the way that we have those workplaces now too, it would be made package less requesting in light of the ability to connect with different contraptions. There will be machine to machine interfaces, where devices chat with each other. We may well watch wearable devices and sensors that can reveal an extensive measure of changes in our lifestyle and give early acknowledgment to illness perils. IoT will prompt extended care about environmental and social issues as more people go ahead with the web and they have more access to new frameworks and answers for guidelines, human rights, regular dangers and preparation. From ecological change to illness evasion, from sagacious ceasing to development organization, from water safeguarding to misuse organization all answers will lie in the IoT, and we will find a better technique than oversee it.

References

Alam, M. R., Reaz, M. B. I., & Ali, M. A. M. (2012). A review of smart homes—Past, present, and future. *IEEE transactions on systems, man, and cybernetics, part C (applications and reviews)*, *42*(6), 1190-1203.

Belfo, F., & Trigo, A. (2013). Accounting information systems: Tradition and future directions. *Procedia Technology*, *9*, 536-546.

Buckingham, M., & Coffman, C. (2014). *First, break all the rules: What the world's greatest managers do differently*. Simon and Schuster.

Ferrulli, P. (2016). Green airport design evaluation (GrADE)–methods and tools improving infrastructure planning. *Transportation Research Procedia*, *14*, 3781-3790.

Geels, F. W. (2004). From sectoral systems of innovation to socio-technical systems: Insights about dynamics and change from sociology and institutional theory. *Research policy*, *33*(6-7), 897-920.

Grant, R. M., & Baden-Fuller, C. (2004). A knowledge accessing theory of strategic alliances. *Journal of management studies*, *41*(1), 61-84.

Jaiganesh, S., Gunaseelan, K., & Ellappan, V. (2017, March). IOT agriculture to improve food and farming technology. In *2017 Conference on Emerging Devices and Smart Systems (ICEDSS)* (pp. 260-266). IEEE.

Jin, J., Gubbi, J., Marusic, S., & Palaniswami, M. (2014). An information framework for creating a smart city through internet of things. *IEEE Internet of Things journal*, *1*(2), 112-121.

Kalimuthu, S., Naït-Abdesselam, F., & Jaishankar, B. (2021). Multimedia Data Protection Using Hybridized Crystal Payload Algorithm with Chicken Swarm Optimization. In *Multidisciplinary Approach to Modern Digital Steganography* (pp. 235-257). IGI Global.

Laoudias, C., Moreira, A., Kim, S., Lee, S., Wirola, L., & Fischione, C. (2018). A survey of enabling technologies for network localization, tracking, and navigation. *IEEE Communications Surveys & Tutorials*, *20*(4), 3607-3644.

Li, S., Xu, L. D., & Zhao, S. (2015). The internet of things: a survey. *Information systems frontiers*, *17*(2), 243-259.

Madupu, P. K., & Karthikeyan, B. (2018, March). Automatic service request system for security in smart home using IoT. In *2018 Second International Conference on Electronics, Communication and Aerospace Technology (ICECA)* (pp. 1413-1418). IEEE.

Mahapatra, B., & Nayyar, A. (2019). Home energy management system (HEMS): Concept, architecture, infrastructure, challenges and energy management schemes. *Energy Systems*, 1-27.

Miorandi, D., Sicari, S., De Pellegrini, F., & Chlamtac, I. (2012). Internet of things: Vision, applications and research challenges. *Ad hoc networks*, *10*(7), 1497-1516.

Moin, S., Karim, A., Safdar, Z., Safdar, K., Ahmed, E., & Imran, M. (2019). Securing IoTs in distributed blockchain: Analysis, requirements and open issues. *Future Generation Computer Systems*, *100*, 325-343.

Obuchi, Y., Yamasaki, T., Aizawa, K., Toriumi, S., & Hayashi, M. (2018, January). Measurement and evaluation of comfort levels of apartments using IoT sensors. In *2018 IEEE International Conference on Consumer Electronics (ICCE)* (pp. 1-6). IEEE.

Quelch, J. A., & Klein, L. R. (2007). The Internet and international marketing. *Readings in modern marketing*, 381-385.

Ransing, R. S., & Rajput, M. (2015, January). Smart home for elderly care, based on Wireless Sensor Network. In *2015 International Conference on Nascent Technologies in the Engineering Field (ICNTE)* (pp. 1-5). IEEE.

Singleton, A. D., Spielman, S., & Brunsdon, C. (2016). Establishing a framework for Open Geographic Information science. *International Journal of Geographical Information Science*, *30*(8), 1507-1521.

Sivanantham, K. (2021). Sentiment Analysis on social media for Emotional Prediction During COVID-19 Pandemic Using Efficient Machine Learning Approach. *Computational Intelligence and Healthcare Informatics*, 215-233.

Soliman, M., Abiodun, T., Hamouda, T., Zhou, J., & Lung, C. H. (2013, December). Smart home: Integrating internet of things with web services and cloud computing. In *2013 IEEE 5th international conference on cloud computing technology and science* (Vol. 2, pp. 317-320). IEEE.

Strengers, Y., & Kennedy, J. (2020). *The smart wife: Why Siri, Alexa, and other smart home devices need a feminist reboot*. MIT Press.

Toseland, R. W., McCallion, P., Gerber, T., & Banks, S. (2002). Predictors of health and human services use by persons with dementia and their family caregivers. *Social science & medicine*, *55*(7), 1255-1266.

Tripathi, M. M., Haroon, M., & Ahmad, F. (2022). A Survey on Multimedia Technology and Internet of Things. In *Multimedia Technologies on the Internet of Things Environment, Volume 2* (pp. 69-87). Springer, Singapore.

Van Nguyen, T., Kim, J. G., & Choi, D. (2009, February). ISS: the interactive smart home simulator. In *2009 11th international conference on advanced communication technology* (Vol. 3, pp. 1828-1833). IEEE.

Vollaro, R. D. L., Evangelisti, L., Carnielo, E., Battista, G., Gori, P., Guattari, C., & Fanchiotti, A. (2014). An integrated approach for an historical building's energy analysis in a smart cities perspective. *Energy Procedia*, *45*, 372-378.

Wang, J., Zhang, H., & Zhou, Y. (2016). Intelligent under frequency and under voltage load shedding method based on the active participation of smart appliances. *IEEE Transactions on Smart Grid*, *8*(1), 353-361.

Chapter 3

Federated Learning for Secure Smart Cities Using IoT

Apurba Koirala[1]
Aaryan Shrestha[2]
Rajeshkannan Regunathan[3,*]
Hrigdev Singh[4]
and Jonathan Atrey[5]
School of Computer Science and Engineering, Vellore Institute of Technology, Vellore, Tamil Nadu, India

Abstract

Limitations of data storing and processing in traditional machine learning have led to constraints in creating high-scalability models. In addition, several security areas may be compromised when accessing and utilizing data. To address these difficulties, recent advances in machine learning have resulted in the conception of Federated machine learning. There are significant issues regarding security, such as the disclosure of private or sensitive information. At times, centralized databases may experience a breach, putting the stored data in danger, or there can be interceptions by hackers during data transmission. As a result, we use Federated Learning to address these issues. The decentralized architecture offered by Federated Learning has allowed us to create more scalable models with definitive architectures while incrementing various security aspects of training data.

The combination of Federated Learning along with data gathered from today's Internet-of-Things has resulted in several studies that may

* Corresponding Author's Email: rajeshkannan.r@vit.ac.in.

In: Blockchain Enabled Secure Big Data Computing …
Editors: L. Ashok Kumar, D. Karthika Renuka, V. Chandrasekar et al.
ISBN: 979-8-89113-243-6

be beneficial concerning different research perspectives in smart cities. In the proposed book chapter, an exhaustive inquiry is carried out to summarize the most recent research on using Federated Learning in various smart city-related domains. A comprehensive knowledge of the recent advances in Federated Learning across multiple industries. A brief history, explanation, and overview of the most critical technologies crucial to Federated Learning come next, along with applications in the context of smart cities.

Federated Learning has the potential to transform how we live, work, and interact in our urban surroundings by combining the power of advanced technology with a privacy-preserving approach, ushering in a new era of collaborative intelligence and invention.

Keywords: smart cities, decentralized environment, federated learning, IoT

Introduction

In the twenty-first century, urbanization has surged, with cities now housing more than half of the world's population. As these urban areas continue to grow, there is a greater need for efficient, sustainable, and intelligent solutions to manage resources and improve the quality of life for their residents. Smart cities are a technologically driven approach that merges advanced digital technologies, big data analytics, and networked systems to produce more efficient, sustainable, and citizen-centric urban settings.

Smart cities use the Internet of Things (IoT), artificial intelligence (AI), and other emerging technologies to improve public services, improve infrastructure, and promote innovation. These advances not only improve transportation, energy management, and trash disposal systems, but they also improve urban people's safety, health, and overall well-being. Yet, the data privacy and security risks that these breakthroughs raise must not be overlooked. Because the smart city ecosystem is heavily reliant on data collection and processing, there is an increasing demand for technologies that can manage sensitive data while retaining the benefits of collaborative intelligence. This is where Federated Learning comes into play.

Federated Learning is a decentralized machine learning technique that allows AI models to learn from distributed datasets while maintaining data privacy. Local devices or edge nodes do model training using their data in this system, and only model updates are exchanged with a central server. This reduces the need for raw data transfer, lowering privacy concerns and bandwidth costs. Smart cities may harness the power of AI while protecting the privacy and security of their inhabitants' data by employing Federated Learning.

Federated Learning has numerous advantages in the context of smart cities. It fosters the development of customized, context-aware solutions by enabling smooth collaboration among multiple stakeholders, including government agencies, private firms, and citizens. This shared intelligence enables city planners and policymakers to make data-driven decisions, improving resource allocation and assuring the city's resilience to challenges like climate change, population expansion, and economic volatility.

Furthermore, Federated Learning makes it possible to incorporate a wide range of data sources, from traffic sensors to healthcare records, without disclosing sensitive information. This enables smart cities to fully utilize their data, resulting in more accurate and robust AI models. Improvements in predictive analytics and real-time decision-making capabilities have improved many elements of urban living, ranging from traffic management and public safety to environmental monitoring and healthcare services.

Research Contributions

The authors' primary goal in the proposed book chapter is to investigate the various uses of Federated Learning in the context of smart cities, with a particular emphasis on its integration with Internet of Things (IoT) devices. The major goal is to evaluate the dependability of Federated Learning in terms of time, cost, scalability, and security. Following an introduction that establishes the significance of Federated Learning, the proposed study goes into the numerous types of Federated Learning, their requirements, and their restrictions. This basic understanding prepares readers for the chapter's focus, which is the use of Federated Learning in IoT and smart cities. Before reaching their conclusions, the authors discuss the challenges that must be overcome as well as the future scope of Federated Learning in smart cities, providing readers with a thorough understanding of the subject. Finally, the authors give their conclusions based on the many findings from the various use cases and recent empirical research discussed throughout the chapter, shining light on the consequences and prospects of Federated Learning in smart city applications.

Incorporating Federated Learning in Smart Cities

As we move deeper into the world of smart cities, Federated Learning will continue to open new avenues for innovation and growth. By encouraging a synergistic link between technology and urban development, we can ensure that future cities are prepared to face the difficult problems that lie ahead.

Transportation systems are one crucial area where Federated Learning may have a substantial influence. With the growth of self-driving cars and intelligent traffic management, there is a greater demand for real-time data processing and decision-making. Federated Learning can aid in the development of a cooperative network of automobiles, traffic lights, and infrastructure, allowing them to learn from each other's experiences while retaining privacy. This collaborative strategy can result in more efficient traffic flow, less congestion, and lower emissions, eventually leading to a cleaner and more sustainable urban environment.

Federated Learning has the potential to transform the way medical data is processed and evaluated in the field of healthcare. Federated Learning can significantly improve the accuracy of disease prediction, early diagnosis, and personalized treatment plans by allowing hospitals, clinics, and research institutions to share AI model updates without exposing sensitive patient information. This approach can also help cities better prepare for and respond to outbreaks, epidemics, and other health emergencies.

Another domain in which Federated Learning can be useful is energy management. Federated Learning can assist improve energy distribution and consumption by combining data from several sources, such as smart meters, weather trends, and energy usage behaviors. This might result in more efficient energy consumption, less strain on the power system, and a smaller carbon impact on the city.

Finally, Federated Learning can help improve public safety and security in smart cities. Federated Learning can help build a safer and more resilient urban environment by enabling multiple stakeholders, such as law enforcement agencies, private security corporations, and people, to share information and work on AI models without risking privacy violations. This can lead to more successful crime prevention initiatives, emergency response plans, and overall increased citizen safety.

Preliminary Concepts about Federated Learning

Aggregation Cycle

When working with a Federated Learning network, the following are the major steps:

Network Configuration

First, specific nodes are assigned to collect all local outputs and updates from each participating node. Following this, a central point for data collection is chosen, and the rest of the network relies on local nodes to process their training data using local models. Consequently, the central data collection point selects several sub-nodes that are suitable for the given task and integrates them into the Federated Learning network.

Local Node Training

Once the local nodes and central collection point have been established, the central server disseminates a model containing all necessary parameters to initiate local data training within each node. The individual nodes then train their respective models, generating updates that may include values such as weights or other forms of unprocessed data. These updates are subsequently uploaded to the network and forwarded to the central collection point.

Aggregating Updates

The central collection point receives updates from all the trained local nodes and proceeds to combine these updates to produce a consolidated model. This final model is generated by averaging the parameters obtained from the local nodes within the federated network and updating the weights based on the data associated with the central server.

Iterative Process until Optimal Model

After the first network iteration, the central server updates the final model and redistributes the resulting combined parameters back to the local nodes for the subsequent iteration of local data training. This cycle of local training and aggregation continues through multiple iterations, with the updated model being shared and refined until an optimal model is achieved. Each iteration allows the local nodes to further refine their models based on the aggregated parameters, gradually improving the performance and accuracy of the global model.

Throughout this iterative process, local nodes maintain their data privacy, as only model updates and not raw data are shared with the central server. This ensures that sensitive information remains protected, while still allowing the network to benefit from the collective intelligence of all participating nodes.

Once the optimal model has been attained, the central server can halt the iterative process, and the final model can be deployed for the desired application. This Federated Learning approach enables the development of robust and accurate models without compromising data privacy and security, making it particularly suitable for scenarios where data protection is a top priority.

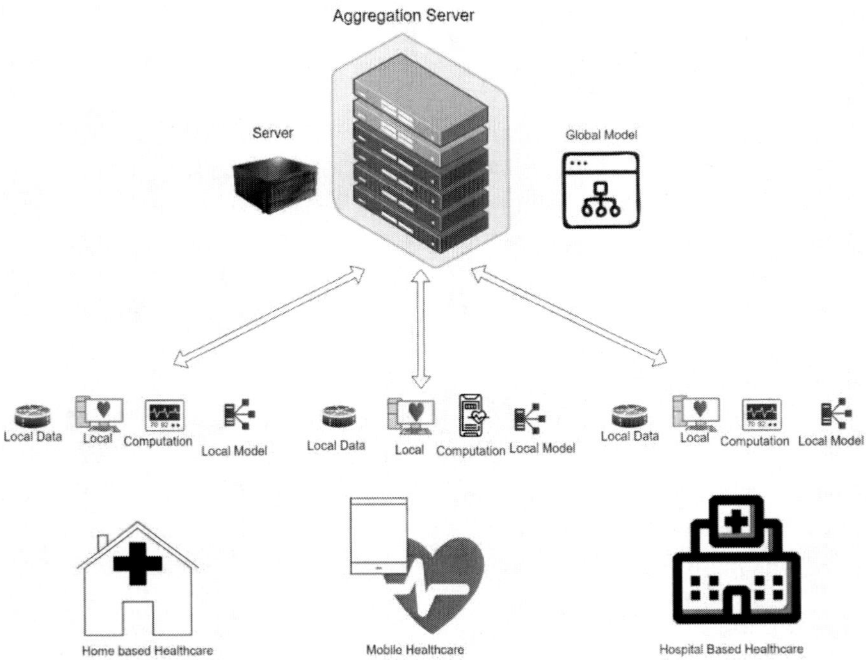

Figure 3.1. FL Architecture.

Figure 3.1 illustrates a federated model that communicates with the central server by distributing a model containing global parameters for every local node to train with, which are then fine-tuned based on local data processing. The figure also demonstrates the transfer of each local node's modifications to the central server and the distribution of a refreshed global model after the central server has combined the incoming parameters.

FedAvg

Federated averaging (FedAvg) is a communication strategy that provides efficiency for a large sample space of clients holding distributed data for training models. Clients in FedAvg secure their data by training them locally and not sharing the actual data while generating a global model by submitting their parameters to an aggregation server. Most deep learning applications rely heavily on the variation of their related stochastic gradient descent (SGD). Hence, while developing FedAvg, or federated averaging technique, McMahan et al. (2017) used federated optimization to apply SGD to an instance. Every round of communication in this example employs a single batch of gradient computation. This method is efficient; nevertheless, training a good model requires a significant number of iterations. Hence, McMahan et al. (2017) used large batch synchronous SGD, similar to the trials used by Chen et al. (2016). To execute this strategy, a group of clients is chosen each round, and the gradient of loss is computed depending on the data that these clients have. This is known as FederatedSGD or FedSGD. Given such a set of FedSGD, the proportion of clients is regarded as a variable C equivalent to 1, and a fixed learning rate 'η' is utilized by each client for computing the average gradient 'g' at the local models 'w.' The Central server combines gradients and performs an update. Each locally trained client performs a single gradient descent step on the current model using local data, and the server then uses the weighted average of the resulting model as its parameter. Following that, the local clients can be repeated several times before proceeding to the next stage, which entails determining the weighted average. FedAvg, or the federated averaging algorithm, is the name given to this approach. Figure 3.2 is an illustration of the FedAvg algorithm.

Drawbacks of Traditional Machine Learning

Traditional Machine Learning has various constraints, some of which are not suitable for training data in a smart environment. They are as follows:

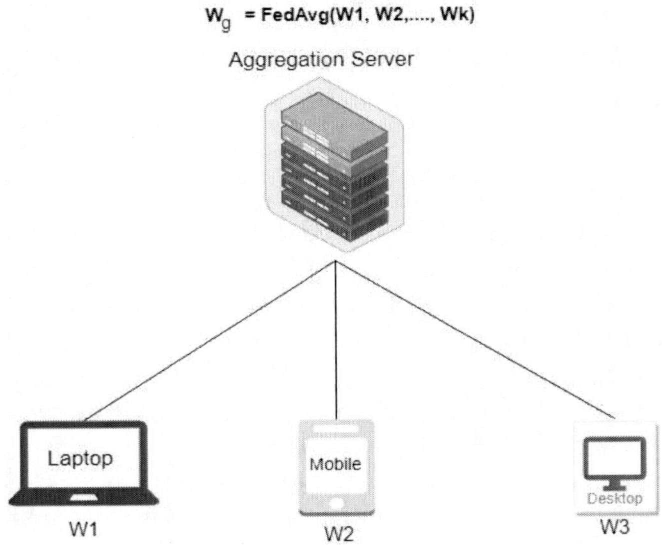

Figure 3.2. FedAvg.

Adversarial Attacks

Traditional machine learning algorithms can be vulnerable to adversarial attacks, where an attacker can manipulate the input data in a way that the model misclassifies the data. For example, an attacker can modify the pixels in an image slightly to make it look different from the original but still be misclassified by the machine learning model.

Data Poisoning

Traditional machine learning algorithms can be susceptible to data poisoning attacks, where an attacker injects malicious or incorrect data into the training dataset to influence the model's behavior. This can result in the model making incorrect predictions or decisions.

Model Inversion

Traditional machine learning models can also be vulnerable to model inversion attacks, where an attacker can use the output of the model to reconstruct sensitive information about the input data. This can be particularly problematic in applications such as credit scoring or medical diagnosis, where the input data is sensitive.

Model Stealing

Attackers can also steal the machine learning model itself, which can be valuable intellectual property. They can do this by reverse-engineering the model's architecture or by exploiting weaknesses in the model's training process.

Lack of Explainability

Traditional machine learning models can sometimes lack transparency and explainability, which can make it difficult to identify and mitigate vulnerabilities or biases in the model.

Motivation for Federated Learning

To eradicate the above-listed issues raised by traditional machine learning that are not suitable for highly scalable modelling training such as smart cities, we utilize federated machine learning. Federated machine learning offers a distributed and secure training environment to train data at local entities. To combat these drawbacks Federated Learning can be used. In Federated Learning the model is trained locally in the user's device. Then the model updates are sent to the aggregation server where a global model is created. The advantage of Federated Learning is that it can allow multiple parties to collaborate on training a machine learning model, without compromising the privacy and security of their data. This can be particularly useful in applications such as healthcare or finance, where sensitive data must be protected. Federated Learning can also be more efficient and scalable than traditional machine learning approaches since the data is distributed across multiple devices or servers, reducing the need to transfer large amounts of data to a central location. In addition, Federated Learning can help to mitigate issues such as bias or overfitting that can arise when training machine learning models on centralized datasets.

Federated Learning (FL) algorithms can overcome some of the security drawbacks of traditional machine learning algorithms.

Data Privacy

FL allows for the training of a machine learning model using data that is distributed across multiple devices without requiring the data to be shared with a third-party entity. This means that sensitive data can be kept private, reducing the risk of data breaches or leaks.

Data Security

FL can also help to enhance data security by ensuring that the data is kept on the local device or server and only model updates are sent to the central server. This can reduce the risk of data theft or data tampering by malicious actors.

Adversarial Attacks

FL can also be used to defend against adversarial attacks, where attackers attempt to manipulate the input data in a way that the model produces incorrect predictions. In FL, each local device or server trains the model on its data and sends only the model updates to the central server. This can make it more difficult for attackers to launch adversarial attacks as they would need to compromise multiple devices or servers simultaneously.

Data Poisoning

FL can also mitigate the risk of data poisoning attacks by utilizing different security protocols on each local node. By implementing varying security measures, the risk of an attacker injecting malicious or incorrect data into the training dataset to influence the model's behavior is reduced significantly.

Explainability

FL can also help to improve the explainability of the machine learning model by allowing for the analysis of the model updates and training data on each device or server. This can help to identify potential biases or errors in the model, as well as improve the transparency of the model's decision-making process.

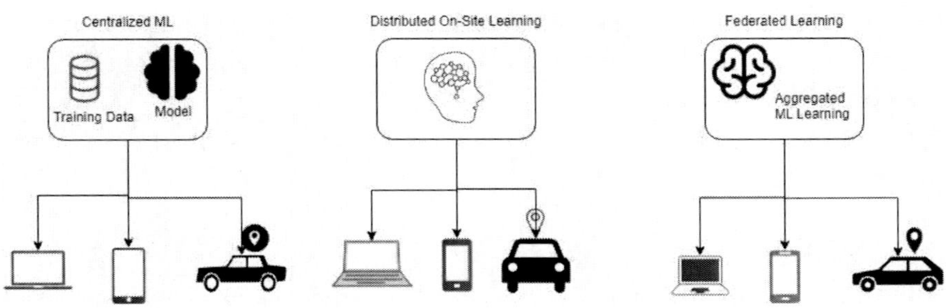

Figure 3.3. Traditional vs Federated Machine Learning.

Figure 3.3 is a diagrammatic representation of the difference between traditional machine learning and federated machine learning.

Literature Survey

Federated Learning has emerged as a promising approach for enabling machine learning on dispersed data sources while maintaining the confidentiality and safety of the nodes that take part in the process. This is especially important in smart cities, which are characterized by the production of massive amounts of sensitive data by a wide variety of IoT devices, such as sensors, cameras, and drones. Federated Learning makes it possible for various devices to work together on training machine learning models without having to share their raw data. As a result, the risk of data breaches is reduced, and the privacy of the persons and organizations involved is preserved. Table 3.1. offers summaries of some relevant papers.

Rasha et al. (2023), offer a comprehensive review of the difficulties associated with maintaining privacy and security when federating learning is used in smart cities. The authors examine a variety of privacy and security problems, including data leakage, model poisoning, and adversarial attacks, and present a set of recommended procedures and potential solutions as a means of mitigating these dangers. In addition to this, they offer an overview of the most recent Federated Learning algorithms and frameworks for smart cities, highlighting both the virtues and limitations of each.

Utomo et al. (2022) present their idea for a federated and trustworthy AI architecture for smart cities. The authors concentrate on the difficulties associated with assuring the trustworthiness of Federated Learning models and present a set of design principles that can be used to construct trustworthy Federated Learning systems. In addition to this, they emphasize the significance of governance and legislation in guaranteeing the ethical application of Federated Learning in smart cities.

Pothole detection is another important application of Federated Learning in smart transportation, as discussed by Alshammari et al. (2022) in their paper. The authors present a Federated Learning-based, three-dimensional pothole detection system that makes use of data collected by numerous sensors and cameras that are mounted on automobiles. They show that their method is effective in detecting potholes and demonstrate how Federated Learning can be used to overcome the difficulties of data heterogeneity and privacy protection in the context of smart transportation. Liu et al. (2020) present a

Federated Learning-based architecture for aerial-ground air quality sensing in smart cities. This framework is described in more detail in the study. The authors offer a Federated Learning strategy to train a machine learning model that predicts the air quality index. The authors deploy swarms of unmanned aerial vehicles (UAVs) to collect data on the air quality in various areas. They show that their strategy is effective in lowering the amount of overhead communication and maintaining the privacy of the participating nodes in the network.

Alsamhi et al. (2019) present an assessment of collaborative smart drones and IoT for improving the smartness of smart cities. The authors examine a variety of uses of drones and the Internet of Things in smart cities, such as traffic control, emergency response, and environmental monitoring. They also highlight the issues of data privacy and security that are present in these applications. In addition to this, they highlight the potential of Federated Learning for enabling collaborative drone and IoT applications and propose a set of research goals for future work.

Rieke et al. (2020) investigate the potential of Federated Learning to shape the future of digital health. The authors highlight the possibility of Federated Learning to enable collaborative machine learning on dispersed health data while maintaining the patients' right to privacy and keeping their data secure. They also highlight the difficulties associated with the heterogeneity of data and the lack of interoperability in digital health, and they suggest several methods to overcome these problems.

A Federated Learning strategy is proposed by Choudhury et al. (2019) as a method for predicting adverse medication reactions based on dispersed health data. The authors present evidence that their method is effective in demonstrating an improvement in the accuracy of the prediction of adverse medication reactions while maintaining the confidentiality of the participating institutions and patients.

The authors of the paper "Privacy-Preserving Traffic Flow Prediction: A Federated Learning Method" Liu et al. (2020) suggest a privacy-preserving Federated Learning approach for traffic flow prediction. In this research, a novel FL technique for model training and aggregation is presented. This approach protects both the privacy and the security of the data. The paper written by Zhang et al. (2021) and titled "FASTGNN: A Topological Information Protected Federated Learning Method for Traffic Speed Forecasting", presents a Federated Learning method that keeps topological information intact and is used for traffic speed forecasting. The authors offer an FL algorithm that aims to protect the privacy of the participating nodes.

Table 3.1. Summary of a few included articles

Reference	Observations	Drawbacks
Jiang et al. (2020)	The paper highlights the challenges and opportunities of using Federated Learning in smart city sensing.	No specific drawbacks were mentioned in the paper.
Li et al. (2020)	The paper provides a comprehensive review of the applications of Federated Learning in various domains.	The paper lacks specific validation of Federate Learning for Smart Cities.
Zheng et al. (2022)	The paper presents recent advances, taxonomy, and open challenges in the application of Federated Learning in smart cities.	The paper lacks a detailed discussion of privacy and security challenges.
Xu et al. (2021)	The paper discusses the application of Federated Learning in healthcare informatics.	The paper lacks a specific focus on smart cities and IoT.
Pandya et al. (2023)	The paper provides a comprehensive survey of the application of Federated Learning in smart cities.	The paper lacks a detailed discussion of the challenges in implementing Federated Learning in smart cities.
Zhang et al. (2022)	The paper discusses the application of Federated Learning in IoT and identifies its challenges and opportunities.	The paper lacks a specific focus on smart cities.
Albaseer et al. (2020)	The paper presents a method for exploiting unlabeled data in smart cities using Federated Learning.	The paper lacks a comprehensive discussion of privacy and security challenges.
Hanjri et al. (2021)	The paper presents a method for water consumption forecasting in smart cities using Federated Learning.	The paper lacks a detailed discussion of the implementation challenges.
Zhou et al. (2021)	The paper discusses the impact of Federated Learning on smart buildings	The paper lacks a specific focus on IoT and smart cities.
Zhao et al. (2020)	The paper provides a comprehensive survey of the applications of Federated Learning in accelerating the Industrial Internet of Things	No specific drawbacks were mentioned in the paper
Jararweh et al. (2020)	The paper discusses the challenges and opportunities of providing trustworthy and sustainable smart city services at the edge.	No specific drawbacks were mentioned in the paper.
Rahman et al. (2022)	The paper proposes a privacy-preserving blockchain-based Federated Learning framework for IoT devices.	The proposed framework may require significant computational resources for blockchain operations.
Boopalan et al. (2022)	The paper provides an overview of the concepts, taxonomies, challenges, and open issues in using Federated Learning-based AI approaches in smart healthcare.	No specific drawbacks were mentioned in the paper.
Ramu et al. (2022)	The paper presents a survey on the fusion of Federated Learning and the Industrial Internet of Things.	No specific drawbacks were mentioned in the paper.

According to the findings presented by Le et al. (2022) Federated Learning has several key applications in smart cities. One of these applications is the prediction of air quality. The authors provide an overview of Federated Learning approaches for air quality index prediction and compare their performance with centralized learning approaches. They also explore the difficulties associated with predicting air quality, such as data heterogeneity and protecting individuals' privacy, and they provide a series of ways to overcome these problems. In their article, Saputra et al. (2019) applied Federated Learning to predict energy demand in electric car networks. The authors used a centralized learning architecture with secure aggregation to address the challenges of data heterogeneity and privacy protection. Zhang et al. (2023) developed a privacy-protecting hybrid Federated Learning architecture for the identification of financial crime. The authors struck a balance between the competing values of data privacy and model accuracy by employing a hybrid approach that combined centralized and Federated Learning.

Rahman et al. (2020) studied the use of Federated Learning as an intrusion detection method in IoT networks. The authors examined the efficacy of three distinct methods, namely centralized learning, on-device learning, and Federated Learning, and compared their results regarding accuracy and privacy.

In general, the studies that were examined for this literature review illustrate the promise of Federated Learning for making smart cities that use IoT more secure while also protecting residents' privacy. They demonstrate how Federated Learning may be used to meet the difficulties of data heterogeneity, privacy protection, and security in a variety of smart city applications. Some examples of these applications include predicting air quality, traffic flow, energy demand, and financial crime detection.

Future Scope and Challenges

Communication Issues in FL-Based Environment

Communication, which is important for FL users, and embedded servers are critical in FL-supported healthcare services. Effective communication resource allocation increases learning results dramatically. This is especially significant if many IoT devices must be deployed on the aggregation server for prolonged link-template upgrades and low-link sample deployments. In this instance, the aggregation server can choose a suitable collection of IoT devices based on an effective planning policy, as documented in several previous research done by Xu et al. (2021), Xia et al. (2020), Luo et al. (2020), Yang et al. (2019). Another major connection issue is the dynamic and rapid changing of the radio channel between the IoT device and the aggregation server, which impacts the reliability and quality of training updates. One possible approach is to address the impact of user disruption, as presented by Bonawitz et al. (2016) and So et al. (2021) as well as more dependable design goals such as device downtime and availability.

Specifications for Deployment

Although the literature has numerous promising health outcomes supported by FL, there are no standard and universal results for comparing the efficacy of different methods to the same condition. Various blockchain frameworks have been proposed for FL systems, such as removing the need for a central server and handling IoT device update locally. Nevertheless, comparing these techniques is challenging since they are advised for various (status) circumstances and utilize different network parameters/datasets to evaluate performance. There are additional critical concerns to consider about the universality and standardization of communication protocols, device hardware, deployment situations, and integration approaches. Recently, IEEE Std 3652.1-2020 gave guidelines on the architecture and design views of FL as given by Qiang et al. (2021), as well as the FL rating table and performance data system.

Quality of Federated Training Data

Variations in computer power and data quality between hospitals can significantly impact educational quality. In this scenario, one viable option is to provide incentives for hospitals and medical institutions to use high-quality data for teaching and to submit trustworthy changes to a single server. The blockchain and game theory are two essential instruments for building stimulation mechanisms as displayed by Xu et al. (2021), Kang et al. (2019) and Sarikaya et al. (2019). By flexibly customizing learning demands, the FL format should be easily adaptable and adaptive for nurses, physicians, and patients (e.g., changing data types, changing learning rates, changing, or regressing the classification of learning objectives). These modifications may have an impact on FL models and default learning patterns (in FL users and embedded servers), necessitating the creation of an adaptable FL methodology. AI is a potential approach since it can increase the flexibility of FL models to future occurrences by using previous data. As a result, DRL employs a classic way to generate an incentive mechanism in the event of poor performance when the opportunity is significant enough (Zhao et al. 2021).

Confidential Dataset Concerns

Different customers may have distinct data types, such as text, photos, sound, and time series, and varied data content, such as blood type, heart rate, facial image, and body temperature, in some treatment situations. Almost all FL approaches in the literature, as is customary, are based on a single data set with a restricted number of characteristics. A novel heterogeneous FL strategy should be devised, with a central server exploring this heterogeneity via the unique construction of the entire paper (Choquette-Choo et al. 2021). Participants in this activity must give a final approach that allows them to employ a diverse collection of models without having to integrate data in one spot.

FL Integrated Advanced Networks

Although 5G networks have not yet reached full availability or commercial deployment, they do exist. Numerous research and development efforts are in progress for emerging 6G wireless technologies (De Alwis et al. 2021). With 6G and Industry 5.0, a variety of applications become possible, such as intelligent healthcare, smart grids, holographic television, and personalized body area networks. Furthermore, several cutting-edge technologies, including blockchain, compression sensors, terahertz and visible light communication, 3D networking, quantum communication, and large-scale intelligent surfaces, are being introduced to meet the stringent requirements of 6G.

Questions arise regarding how Federated Learning capabilities will be incorporated into future 5G/6G medical devices and how 6G devices, like smart implants and wearable devices, will be utilized in various FL-based healthcare and novel health services that 6G will enable. For instance, future e-health services will be enhanced by AI and Federated Learning capabilities to improve quality of life and reduce hospital admissions (Mucchi et al. 2020). These questions provide an exciting avenue for further research.

FL's Provable Privacy

Although FL offers great promise to secure users' data privacy, the high sensitivity of health data raises a variety of privacy concerns that must be addressed carefully, particularly in the context of smart healthcare. Federated Learning privacy problems, according to Mothukuri et al. (2021), may be classed as the member's last attack, inadvertent information leaking, and generative antagonism. An attacker, for example, may utilize the global FL template to see if a sample of data exists in the FL health dataset. When a patient's gadget communicates local model updates to central servers in hospitals and healthcare institutions, patient information is also acquired. There is also a sophisticated encryption technology being developed for securing personal information for medical Care systems and ensuring differential confidentiality. Additional research into the use of differential confidence to increase the secrecy of the FL system is being considered (Wu et al. 2021) (Kerkouche et al. 2021).

Cyber Defence Aspects

In an FL-based smart care system, multiple client participants can behave as attackers, sending toxic model updates or disinformation to undermine model integration. When learning local data, an attacker might infect data attribute information or modify local updates while moving templates between local clients and central servers. Remote attackers can utilize their assaults on the server to obtain information about the accumulated global pattern, resulting in serious privacy problems such as information breaches. Fixing these security flaws is a significant concern for smart FL-based healthcare systems. Various options, such as the use of differential confidentiality, should be investigated to safeguard the training dataset against data corruption (Hu et al. 2020). Furthermore, by encrypting local updates and sharing keys between the client and a central server, the creation of a secure integration technique is a viable option to create a double-mask structure to protect clients from data fabrication and assaults (Fereidooni et al. 2021).

IIDness Issue of Data

Inconsistencies in medical data sets that might lead to learning FL in training are major challenges that must be addressed to obtain the necessary learning performance in FL-based medical systems. A hospital, for example, may have a higher frequency of a certain type of regional disease than another hospital in a different geographical location. Participation in federal data education is challenging in this scenario since label distribution differs by the medical institution. If this does not resolve the inconsistency problem, it may worsen or possibly depart from learning the data. As a result, solutions to non-IID challenges, such as producing new subsets of data sets for fair distribution to consumers, should be created to successfully shape data in FL-based smart healthcare. Another intriguing technique is to execute the mobility of characteristics amongst diverse consumers by modifying the distribution of characteristics on the customer side before averaging local models. To examine non-IID data in the FL-based smart care sector, quantitative measures such as standard deviation, a curvature of label/function distribution, and accuracy and accuracy for homogenous sections are required (Li et al. 2021).

Conclusion

Federated Learning is a rapidly growing area of research, and its combination with the medical Internet of Things could significantly transform the safety of healthcare data. The proposed book chapter aims to assess whether Federated Learning is a highly reliable topic in terms of enhancing the confidentiality and security of data for intelligent urban environments and if it can contribute to developing a more scalable model for handling sensitive information. The proposed book chapter explores some of the most crucial real-world implementations that have been examined and simulated for research purposes. In concluding the book chapter, vital discussions regarding the challenges faced by Federated Learning frameworks, as well as the prospects of Federated Learning in healthcare settings, are presented. In summary, this topic possesses substantial research potential, with the capability to develop advanced AI methodologies in the future

References

Albaseer, Abdullatif, Bekir Sait Ciftler, Mohamed Abdallah, and Ala Al-Fuqaha. "Exploiting unlabeled data in smart cities using federated edge learning." In *2020 International Wireless Communications and Mobile Computing (IWCMC)*, pp. 1666-1671. IEEE, 2020.

Alsamhi, Saeed H., Ou Ma, Mohammad Samar Ansari, and Faris A. Almalki. "Survey on collaborative smart drones and internet of things for improving smartness of smart cities." *Ieee Access* 7 (2019): 128125-128152.

Alshammari, Sami, and Sejun Song. "3Pod: Federated Learning-based 3 Dimensional Pothole Detection for Smart Transportation." In *2022 IEEE International Smart Cities Conference (ISC2)*, pp. 1-7. IEEE, 2022.

Bonawitz, Keith, Vladimir Ivanov, Ben Kreuter, Antonio Marcedone, H. Brendan McMahan, Sarvar Patel, Daniel Ramage, Aaron Segal, and Karn Seth. *"Practical secure aggregation for Federated Learning on user-held data."* arXiv preprint arXiv:1611.04482 (2016).

Boopalan, Parimala, Swarna Priya Ramu, Quoc-Viet Pham, Kapal Dev, Praveen Kumar Reddy Maddikunta, Thippa Reddy Gadekallu, and Thien Huynh-The. "Fusion of Federated Learning and industrial Internet of Things: A survey." *Computer Networks* (2022): 109048.

Chen, Jianmin, Xinghao Pan, Rajat Monga, Samy Bengio, and Rafal Jozefowicz. *"Revisiting distributed synchronous SGD."* arXiv preprint arXiv:1604.00981 (2016).

Choquette-Choo, Christopher A., Natalie Dullerud, Adam Dziedzic, Yunxiang Zhang, Somesh Jha, Nicolas Papernot, and Xiao Wang. *"CaPC learning: Confidential and private collaborative learning."* arXiv preprint arXiv:2102.05188 (2021).

Choudhury, Olivia, Yoonyoung Park, Theodoros Salonidis, Aris Gkoulalas-Divanis, and Issa Sylla. "Predicting adverse drug reactions on distributed health data using Federated Learning." In *AMIA Annual Symposium proceedings*, vol. 2019, p. 313. American Medical Informatics Association, 2019.

De Alwis, Chamitha, Anshuman Kalla, Quoc-Viet Pham, Pardeep Kumar, Kapal Dev, Won-Joo Hwang, and Madhusanka Liyanage. "Survey on 6G frontiers: Trends, applications, requirements, technologies and future research." *IEEE Open Journal of the Communications Society* 2 (2021): 836-886.

Fereidooni, Hossein, Samuel Marchal, Markus Miettinen, Azalia Mirhoseini, Helen Möllering, Thien Duc Nguyen, Phillip Rieger et al. "SAFELearn: Secure aggregation for private Federated Learning." In *2021 IEEE Security and Privacy Workshops* (SPW), pp. 56-62. IEEE, 2021.

Hanjri, Mohammed El, Hibatallah Kabbaj, Abdellatif Kobbane, and Amine Abouaomar. "Federated Learning for Wa Mitra, Angan, Yanik Ngoko, and Denis Trystram. "Impact of Federated Learning on smart buildings." In *2021 International Conference on Artificial Intelligence and Smart Systems (ICAIS)*, pp. 93-99. IEEE, 2021.ter Consumption Forecasting in Smart Cities." *arXiv preprint arXiv: 2301.13036* (2023).

Hu, Rui, Yuanxiong Guo, Hongning Li, Qingqi Pei, and Yanmin Gong. "Personalized Federated Learning with differential privacy." *IEEE Internet of Things Journal* 7, no. 10 (2020): 9530-9539.

Jararweh, Yaser, Safa Otoum, and Ismaeel Al Ridhawi. "Trustworthy and sustainable smart city services at the edge." *Sustainable Cities and Society* 62 (2020): 102394.

Jiang, Ji Chu, Burak Kantarci, Sema Oktug, and Tolga Soyata. "Federated Learning in smart city sensing: Challenges and opportunities." *Sensors* 20, no. 21 (2020): 6230.

Kang, Jiawen, Zehui Xiong, Dusit Niyato, Shengli Xie, and Junshan Zhang. "Incentive mechanism for reliable Federated Learning: A joint optimization approach to combining reputation and contract theory." I*EEE Internet of Things Journal* 6, no. 6 (2019): 10700-10714.

Kerkouche, Raouf, Gergely Acs, Claude Castelluccia, and Pierre Genevès. "Privacy-preserving and bandwidth-efficient Federated Learning: an application to in-hospital mortality prediction." In *Proceedings of the Conference on Health, Inference, and Learning,* pp. 25-35. 2021.

Le, Duy-Dong, Anh-Khoa Tran, Minh-Son Dao, Mohamed Saleem Haja Nazmudeen, Viet-Tiep Mai, and Nhat-Ha Su. "Federated Learning for Air Quality Index Prediction: An Overview." In *2022 14th International Conference on Knowledge and Systems Engineering (KSE)*, pp. 1-8. IEEE, 2022.

Li, Li, Yuxi Fan, Mike Tse, and Kuo-Yi Lin. "A review of applications in Federated Learning." *Computers & Industrial Engineering* 149 (2020): 106854.

Li, Qinbin, Yiqun Diao, Quan Chen, and Bingsheng He. *"Federated Learning on non-iid data silos: An experimental study."* arXiv preprint arXiv:2102.02079 (2021).

Liu, Yi, J. Q. James, Jiawen Kang, Dusit Niyato, and Shuyu Zhang. "Privacy-preserving traffic flow prediction: A Federated Learning approach." *IEEE Internet of Things Journal* 7, no. 8 (2020): 7751-7763.

Liu, Yi, Jiangtian Nie, Xuandi Li, Syed Hassan Ahmed, Wei Yang Bryan Lim, and Chunyan Miao. "Federated Learning in the sky: Aerial-ground air quality sensing framework with UAV swarms." *IEEE Internet of Things Journal* 8, no. 12 (2020): 9827-9837.

Luo, Siqi, Xu Chen, Qiong Wu, Zhi Zhou, and Shuai Yu. "HFEL: Joint edge association and resource allocation for cost-efficient hierarchical federated edge learning." *IEEE Transactions on Wireless Communications* 19, no. 10 (2020): 6535-6548.

McMahan, Brendan, Eider Moore, Daniel Ramage, Seth Hampson, and Blaise Aguera y Arcas. "Communication-efficient learning of deep networks from decentralized data." In *Artificial intelligence and statistics,* pp. 1273-1282. PMLR, 2017.

Mothukuri, Viraaji, Reza M. Parizi, Seyedamin Pouriyeh, Yan Huang, Ali Dehghantanha, and Gautam Srivastava. "A survey on security and privacy of Federated Learning." *Future Generation Computer Systems* 115 (2021): 619-640.

Mucchi, Lorenzo, Sara Jayousi, Stefano Caputo, Elisabetta Paoletti, Paolo Zoppi, Simona Geli, and Pietro Dioniso. "How 6G technology can change the future wireless healthcare." In *2020 2nd 6G wireless summit* (6G SUMMIT), pp. 1-6. IEEE, 2020.

Pandya, Sharnil, Gautam Srivastava, Rutvij Jhaveri, M. Rajasekhara Babu, Sweta Bhattacharya, Praveen Kumar Reddy Maddikunta, Spyridon Mastorakis, Md Jalil Piran, and Thippa Reddy Gadekallu. "Federated Learning for smart cities: A comprehensive survey." *Sustainable Energy Technologies and Assessments* 55 (2023): 102987.

Qiang, Fan, Tong Lixin, and Lv Richard. "White paper-IEEE federated machine learning." (2021): 1-18.

Rahman, Anichur, Md Sazzad Hossain, Ghulam Muhammad, Dipanjali Kundu, Tanoy Debnath, Muaz Rahman, Md Saikat Islam Khan, Prayag Tiwari, and Shahab S. Band. "Federated Learning-based AI approaches in smart healthcare: concepts, taxonomies, challenges and open issues." *Cluster Computing* (2022): 1-41.

Rahman, Sawsan Abdul, Hanine Tout, Chamseddine Talhi, and Azzam Mourad. "Internet of things intrusion detection: Centralized, on-device, or Federated Learning?." *IEEE Network* 34, no. 6 (2020): 310-317.

Ramu, Swarna Priya, Parimala Boopalan, Quoc-Viet Pham, Praveen Kumar Reddy Maddikunta, Thien Huynh-The, Mamoun Alazab, Thanh Thi Nguyen, and Thippa Reddy Gadekallu. "Federated Learning enabled digital twins for smart cities: Concepts, recent advances, and future directions." *Sustainable Cities and Society* 79 (2022): 103663.

Rasha, Al-Huthaifi, Tianrui Li, Wei Huang, Jin Gu, and Chongshou Li. "Federated Learning in Smart Cities: Privacy and Security Survey." *Information Sciences* (2023).

Rieke, Nicola, Jonny Hancox, Wenqi Li, Fausto Milletari, Holger R. Roth, Shadi Albarqouni, Spyridon Bakas et al. "The future of digital health with Federated Learning." *NPJ digital medicine* 3, no. 1 (2020): 119.

Saputra, Yuris Mulya, Dinh Thai Hoang, Diep N. Nguyen, Eryk Dutkiewicz, Markus Dominik Mueck, and Srikathyayani Srikanteswara. "Energy demand prediction with Federated Learning for electric vehicle networks." In *2019 IEEE global communications conference (GLOBECOM)*, pp. 1-6. IEEE, 2019.

Sarikaya, Yunus, and Ozgur Ercetin. "Motivating workers in Federated Learning: A stackelberg game perspective." IEEE Networking Letters 2, no. 1 (2019): 23-27.

So, Jinhyun, Başak Güler, and A. Salman Avestimehr. "Turbo-aggregate: Breaking the quadratic aggregation barrier in secure Federated Learning." *IEEE Journal on Selected Areas in Information Theory* 2, no. 1 (2021): 479-489.

Utomo, Sapdo, A. John, Adarsh Rouniyar, Hsiu-Chun Hsu, and Pao-Ann Hsiung. "Federated Trustworthy AI Architecture for Smart Cities." In *2022 IEEE International Smart Cities Conference (ISC2)*, pp. 1-7. IEEE, 2022.

Wu, Maoqiang, Dongdong Ye, Jiahao Ding, Yuanxiong Guo, Rong Yu, and Miao Pan. "Incentivizing Differentially Private Federated Learning: A Multidimensional Contract Approach." *IEEE Internet of Things Journal* 8, no. 13 (2021): 10639-10651.

Xia, Wenchao, Tony QS Quek, Kun Guo, Wanli Wen, Howard H. Yang, and Hongbo Zhu. "Multiarmed bandit-based client scheduling for Federated Learning." *IEEE Transactions on Wireless Communications* 19, no. 11 (2020): 7108-7123.

Xu, Bo, Wenchao Xia, Jun Zhang, Tony QS Quek, and Hongbo Zhu. "Online client scheduling for fast Federated Learning." *IEEE Wireless Communications Letters* 10, no. 7 (2021): 1434-1438.

Xu, Jie, Benjamin S. Glicksberg, Chang Su, Peter Walker, Jiang Bian, and Fei Wang. "Federated Learning for healthcare informatics." *Journal of Healthcare Informatics Research* 5 (2021): 1-19.

Xu, Jie, Heqiang Wang, and Lixing Chen. "Bandwidth allocation for multiple Federated Learning services in wireless edge networks." *IEEE Transactions on Wireless Communications* (2021).

Yang, Howard H., Zuozhu Liu, Tony QS Quek, and H. Vincent Poor. "Scheduling policies for Federated Learning in wireless networks." *IEEE transactions on communications* 68, no. 1 (2019): 317-333.

Zhang, Chenhan, Shuyu Zhang, J. Q. James, and Shui Yu. "FASTGNN: A topological information protected Federated Learning approach for traffic speed forecasting." *IEEE Transactions on Industrial Informatics* 17, no. 12 (2021): 8464-8474.

Zhang, Haobo, Junyuan Hong, Fan Dong, Steve Drew, Liangjie Xue, and Jiayu Zhou. "A Privacy Preserving Hybrid Federated Learning Framework for Financial Crime Detection." *arXiv preprint arXiv:2302.03654* (2023).

Zhang, Tuo, Lei Gao, Chaoyang He, Mi Zhang, Bhaskar Krishnamachari, and A. Salman Avestimehr. "Federated Learning for the internet of things: applications, challenges, and opportunities." *IEEE Internet of Things Magazine* 5, no. 1 (2022): 24-29.

Zhao, Jie, Xinghua Zhu, Jianzong Wang, and Jing Xiao. "Efficient client contribution evaluation for horizontal Federated Learning." In *ICASSP 2021-2021 IEEE International Conference on Acoustics, Speech and Signal Processing* (ICASSP), pp. 3060-3064. IEEE, 2021.

Zhao, Yang, Jun Zhao, Linshan Jiang, Rui Tan, Dusit Niyato, Zengxiang Li, Lingjuan Lyu, and Yingbo Liu. "Privacy-preserving blockchain-based Federated Learning for IoT devices." *IEEE Internet of Things Journal* 8, no. 3 (2020): 1817-1829.

Zheng, Zhaohua, Yize Zhou, Yilong Sun, Zhang Wang, Boyi Liu, and Keqiu Li. "Applications of Federated Learning in smart cities: recent advances, taxonomy, and open challenges." *Connection Science* 34, no. 1 (2022): 1-28.

Zhou, Jiehan, Shouhua Zhang, Qinghua Lu, Wenbin Dai, Min Chen, Xin Liu, Susanna Pirttikangas, Yang Shi, Weishan Zhang, and Enrique Herrera-Viedma. "A survey on Federated Learning and its applications for accelerating industrial internet of things." *arXiv preprint arXiv:2104.10501* (2021).

Chapter 4

Internet of Things for Smart Cities in E-Waste Management: Overview and Challenges

Ashwini, A.[1,*]
and Kavitha, V.[2]

[1]Department of ECE, Vel Tech Rangarajan Dr. Sagunthala R&D Institute of Science and Technology, Chennai. Tamilnadu, India
[2]Department of CSE, University College of Engineering, Kancheepuram, Tamilnadu, India

Abstract

The term "smart" has become more significant as a result of technical advancements in digitalization. Due to its tremendous advancements in innovation and favorable effects on the general public, Internet of Things (IoT) is actually regarded as the primary drivers involving drastic change. IoT technologies' primary objectives are to boost technological effectiveness, streamline procedures across numerous industries, and ultimately raise living standards. But this rapid growth needs to be closely watched and evaluated from an environmental viewpoint in order to avoid negative effects and maintain proper utilization of limited resources globally. The primary goal of this review chapter is to aid the bibliophile in comprehending the different aspects of communities that are smart and reasons to become smart. By examining current articles that were taken from the different database, it was possible to determine the peculiar nature of IoT study. Developing methods for the active management of e-waste streams involving recycling, reuse, and recovery is the primary goal of this chapter. Since disposing of household electronic waste is not beneficial, extensive research is being done in this

* Corresponding Author's Email: a.aswiniur@gmail.com.

In: Blockchain Enabled Secure Big Data Computing …
Editors: L. Ashok Kumar, D. Karthika Renuka, V. Chandrasekar et al.
ISBN: 979-8-89113-243-6

area. To assist people in managing their waste and delivering it to the nearby e-waste collection facilities, many mobile apps are being developed. The majority of communities that are transitioning focus on sustainability. Some of the most crucial IoT applications for smart towns include smart infrastructures, autonomous parking, medical systems, smart power systems, and transit. An indepth evaluation depicts that the internet is used in concurrence with many novel technologies, including wireless systems, artificial intelligence, the block-chain techniques, computation with cloud-based concepts, and analysis on big data. With the increase in consumption rates and shortage in mineral wealth, e-waste generation is considered as the most common type of solid waste. More thorough study has to be conducted in these platforms to create novel, important and favorable conditions for community in waste management to create sustainable cities. It maps global studies within a smart structure to give academics and professionals a broad perspective. At the absolute, smart cities have been viewed as a significant step in this path. IoT technologies should therefore be significant in the future, with their unique utility and importance.

Keywords: analysis on big data, artificial intelligence, e-waste management, Iot, smart cities

Introduction

Technologies built on the Internet of Things (IoT) offer completely different perspectives on how various industries, such as engineering, medicine, agriculture and uncharted regions, change with time (Alulema et al., 2018). This comes into activation by the pervasive improvements in communications network accessibility, software as well as hardware platform development, and data analysis tool development (Maranda et al., 2019). 43.7 million tonnes of e-waste are projected to be produced worldwide in 2016 (Turaga et al., 2019). The idea of managing e-waste in smart cities was initially put forth as a possible remedy for problems caused by development (Ahmed et al., 2020). The idea of a "smart city" refers to towns that were "smart" ten years ago, making it similar to a moving target. It entails utilizing a variety of contemporary technologies to enhance metropolitan living standards while using fewer assets and doing so more economically. The majority of IoT initiatives currently underway involve connected buildings and vehicles, the chemicals sector, and industrial IoT (Gharaibeh et al., 2017).

The ICT convergence of IoT has been predicted to deliver distinctive characteristics in smart cities with the least amount of public participation. There are a lot of contemporary towns with top-notch economies, transportation systems, governments, and environmental aspects. Only a few instances of typical smart solutions include the production of water, electrical grids, businesses, traffic management, consumption and transportation (Martin et al., 2020). By incorporating IoT into these distinctive characteristics, they can advance their development and improve the outcomes of their operations. In the EU and US, fewer projects are finished in the connected vehicle, connected agriculture, and IoT industries (Statista, 2020).

According to IoT Forecasts, there will be over 50 billion IoT devices worldwide by 2050, creating enormous, interconnected gadgets of networks that includes all models from smartphones to household equipment. The primary problems include numerous tools for improvement in network monitoring operations (Kakkavas et al., 2020), the security-based administration, software bugs (Mauro et al., 2020), the consumption of time to upkeep IoT based networking system, and ultimately subjects with IoT. Additionally, effective e-waste management practices were not adopted on Asia (Almusaylim et al., 2020) particularly about managing household e-waste, which forms the reason contributing for successful recycling operations (Ilankoon et al., 2018 and Yong et al., 2019). The concepts of smart cities have already been adopted by a large number of important cities around the world, including Toronto, London, New York, Paris, Seoul and Shanghai. The idea of sustainability is at the heart of most transitioning regions.

This indicates that the requirement for secure e-waste disposal is connected to internet connectivity and that the IoT applications subject that received focus in the field of managing this waste stream for solids. Two of such biggest obstacles are the high lead level of e-waste and the severe environmental harm. Recycling rates increase globally on average by 5% to 6% yearly. Each person produces more than 6 kg of electronic debris yearly (Song et al. 2015). To ensure equitably and sustainably growing IoT, more planned initiatives in the e-waste issue are undeniably preferable. The reality that more than 55% of the worldwide people lack appropriate electronic debris constitutes one of the main issues. It is necessary to more effectively utilize potential in order to guarantee a steady supply of essential basic materials (Krishnaswamy et al., 2019). It has been determined that e-waste collection techniques from households must constantly be improved in order to control this generation (Kang et al., 2020). Finding the material flows and generation of domestic e-waste has proven to be challenging.

Although e-waste management systems have not gained importance in previous research studies, it is anticipated that they will eventually make it simpler to gather domestic e-waste. The study's goal is to put in place e-waste management methods that are appropriate for its effective management. It is used in conventional solid waste management. Overall, smart cities have gained a significant development, and there have been made a substantial global effort in this direction. IoT integration for managing e-waste in future cities, data with multiple applications are produced for a variety of technologies.

Literature Review

If a developed metro area makes use of information and communication technology (ICT) then vast population density is produced to connect and view crucial services and infrastructure, it is referred to as a "smart city" (Mohammed et al., 2018). By 2050, 69% of the world's people, according to the UN, will reside in urban areas. To track disposal details and estimate the quantity of e-waste present, sensors were added to an apartment automation e-waste collection box. As a result of this effort, the general public is anticipated to utilize this mobile application to dispose of their domestic electronic materials (Khan et al., 2022 and Asensio et al, 2015).

K-means clustering, and a simple linear regression are used as correlation strategies. According to this study, there is a connection between the Smart City Index (SCI) ranking, the amount of electronic waste produced (Amiruddin et al., 2019 and Denovais et al., 2021), and the general economic well-being of the smart city community (Tiwari et al., 2019). Smart cities driven by IoT facilitate improved oversight of a range of services, including parking, healthcare, and transportation with security cameras (Osterrieder et al., 2020).

Additionally, the administration of electricity, intelligent energy management in structures, agriculture and transportation (Mekki et al., 2019 and Janssen et al., 2019). The diversity and dispersed nature of IoT technologies bring connection issues, ambiguity regulations, sociological and ethical issues, public approval lacking, and government strategy. A system that uses the cloud is advised for data tracking on electrical goods sold, especially for offline and online purchases (Das et al., 2020).

IoT technology could be used in conjunction with intelligent appliances and a sophisticated management system to reduce e-waste. Although the IoT has produced a substantial body of literature in the smart city context (Luca et al., 2019; Deakin et al., 2018; Fahmidehgholami et al., 2020 and Mekkaoui et al., 2017), only a few investigations have attempted to use bibliometric tools to review current research. WSN-based urban smart city apps are represented with a focus on network performance evaluation using advanced intelligent computational methods (Aiyappan et al., 2023). It is suggested to use a smart trash that manages projects in waste disposal. The system uses devices for trash mass measurement in the bin. By adjusting to the network environment, the system can handle all data related to waste management (Wijaya et al., 2017).

The authors concluded that IoT is a key driver for the advancement of cities after reviewing important, insightful, and informative writings. In order to automate all phases of trash management, (Aithal, 2021) documented the use of IoT sensors based on the cloud trash surveillance system with machine learning.

Zhao et al., (2019) proposed a comprehensive analysis of knowledge space representation methods are used in smart city investigation. The trash container will have more capacity, and alert systems will be used to rapidly empty it. Municipal trash cans are overflowing and not being cleansed on a regular basis. It has a buildup of garbage that is bad for the environment, spreads disease, deteriorates the area, and makes the locals unhygienic (Zavare et al., 2017). Fundamentally, IoT is a key technology that enables sustainable development of urban areas, according to the research. By integrating blockchain technology into an IoT system, this study aims to address the complex issues surrounding management, exchange of data, policy forging, automation, tracking authentication guidelines, and network administrators of IoT nodes (Porru et al., 2020).

The blockchain relies on an IoT system for documenting all business processes and actions taken on electrical equipment during post-production. Five intelligent contracts that monitor user actions on the distributed, immutable database make up the system's core. This makes use of an integrated distributed storage device. The structure is investigated to ascertain smart contracts' functionalities that the gas consumed using Ethereum blockchain. The IoT-smart cities was reviewed by Bellini et al., (2022), noted the important trends and open questions. IoT technologies were used to create effective and sustainable smart cities.

Su et al., (2018) investigated how the absence of 5G networks would prevent the creation of sustainable smart communities. The literature on edge computing and the Web Issues has been thoroughly analyzed in a number of application areas, including smart homes and towns, by Wang et al., (2021). Vishnu et al., (2021) provides information on the scattered and unorganized IoT applications in smart communities that can be organized through IoT research. This research review is specifically interested in assessing the development of IoT study using pertinent articles from Scopus database that were collected between 2015 and 2022. The primary goal of this chapter is to thoroughly explain to bibliophiles the various aspects of smart cities and what constitutes them as truly intelligent. In this volume, we'll look at characteristics of smart cities. The descriptive methods of research come in second. Thirdly, we discuss the significant challenges that smart communities must overcome. After highlighting the importance of the study's results, the chapter is resolved with its limitations.

Characteristics of IoT

This section outlines the methodology that compares smart community ranking in order to pinpoint their comparative advantages. There are various variables that contribute to each attribute as shown in Figure 4.1. The following are traits of networks for smart cities.

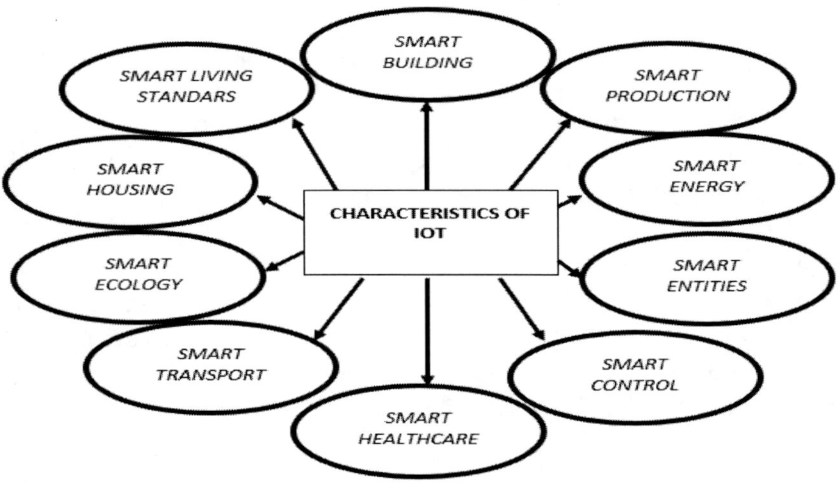

Figure 4.1. IoT Characteristics.

Smart Building

The initial step in developing secure settings is implementing risk management. In order to accomplish this, a smart building must be able to recognize, identify, block access, combat, and recognize entrance at every level. Other protection precautions have to be made to bring an effective measure (Brisson et al., 2019 and Abubaker et al., 2021). Workspaces that support BYOD (Bring Your Own Device) exchanges have to be strengthened at smart structures. They should overcome such constraints of outdated methods.

Smart Production

The wealth and resources of a community affects the economy with respect to production for goods and service consumption. The aim of smart economies is to enhance infrastructure and service delivery by utilizing information, technology, and statistics. Access to utilities also includes healthcare, cheap housing, power, and IT communication. Entrepreneurship, invention, patents, a flexible labor market, productivity, worldwide integration of markets, etc. are all components of a smart economy (Kumar and Bharat, 2017 and Beneicke et al., 2019). The environment must readily adjust to the requirements of the people and support their more responsible resource use. By 2030, it is anticipated that the worldwide cost of new urban infrastructure will be $40 trillion, creating enormous opportunities for innovation (Lombardi et al., 2021).

Smart Energy

Smart energy significantly reduces running costs for both government institutions. Thanks to smart energy sources, cities can understand their profile of energy consumption. Energy use has become more democratic as a result of smart energy (Thornbush et al., 2021). Access to electricity is now possible without any logistical or technical barriers. When energy is spread, it is more resilient. Smart energy laws, which also protect vital supplies, oversee efficiently regulating the development of renewable sources.

Smart Entities

The prime element of a smart city is intelligent individuals. People with intelligence construct cities. Numerous factors, such as inventiveness, schooling, participation, and creativity, can be used to gauge someone's intelligence. The human factor is another consideration in the use of local information. Innovation with creative thinking go hand in hand, and many of these smart communities have promoted private enterprise by providing a safety net for all of these failure cases (Gupta et al., 2017). It also pertains to the initiative support in smart cities. Public participation is the primary element of success for a city. There will be a failure of smart cities if locals are not encouraged to co-create and share their ideas.

Smart Control

E-government tools usage and initiatives have increased citizen involvement in smart communities. They are unique as every smart community contributes to different objectives. Smart cities frequently have their own form of government as a consequence (Mutiara et al., 2018). This enhances the transparency, accountability, and accessibility of governmental services. A few of the many elements of smart governance include choice-based participation, communities and the public services and policy instances, etc. Residents of smart cities frequently interact with the government and offer input through a variety of online platforms (Lopes, 2017).

Smart Transport

Smart mobility has an influence on how much time and money city dwellers spend on their values. All of these concepts are presently being investigated, put to the test, or are just starting to be implemented (Orlowski et al., 2019 and Yong et al., 2019). In many smart cities, real-time data for transportation networks and smart cards are important themes. Smart mobility encompasses the newest commuter look trends. Other manifestations of intelligent movement include technology that links people to transportation and trip booking software that takes digital currencies (Davor et al., 2018).

Smart Ecology

One of the topics that is commonly discussed in smart cities is green communication. Urban regions need to minimize their carbon footprint. There are initiatives to support improved management and more environmentally sound electric vehicles. Additionally, intelligent structures can enhance both background and its attractiveness (Chen et al., 2023). The "smart ecology" is made up of a variety of components, such as resource management, municipal infrastructure, water and use of energy with carbon footprint reduction, and environmental protection.

Smart Living Standards

Smart cities frequently use smart technology to handle urbanization problems inorder to raise living standards. A smart city employs techniques to lessen waste created during building demolition, construction, and also restoration. It also uses waste to produce energy, make fertilizer, and manage limited water resources more responsibly (Dvab et al., 2023).

Smart Housing

Residents have a big influence on how a community develops. Addressing housing conditions, illnesses, personal safety, possibilities for learning, tourist attractions, and a sense of community are essential to improve citizens' quality of life (Anagnostopoulos and heodoros, 2023). The cultural agenda, tourist guides, education, and well-being can all be advanced and supported by these variables.

Monitoring Healthcare

Smart cities are able to gather data and monitor wellbeing by bringing healthcare inside the home. Due to the seamless integration of this data into electronic health records, doctors can virtually monitor patients in real-time. Patients in smart cities are increasingly likely to consult with physicians while unwinding in their living rooms, thanks to telemedicine (Chauhan and Meetali,

2023). Artificial intelligence (AI) in combination with processing of natural languages offers clinical decision support. AI might one day push boundaries which are presently unimaginable.

Research System

The literature for IoT smart cities is evaluated through bibliometric analysis. The evaluation presented in this review are particularly dependent on the stages listed below:

1. Basic evaluation of data concepts
2. Finding keyword with database establishment
3. Data through evaluation
4. Keyword framed network

Examining the general patterns that emerge from the grouping of keyword co-occurrences takes place in the first three steps. BibExcel was used in this study because it is incredibly flexible when editing and adjusting data obtained from scientific datasets that are compatible with analysis programs like VOS viewer, Gephi and SciNet. The VOS viewer software assisted in improving the network presentation and visualization. The four-step methodical approach is fully explained in the following subsections.

Basic Evaluation of Data Concepts

All bibliographic records are analyzed along with the metadata that the authors have provided, including the titles, keywords, documents, and authors. 1,070 works overall, with 6,283 authors contributing, have a cooperation index of 5.16. According to the articles that have been analyzed, the cooperation index values are chosen. The evaluation technique, presented by Ajiferuke et al. (1988), is productivity measurement, where a number with zero closure denotes that majority of research works are written by only one author (Price and Beaver., 1966).

Finding Keyword with Database Establishment

The Scopus database for articles on smart communities and the Internet of Things are well examined. It is popular on comprehensive approaches with precise tracking of books from international editors like Taylor & Francis, Springer, Elsevier, IEEE, IETE and Emerald Insight (Foncubierta et al., 2014 and Thelwall et al., 2015). The papers were selected to guarantee the sources with the highest caliber focused on academics (Ramos et al., 2004). Only peer-reviewed papers were used in this study because it guarantees reliable IoT research on the topic of smart cities. Journal papers were searched using keywords "internet of things" and "smart city" which eliminates false-positive results. Thus, the limitations were eliminated on duration or topic, so all potentially pertinent papers could be included. In total, 1,977 papers were in the initial search findings.

Data through Evaluation

Once the keyword network had been created, the clusters were found and given labels on the subjects denoted by the topmost keywords. The clustering goals in relation to smart cities contributes to providing a reasonably thorough overview of IoT studies. This scientometric tool allows researchers that spots out the research trends with novel developments by statistically examining frequency (Chiu et al., 2007). As a result, the main IoT and smart city study areas are examined.

Keyword Framed Network

A useful metric that provides an in-depth examination of the main research areas in a scientific subject is a term overlap network (Borner et al., 2003). The network is constructed using analysis of co-network techniques, as well as the centralized measures and intensity created by Callon et al., (1986). According to a number of researchers, the ease of use and importance in keyword analysis supports its widespread use in literature with academic trends and emerging areas (Zong et al., 2013). Two terms provide better co-occurrence association if they recurrently show publications.

The purpose of network search term is to provide researchers with knowledge about the substance of publications so they can further develop the structure of scientific models. To build a network of term co-occurrences, we used the program VOS-viewer. Researchers build a network using the keywords as nodes thanks to the mapping VOS-viewer method (Lee et al., 2020). Therefore, a large density number indicates a close nodal distance.

Methods in E-Waste Management – A Descriptive Analysis

Block Chain Systems

After it has been created, electronic waste can be tracked and traced using a system that uses blockchain technology. Stakeholder selection factors based on reputation shorten the supply chain. Certificate issuance as proof of storage device data erasure. Implementation and testing are done on the Ethereum network. The system's viability is supported by studies of its affordability and security. The solution monitors the amount of e-waste produced and uses a reward system to entice users to position in their governmental agencies that can do so effectively and sustainably (Dua et al., 2020). It is feasible to set up this unregulated economic sector so that a public-private partnership (PPP) model is usable, and it has enormous potential that remains untapped.

Deep Learning Strategies

In particular, artificial intelligence can be utilized to quickly and accurately sort a big stream of waste in lieu of conventional techniques. It utilizes transfer learning as a specific AI method by refining AlexNet's output layers as a trained model and placing the execution to use a tiny data set made up of 10 classes from 5 different smartphone manufacturers. With momentum optimizer, a generalized model precision of about 97% is produced. The proposed paradigm includes three separate tasks.

The first step is to collect specific trash. In the succeeding step, web submissions created based on TL method (VGG16) helps in classification of the verified photos, grouping with the same kind dependent on classification categories, and ensure that every bin only contains one particular type of trash. To easily recycle such materials and significantly reduce waste, the appropriate authority is notified. This algorithm has a 97% accuracy rate (Mridha et al., 2021). Keypad, mouse, power supply, and desktop computer clicks are also included in the electronic dataset. The titles' descriptions include things made of paper, cardboard, and plastic. The photos are splitted into groups using deep learning techniques, with a training success rate of 96%. To further enhance the classification, it is beneficial to examine the performance of various filters and kernel sizes.

YOLO Model

In order to return users' deposits, the organizers are focusing on developing a sophisticated e-waste container that identifies the sort of e-waste being disposed. The YOLOv4 technique uses system differentiation on various types of e-waste. The devices that are approximated with precision, reliability, and recall are determined using a variety of tests. On average, 93.33% accuracy is attained (Sampedro et al., 2021). The advanced YOLOv5 model achieves mean accuracy relevant to test dataset (Zhou et al., 2023).

Dynamic CNN Model

It is vital to distinguish between various trash and metals. The DCNN model, also known as electronic trash management, is utilized as a result. The precise mapping of the image features expands classification success (Latha et al., 2022). These interim, waste collection can be improved to decrease distance, travel and time. Smartphone users are continuously monitoring the amount of e-waste in virtual trash that can ensure prompt gathering. Furthermore, it greatly reduces the testing, training, localization, categorization, and verification possibilities.

Analysis Using Material Flow System

Material flow analysis (MFA) is the common and popular technique in industrial ecology for input-output measurement for material and analyzing flux and routes of different material flow throughout the full arrangement along with related regional numbering, national product that is with elemental models (Islam et al., 2019).

Smart Cities and Its Importance

Smart cities address the problems that emerging capitals will experience in the near future. The United Nations estimates that by 2050, 7 million people will live in urban regions. As a consequence, there is a lot of pressure on cities to provide for the needs of their citizens (Choi et al., 2019). Several of advantages of smart communities are listed below:

(a) Better city planning and growth planning
(b) The provision of e-government amenities to people using routes that are optimized
(c) Cost and labor savings
(d) Local growth in the economy
(e) Improved output and service quality
(f) Maintaining environmental credentials through trash minimization, use of minimal energy, recycling, and waste treatment.
(g) Digital energy meters for ecological buildings and renewable energy.

Challenges in Smart Cities

Planning of Resources

The key problem that smart communities face is the obtainability of resources. The shortage of government infrastructure is the most glaring illustration. For all the cities, the price of implementing innovative, clever solutions was quite expensive (Zhou et al., 2020 and Ferreira et al., 2014). Even though the government pays extra attention to the e-waste sector, much work needs to be done. Even the most elementary refuse collection, recycling, and disposal cannot be managed by the government.

E-waste-related duties like collection, shipping, recycling, destruction, and other activities are done by unlicensed scrap dealers with untrained laborers. The absence of an adequate system, however, makes it promising for a greater quantity of e-waste for recycling the disposal, which creates various issues.

Infrastructure

In a smart object, the infrastructures for electrical power, healthcare, schooling, and movement all need to be more robust. Improvement in e-waste management, reuse and recycling, employ qualified workforces and use fewer industrial tools. Hence, the cities must be outfitted with the newest technology in order to execute an effective plan to build the smart city infrastructures.

Innovation Standards

Utilization of smart technologies, initiatives in smart cities need skill varieties. More people with necessary skills to work on these projects are hard to come by, both now and in the future (Hawryszkiewycz and Igor, 2014). Groups in India lack supply chains, so smart societies will need to use innovative tactics. On the basis of the data documentation, the principal editor is updated.

Smart Cities – The Future Idea

Smart cities can help us handle the problem of our rapidly expanding urbanization. The E-waste (Management) Rules, 2016 increased the number of products covered by the rule to over 21 as of the date they took force on October 1, 2017. (Schedule-I). The United Nations (UN) predicts that by 2050, 6 billion people will reside in urban areas. In addition, it's believed that cities use about 72% of the energy produced worldwide. This has resulted in a substantial demand for resources and numerous challenges for cities. There is a widespread and growing level of worry about the challenges that cities may encounter given the present growth rate. Not only does it help with e-waste dumping issues, but it also generates employment. On the basis of the demand for employees with all levels of education and experience, the e-waste recycling facilities have improved.

Conclusion

A more sustainable future can be pioneered by IoT-enabled smart cities, supporting the concept of efficient waste management in these neighborhoods.

The concepts discussed in this chapter clearly demonstrate how towns have dealt with the issues to raise the standard of living for their citizens. Effective waste management techniques are advantageous when creating sustainable communities. Since the generation of e-waste is the quickest in the field of solid refuse management, e-trash collection is efficient. It is ideal for places where there is a lot of foot traffic, such as schools, hospitals, parks, markets, etc. Cities around the world are working to change themselves intelligently. It is frequently necessary to carefully fund, organize, and sustain smart cities. Cities have no obligation to abandon the traditional methods and switch to smart bins in order to improve the efficiency of waste collection and recycling.

References

Abubaker, Nermeen, Paul Szabo-Müller, and Uwe Handmann. (2021). "Transfer learning-based method for automated e-waste recycling in smart cities." *EAI Endorsed Transactions on Smart Cities*, 5, no. 16.

Ahmed, Vian, Karam Abu and Sara. (2020). "An investigation into stakeholders' perception of smart campus criteria: the American university of Sharjah a case study." *Sustainability*, 12, no. 12, 5187.

Aithal, P. S. (2021). "Smart city waste management through ICT and IoT driven solution." *International Journal of Applied Engineering and Management Letters (IJAEML)*, 5, no. 1, 51-65.

Aiyappan, Asha, A. Rajesh, Ilango, Urooj, and Salem Alelyani. (2023). "Optimized RNN-based Performance Prediction of IoT and WSN-oriented Smart City Application using Improved Honey Badger Algorithm." *Measurement*, 112505.

Ajiferuke, Isola, Q. Burell, and Jean Tague. (1988). "Collaborative coefficient : A single measure of the degree of collaboration in research." *Scientometrics*, 14, no. 5-6, 421-433.

Almusaylim, Zahrah, Alhumam, and Jhanjhi. (2020). "Proposing a secure RPL based internet of things routing protocol: A review." *Ad Hoc Networks*, 101, 102096.

Alulema, Javier, and Luis. (2018). "A Cross-Device Architecture for Modelling Authentication Features in IoT Applications." *J. Univers. Comput. Sci.*, 24, no. 12, 1758-1775.

Amiruddin, Anak Agung Putri Ratna, and Riri Sari. (2019). "Systematic review of internet of things security." *International Journal of Communication Networks and Information Security*, 11, no. 2, 248-255.

Anagnostopoulos, Theodoros. (2023). "Smart City.", 7-15.

Asensio, Ángel, Teresa Blanco, Rubén Blasco, Alvaro Marco, and Roberto Casas. (2015). "Managing emergency situations in the smart city : The smart signal." *Sensors*, 15, no. 6, 14370-14396.

Bellini, Pier Francesco, Paolo Nesi, and Gianni Pantaleo. (2022). "IoT-enabled smart cities: A review of concepts, Framework and key technologies." *Applied Sciences*, 12, no. 3, 1607.

Beneicke, Julia, Angel A. Juan, Fatos Xhafa, David Lopez-Lopez, and Alfons Freixes. (2019). "Empowering citizens' cognition and decision making in smart sustainable cities." *IEEE Consumer Electronics Magazine*, 9, no. 1, 102-108.

Borner, Katy, Chaomei Chen, and Kevin W. Boyack. (2003). "Visualizing knowledge domains." *Annual review of information science and technology*, 37, no. 1, 179-255.

Brisson, C. R. E., Dan Doggendorf, and Michael J. Savoie. (2019). "Cybersecurity of building technology: Smart cities and smart buildings require smart protection." *CRE Real Estate Issues*, 43, no. 1, 1-9.

Callon, Michel, Arie Rip, and John Law, eds. (1986). *Mapping the dynamics of science and technology: Sociology of science in the real world*. Springer.

Chauhan, Meetali. (2023). "Smart Healthcare Solutions for Smart Cities." In *AI-Centric Smart City Ecosystems*, pp. 247-260. CRC Press.

Chen, Chong-Wen. (2023). "Can smart cities bring happiness to promote sustainable development? Contexts and clues of subjective well-being and urban livability." *Developments in the Built Environment*, 13, 100108.

Chiu, Wen-Ta, and Yuh-Shan Ho. (2007). "Bibliometric analysis of tsunami research." *Scientometrics*, 73, no. 1, 3-17.

Choi, Jonghwa, and Sanghyun Ahn. (2019). "Scalable service placement in the fog computing environment for the IoT-based smart city." *Journal of Information Processing Systems*, 15, no. 2, 440-448.

Das, Ankita, Debnath, Modak, Abhijit, and Debasish De. (2020). "E-waste inventorisation for sustainable smart cities in India: A cloud-based framework." In *2020 IEEE International Women in Engineering (WIE) Conference on Electrical and Computer Engineering (WIECON-ECE)*, pp. 332-335. IEEE.

Davor, Marko Slavulj, Dino Šojat, and Julian Jurak. (2018). "The role of smart mobility in smart cities." In *Fifth International Conference on Road and Rail Infrastructure (CETRA 2018)*, pp. 17-19.

Denovais, Marcos Pereira, Victor De Almeida Xavier, Lúcia Helena Xavier, and Junseok Hwang. (2021). "Modelling e-waste management data in smart cities." In *2021 2nd Sustainable Cities Latin America Conference (SCLA)*, pp. 1-4. IEEE.

Deakin, Mark, and Alasdair Reid. (2018). "Smart cities: Under-gridding the sustainability of city-districts as energy efficient-low carbon zones." *Journal of Cleaner Production*, 173, 39-48.

Dua, Amit, Akash Dutta, Nishat Zaman, and Neeraj Kumar. (2020). "Blockchain-based E-waste management in 5G smart communities." In *IEEE INFOCOM 2020-IEEE conference on computer communications workshops (INFOCOM WKSHPS)*, pp. 195-200. IEEE.

Fahmidehgholami, Mahdi, and Didar Zowghi. (2020). "An exploration of IoT platform development."

Ferreira, Manuel Portugal, Cláudia Frais Pinto, and Fernando Ribeiro Serra. (2014). "The transaction costs theory in international business research: a bibliometric study over three decades." *Scientometrics* 98, 1899-1922.

Foncubierta, Antonio, Henning Müller, and Adrien Depeursinge. (2014). "Retrieval of high-dimensional visual data : current state, trends and challenges ahead." *Multimedia tools and applications* 69, 539-567.

Gharaibeh, Mohammad A. Salahuddin, Sayed, Abdallah, Issa, Mohsen and Fuqaha. (2017). "Smart cities: A survey on data management, security, and enabling technologies." *IEEE Communications Surveys & Tutorials* 19, no. 4, 2456-2501.

Gupta, Shristi, Syed Ziaul Mustafa, and Harish Kumar. (2017). "Smart people for smart cities: A behavioral framework for personality and roles." In *Advances in Smart Cities*, pp. 23-30. Chapman and Hall/CRC.

Hawryszkiewycz, Igor T. (2014). "Cloud requirements for facilitating business collaboration : a modeling perspective." *Journal of organizational computing and electronic commerce* 24, no. 2-3, 174-185.

Ilankoon, Ghorbani, Chong, Gamini Herath, Moya, and Jochen Petersen. (2018). "E-waste in the international context–A review of trade flows, regulations, hazards, waste management strategies and technologies for value recovery." *Waste management*, 82, 258-275.

Islam, Md Tasbirul, and Nazmul Huda. (2019). "Material flow analysis (MFA) as a strategic tool in E-waste management : Applications, trends and future direction." *Journal of environmental management*, 244, 344-361.

Janssen, M., Luthra, S., Mangla, S., Rana, N. P., and Dwivedi, Y. K. (2019). Challenges for adopting and implementing IoT in smart cities." *Internet Research*, 29, 1589-616.

Kakkavas, Grigorios, Despoina Gkatzioura, Vasileios Karyotis, and Symeon Papavassiliou. (2020). "A review of advanced algebraic approches enabling network tomography for future network infrastructures." *Future Internet*, 12, no. 2, 20.

Kang, Kai Dean, Harnyi Kang, I. M. S. K. Ilankoon, and Chun Yong Chong. (2020). "Electronic waste collection systems using Internet of Things (IoT): Household electronic waste management in Malaysia." *Journal of cleaner production*, 252, 119801.

Khan, Atta Ur Rehman, and Raja Wasim Ahmad. (2022). "A blockchain-based IoT-enabled E-Waste tracking and tracing system for smart cities." *IEEE Access*, 10, 86256-86269.

Krishnaswamy, Jayaraman, Vejayon, Raman, and Imtiaz Mostafiz. (2019). "The proposed e-waste management model from the conviction of individual laptop disposal practices-An empirical study in Malaysia." *Journal of Cleaner Production*, 208, 688-696.

Kumar, T.V. and Bharat Dahiya. (2017). "Smart economy in smart cities." *Smart economy in smart cities*, 3-76.

Latha, C. Jenifa, Kalaiselvi, K., Ramanarayan, S., Srivel, R., Vani, S., and Sairam, T. V. M. (2022). "Dynamic convolutional neural network based e-waste management and optimized collection planning." *Concurrency and Computation : Practice and Experience*, 34, no. 17, e6941.

Lee, Doyeon, Jongseok Kang, and Keunhwan Kim. (2020). "Global collaboration research strategies for sustainability in the post COVID-19 era: Analyzing virology-related national-funded projects." *Sustainability*, 12, no. 16, 6561.

Lombardi, Patrizia, Silvia Giordano, Hend Farouh, and Wael Yousef. (2012). "Modelling the smart city performance." *Innovation : The European Journal of Social Science Research*, 25, no. 2, 137-149.

Lopes, Vasco, (2017). "Smart governance: A key factor for smart cities implementation." In *2017 IEEE International Conference on Smart Grid and Smart Cities (ICSGSC)*, pp. 277-282. IEEE.

Luca, Mora, Deakin Mark, and Roberto Bolici. (2017). "*The First Two Decades of Smart-City Research: A Bibliometric Analysis Journal of Urban Technology*." Vol. 24.

Maranda, Aneta, Daniel Kaczmarek, Natalia Kryvinska, and Fatos Xhafa. (2019). "Studying usability of AI in the IoT systems/paradigm through embedding NN techniques into mobile smart service system." *Computing*, 101, 1661-1685.

Conti, Mauro, Pallavi Kaliyar, Md Masoom Rabbani, and Silvio Ranise (2020). "Attestation-enabled secure and scalable routing protocol for IoT networks." *Ad Hoc Networks*, 98, 102054.

Mekkaoui, Ali, Laouer, M., and Mimoun, Y. (2017). "Modeling and simulation for smart grid integration of solar/wind energy." *Leonardo Journal of Sciences*, 30, 31-46.

Mekki, Kais, Eddy Bajic, Frederic Chaxel, and Fernand Meyer. (2019). "A comparative study of LPWAN technologies for large-scale IoT deployment." *ICT express*, 5, no. 1, 1-7.

Mohammed, Ali Abdul-Jabbar, M. A. Burhanuddin, Halizah Basiron, and Durian Tunggal. (2018). "Key enablers of IoT strategies in the context of smart city innovation." *Jour Adv Res. Dyn. Control Syst*, 10, no. 04, 582-589.

Mridha, Krishna, Rabindra Nath Shaw, and Ankush Ghosh. (2021). "Intelligent based waste management awareness developed by transfer learning." In *2021 IEEE 4th International Conference on Computing, Power and Communication Technologies (GUCON)*, pp. 1-5. IEEE.

Mutiara, Dewi, Siti Yuniarti, and Bambang Pratama. (2018). "Smart governance for smart city." In *IOP Conference Series: Earth and Environmental Science*, vol. 126, no. 1, p. 012073. IOP Publishing.

Orlowski, Aleksander, and Patrycja Romanowska. (2019). "Smart cities concept : Smart mobility indicator." *Cybernetics and Systems*, 50, no. 2, 118-131.

Osterrieder, Philipp, Lukas Budde, and Thomas Friedli. (2020). "The smart factory as a key construct of industry 4.0: A systematic literature review." *International Journal of Production Economics*, 221, 107476.

Porru, Simone, Francesco Edoardo Misso, Filippo Eros Pani, and Cino Repetto. (2020). "Smart mobility and public transport : Opportunities and challenges in rural and urban areas." *Journal of traffic and transportation engineering (English edition)* 7, no. 1, 88-97.

De Solla Price, Derek J., and Donald Beaver (1966). "Collaboration in an invisible college." *American psychologist*, 21, no. 11, 1011-1018.

Ramos, Antonio-Rafael, and José Ruíz-Navarro. (2004). "Changes in the intellectual structure of strategic management research: A bibliometric study of the Strategic Management Journal, 1980–2000." *Strategic management journal*, 25, no. 10, 981-1004.

Sampedro, Gabriel Avelino, Ryanne Gail C. Kim, Yohana Jayanti Aruan, Dong-Seong Kim, and Jae-Min Lee. (2021). "Smart e-waste bin development based on yolov4 model." In *2021 1st International Conference in Information and Computing Research (iCORE)*, pp. 125-128. IEEE.

Song, Qingbin, and Jinhui Li. (2015). "A review on human health consequences of metals exposure to e-waste in China." *Environmental Pollution*, 196, 450-461.

Statista, Number of connected devices worldwide 2030, Statista, 2020. https://www.statista.com/statistics/802690/worldwide-connected-devices-by-accesstechnology/. Accessed 14 November 2020.

Su, Haonan, and Cheolkon Jung. (2018). "Perceptual enhancement of low light images based on two-step noise suppression." *IEEE Access*, 6, 7005-7018.

Thelwall, Mike, and Nabeil Maflahi. (2015). "When are readership counts as useful as citation counts? Scopus versus Mendeley for LIS journals."

Thornbush, Mary, and Oleg Golubchikov. (2021). "Smart energy cities: The evolution of the city-energy-sustainability nexus." *Environmental Development*, 39, 100626.

Tiwari, Pulkit, P. Vigneswara Ilavarasan, and Sushil Punia. (2019). "Content analysis of literature on big data in smart cities." *Benchmarking : An International Journal*, 28, no. 5, 1837-1857.

Turaga, Mohana R., Bhaskar, Satish Sinha, Daniel, Morton, Rachna Arora, Sandip et al. (2019). "E-waste management in India: issues and strategies." *Vikalpa*, 44, no. 3, 127-162.

Vishnu, S., Ramson, Samson, Theodoros, Adnan, Xiaozhe Fan, S. Srinivasan, and Kirubaraj. (2021). "IoT-Enabled solid waste management in smart cities." *Smart Cities*, 4, no. 3, 1004-1017.

Wang, Yiou, Fuquan Zhang, Junfeng Wang, Laiyang Liu, and Bo Wang. (2021). "A bibliometric analysis of edge computing for internet of things." *Security and Communication Networks*, 2021, 1-10.

Web source : Globalewaste. Globalewaste.org/wp-content/uploads/2018/10/GlobalE-waste-Monitor-2017.pdf, (accessed, March 25, 2020).

Web source : Natalie Gagliordi. Zdnet.com/article/iot-to-drive-growth-inconnected-devices-through-2022-cisco/, (accessed, March 24, 2020).

Wijaya, Aksan Surya, Zahir Zainuddin, and Muhammad Niswar. (2017). "Design a smart waste bin for smart waste management." In *2017 5th International Conference on Instrumentation, Control, and Automation (ICA)*, pp. 62-66. IEEE.

Yong, Lim, Ilankoon, I. M. S. K. (2019). An analysis of electronic waste management Strategies and recycling operations in Malaysia : Challenges and futurs prospects. *Journal of Cleaner Production*, 224, 151-166.

Yong, Yun Siew, Yi An Lim, and Ilankoon, I. M. S. K. (2019). "An analysis of electronic waste management strategies and recycling operations in Malaysia : Challenges and futurs prospects." *Journal of Cleaner Production*, 224, 151-166.

Zavare, Shraddha, Rashmi Parashare, Shivani Patil, Pooja Rathod, and Vanita Babanne. (2017). "Smart City waste management system using GSM." *Int. J. Comput. Sci. Trends Technol*, 5, no. 3, 74-78.

Zhao, Li. (2019). "TANG, Zhi-ying; ZOU, Xin." *Mapping the Knowledge Domain of Smart-City Research: A Bibliometric and Scientometric Analysis. Sustainability, MDPI AG*, 11, no. 23, 6648.

Zhou, Haoshi, Qiu, and Zheng. (2023). "Object Detection for Construction Waste Based on an Improved YOLOv5 Model." *Sustainability*, 15, no. 1, 681.

Zhou, Shuling, Xi, Z., Liu, J., Zhang, K., and Zhao, Y. (2020). "Exploring development of smart city research through perspectives of governance and information systems: A scientometric analysis using CiteSpace." *Journal of Science and Technology Policy Management*, 11, no. 4, 431-454.

Zong, Qian-Jin, Hong-Zhou Shen, Qin-Jian Yuan, Xiao-Wei Hu, Zhi-Ping Hou, and Shun-Guo Deng. (2013). "Doctoral dissertations of Library and Information Science in China : A co-word analysis." *Scientometrics*, 94, 781-799.

Chapter 5

Blockchain for Privacy and Security in Data Sharing in Industries

S. Swetha[*]
and P. M. Joe Prathap
Department of Computer Science and Engineering,
R.M.D. Engineering College, India

Abstract

Recent developments in communication technology and data-sharing techniques have spread quickly in a variety of industries. Security and data sharing are therefore seen as being crucial for maximizing the value of data. Also, the early data-sharing methods could not keep track of how electronic data was being used. The unwillingness of data providers to provide their information is regarded as a challenging task. Data integrity, data privacy, and infrastructure security are among the issues with big data safety that need to be addressed. Using encryption techniques that are appropriate for big data security over the cloud ensures thehighest level of security for big data analysis, storage, and processing. Transparency in transmission applications is provided by a blockchain. As a result, this study investigates a novel, three-phase secure dataexchange paradigm. the administration, sharing, and outsourcing of large data sets. With the help of blockchain, smart contracts are created. By fusing smart contracts, which automatically carry out default instructions, with traceable, non-tamper-evident features, blockchain-based schemes provide a fresh solutionto this security dilemma. Some tactic aids in advancing big data's ultimate security and data exchange. The results of the experimental research showed that the Blockchain for

[*] Corresponding Author's Email: swethasundar2611@gmail.com.

In: Blockchain Enabled Secure Big Data Computing …
Editors: L. Ashok Kumar, D. Karthika Renuka, V. Chandrasekar et al.
ISBN: 979-8-89113-243-6

big data sharing framework was more effective at securing data transactions.

Keywords: blockchain, data-sharing, privacy, security

Introduction

Managing large amounts of data remains a significant challenge for data managers due to the constantly increasing volume of data produced, which is an essential component of the operation of many organisations. Businesses can now acquire and retrieve data in a more timely and cost-effective manner thanks to the development of new and affordable computing devices, as well as cloud computing platforms. Zheng et al., (2020), it is, however, difficult to securely transmit data across a network while lowering overhead costs. The management of large amounts of data continues to be a significant challenge for data managers due to the ever-increasing volume of data produced, which is an essential component of the operation of many organizations. The invention of novel and affordable computing devices, as well as cloud computing platforms, has enabled businesses to acquire and retrieve data in a more timely and cost-effective manner. However, securely transmitting and receiving data over a network while lowering overhead costs is a difficult task. Shrestha et al., (2020), The exchange of large amounts of data using a decentralised design not only puts strain on the nodes in the path between the sender and the recipient, but it also causes network congestion. Internal nodes that can reroute data may not have enough buffer capacity to temporarily retain data and retransmit it to subsequent nodes. Zhu et al., (2019), This has the potential to cause sporadic data transmission failures as well as frequent thrashing. As a result, selecting the next node to transmit data can be a time-consuming process, increasing the time required to receive and analyze data in its entirety. Blockchain has significantly aided the provision of transparency required for a transmission application. An architecture in Figure 1 that supports concurrent multithreading ensures that data is quickly routed to various recipients across a network. This research focuses on methodologies and procedures for the timely and rapid transfer of data fragments from large amounts of data that a user wishes to convey. The preliminarymodel data is sent to a remote user who is not physically close to the sender via a secure route made possibleby blockchain technology.

Related Works

Over the last decade, research consortiums have made technical breakthroughs in adapting data-sharing methodologies. Research-based initiatives may benefit from a collaborative effort and sound judgment in this approach. Anjana P.S et al., (2019), The open exchange of relevant data is the first step towards fully capitalizing on advances in scientific research.

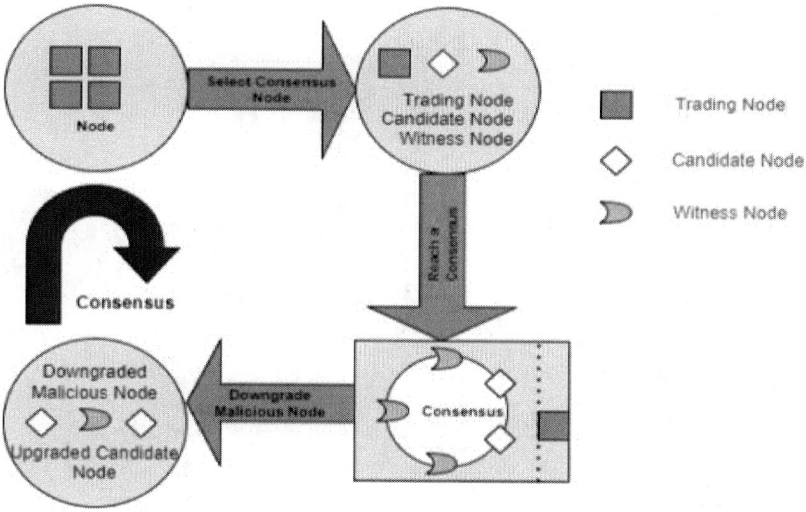

Figure 1. Ripple of Consensus Ledger.

Nonetheless, it is critical to understand the three W's: what, when, and where. These questions must be thoroughly answered before proceeding with the data exchange procedure. The owner of the data set should be rewarded or incentivized in some way, which could be improved. This study makes use of blockchain's security features to allow for the safe exchange of personal information. Blockchain, or distributed ledger technology architecture in Figure 2 is the latest buzz in the field of information technology. Blockchain technology has been used in a wide range of financial and non-financial applications. According to one Silicon Valley expert, the consensus process is the most fundamental and important invention ever. Biswas et al., (2017), The current paper currency's reliance on an external authority jeopardizes security, trust, and privacy. The main selling point of blockchain is its dependability, which is achieved by eliminating intermediaries. Nakamoto founded Bitcoin, the first decentralised digital currency, in 2008.

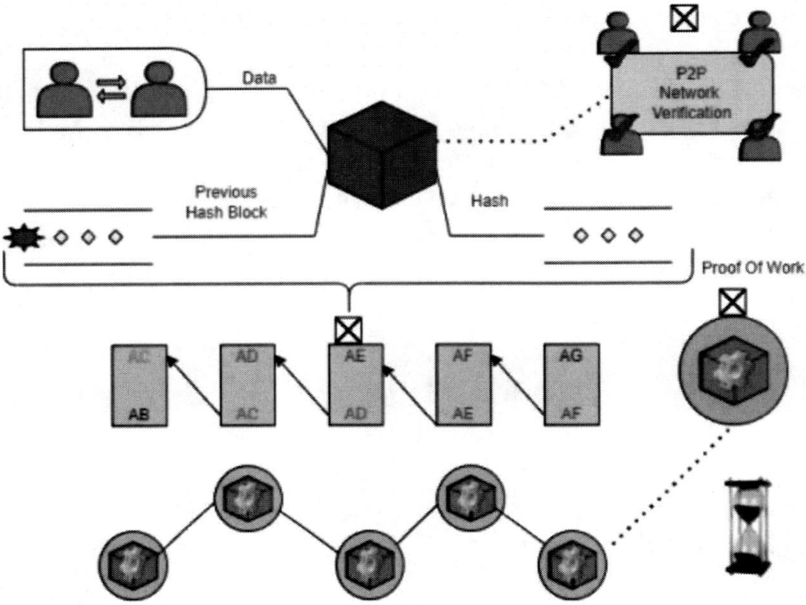

Figure 2. Block Chain Distributed Ledger.

This allows money to be transferred instantly from one person to another. Nakamoto later published the protocol for the genesis block's first 50 coins in a paper. Blockchain is now being used in a wide range of fields due to its remarkable benefits, including the cloud, IoT, data trade, information security, healthcare, and many more. The fear of researchers' data being misappropriated and misinterpreted is at the heart of the data-sharing problem. This is because the research community's trust in one another is still developing, making it difficult to implement mature data-sharing techniques. Rather than making all data publicly available, solutions include safeguarding all individuals' identities and limiting access to data. However, these solutions do not ensure data privacy, data integrity, or data usage audit trails. Cloud servers house themajority of the world's information, making them a single point of failure. One of several risks associated with relying on a centralized body is a single-point failure. Having other parties provide data backups is a preventative measure against this type of disaster. Crossby et al., (2016), We can eliminate the need for a trusted third party when developing a trust-based paradigm by introducing a distributed ledger system (blockchain) that is both trustworthy and transparent. Decentralized storage, which uses a distributed ledger to allow data to be stored independently on different computers across a network, is one solution. The problem stems from network nodes' inability to adequately store or process data. To accomplish this, the interplanetary file system (IPFS) peer-to-peer architecture is adapted. Existing systems do not give data owners complete control over who can see and use their information. The owner of the data rarely participates in the exchange. When it comes to distributing information and settling payments, for example, the owner takes a back seat while escrow handles everything. Panescu et al., (2018), The owner receives a royalty payment at the end of the day. Keeping financial transactions honest is a difficult task in the absence of blockchain technology. This means that not everyone will receive their fair share of the funds. In this case, the blockchain's trust and openness among network nodes would be invaluable in ensuring that any funds received from a data requester are distributed equitably. When a consumer purchases material from third parties, the quality and integrity of that data must be preserved at all costs. Park et al., (2018), Instead of several separate approaches, one method is provided to address these difficulties. The owner is in charge of managing the transfer of digital data to other users in the scenario presented. The data's rightful owner can limit access to it by imposing a set of constraints. When a request for data is made, the data's owner contacts an IPFS server. When a user registers for a data request, digital signatures are used for authentication. If the request is made before that time, the transaction on the blockchain will be cancelled. In By incorporating an encryption technique into IPFS hashes, data stored there is protected. Shamir secret sharing (SSS) Dhinakaran et al., (2022), is a technique for encrypting hashes by dividing them into a set of shares, each of which is kept secret until the owner requires it. The smart contract protects the encrypted stock certificates. A consumer initiates data access by sending a request, which is validated with digital signatures. After obtaining the material via encrypted shares, the consumer is expected to register reviews on the review system. Customers are encouraged to re-register by offering a 10% return on their real money deposit. Digital Media Security Dias et al., (2018), proposed a technique in the blockchain that ensures the validity and nonrepudiation of digital material while also protecting the privacy of traceable encryption. The authors address the issue of disclosing sensitive information by revealing the user's secret key, which is useless to the recipient. If the shared key is compromised, tracing the origin of the secret key becomes more difficult. Furthermore, current systems have a problem with private information leaking during the access control process. To keep the kingdom's keys safe, the authors used a privacy-protecting method such as attribute- based encryption (ABE). However, the decryption technique's effectiveness has not improved. Proper management of associated rights is required to secure digital information. Current methods for protecting data rights fall short in three key areas: openness, decentralisation, and trust. G. Georgakakos et al., (2019), proposed a decentralised approach based on blockchain to address the issues outlined above. Everyone has access to the licences and transactions associated with digital content usage. One of the primary goals of smart contract architecture is automatic licence assignment. The licensee is free to charge whatever they want for the license's resale market. However, in order to complete critical acquisitions, network peers require powerful computing resources. Hasan et al., (2018), focus on digital rights management with blockchain to prevent the misuseof sensitive digital content. To address these issues, DRMchain technology is being offered. This method ensures proper use by allowing only verified people access to protected digital material. For this purpose, two blockchains have been developed; one keeps the original content and its cypher summary, while the other keeps the cypher summary of encrypted data. DRMchain provides users with both an audit trail of any violations and a secure environment. The proposed system protects digital content, ensures secure authorization, and employs multiple signatures to regulate usage. Selvi et al., (2017), However, using the Ethereum token to secure digital data may lead researchers down an unexpected path. Sharing data is the most effective way to reap the benefits of research. Various information-exchange strategies have been developed and addressed in published works. There is insufficient literature addressing the incentive mechanism to promote data sharing. To fill these gaps, Joe et al., (2008), examined medical and health records in order to identify incentive mechanisms, comparing pre- and post-analysis results. The study examines the rate of data sharing for medical and health information using a single incentive. As a result, additional research with financial incentives to share data is required.

Big Data Sharing Architecture Proposal

In today's business and scientific environments, big data sharing is critical. Most organisations use a large number of cutting-edge technologies and applications to enable billions of users around the world to generate more data in their daily lives via smartphones, social media sites, and other smart devices. The advancement of various technologies such as IoT, AI, and cloud computing increased the use of big data. Naz et al., (2019), The use of more big data provides more benefits to organisations and businesses, as well as creating a link between generations and utilising data to achieve fair data trading. GBDEX, Terbine, and Data trading are three platforms that provide big data services. The data is obtained from big data curators by the big data sharing platform. Li et al., (2018), Most individual data provider data quantities are insufficient to meet the needs of organisations, and it is also difficult to provide data to end users directly. In the current big data era, developing and providing highly secure as well as unique data-sharing platforms to acquire information from entity information resources and supply data to data users has secured a high significance rate.

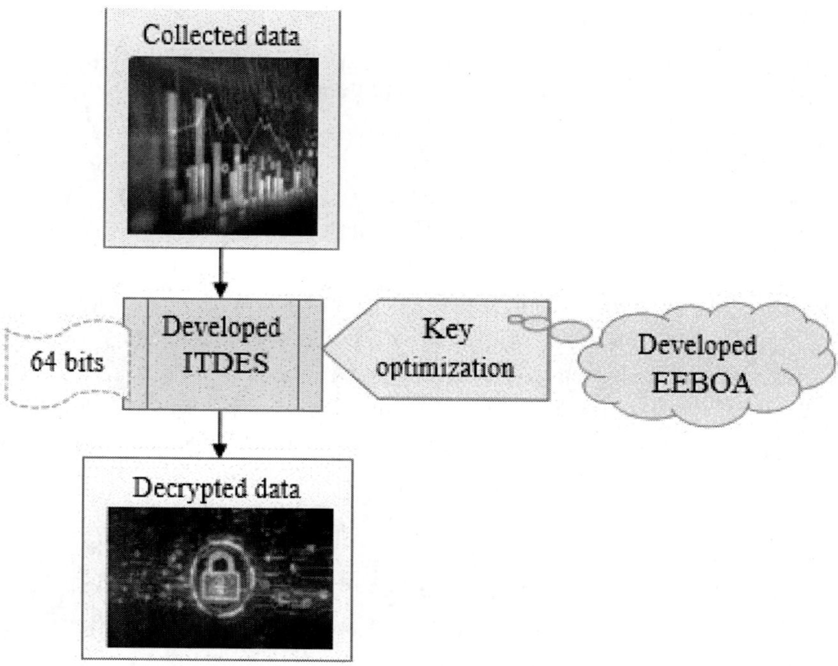

Figure 3. Architecture Presentation of Big Data Sharing Model using Blockchain Technology.

The majority of data-sharing platforms are used in industry, and large data-sharing frameworks are developed in the research community. Furthermore, various solutions derived from cutting-edge models are primarily related to the centralised third-party data-sharing platform, which faces additional issues such as intractability, uncontrollability, privacy disclosure, and opaque transactions. Menon et al., (2016), The limitations presented in existing big data sharing approaches must be addressed, so a novel multi-hop concurrent big data sharing framework based on blockchain techniques and encryption approaches is designed. The developed multi- hop concurrent big data sharing model is architecturally represented in Figure 3 using encryption techniques and blockchain technology.

Inaccuracy and Validation

A transmission in progress can fail for a variety of reasons, including:

a. No connecting nodes

Task: Transferring data fragments between the transmitter and the recipient can be difficult when they are geographically separated and there is no intermediary node between them.

Action: The only solution to this issue is to have a transmitter that can send data in chunks to the receiver without interfering with it.

b. Intermediate nodes with insufficient resources

Task: When intermediary nodes are available for transmission but lack the capability to send data fragmentsin their entirety, this problem arises. Storage capacity, memory, bandwidth, and a high bitrate are some of the resources required for transmission.

Action: Underpowered nodes may retry processing previously rejected data packets as a workaround. This will further divide the packages into manageable chunks. This method allows you to transmit dynamic data passively.

c. Inefficient transitional nodes

Task: Intermediate nodes frequently fail to relay received packets due to limited bandwidth or other factors.As a result, the role of intermediate nodes in the transmission chain is significantly reduced.

Action: The practical solution is for the smooth transition nodes to broadcast a request for information to all linked nodes, hoping that one of those nodes will send the data to the intended receiver. As a result, even if a node is out of order, the system will continue to deliver data. Menon et al., (2019), If the transmitter is located a long distance from the receiver and the intermediate nodes are out of order, the recipient may not receive any data. The system works flawlessly whenever at least one of the intermediary nodes on the chain can recover from an out-of-order or temporarily disabled state.

d. Nodes that are harmful in the middle

Task: This is far more pressing than the issue of out-of-order nodes. Both the though nodes and malicious nodes problems involve independent third-party nodes that are either useless or intended by a malicious userto operate incorrectly. [21]When malicious people forward information, they want the recipient to believe it came directly from the source.

Action: Because blockchain technology is being used, the data checksum was previously determined, and the shared information fragment has been added to the existing chain. As a result, if the checksum suppliedby the sender does not match the checksum of the received data fragment, which is impossible if the networkis secure, the receiver will reject any faulty data fragments.

Conclusion

This chapter explores a blockchain-based infrastructure for the secure transfer of digital assets and the distribution of associated data. The primary goal of the proposed scenario is to provide the client with accurate and high-quality information while also providing the proprietor with a solid foundation on which to build a successful business. Swetha et al., 2022, A decentralised storage network, such as IPFS, can assist with reducing owner-side swelling. A blockchain-based method of sending massive amounts of data over a network without exposing the data to potential threats. The immutability and public nature of a blockchain are assets to the system. Because of the peer-to-peer network, large amounts of data can now be transmitted quickly and without relying on a central server. The proposed system can reliably send data to any size receiver. It'sa divide-and-conquer algorithm in that data is encrypted at each level before being sent to the next set of nodes. Because of the adaptability of the system's network, it can be used in a wide range of practical settings. In its current state, the system allows for a consensus recovery ratio of more than 80%. To save time and money on retransmissions while maintaining fault tolerance with fewer nodes, a variant of Byzantine Fault Tolerance known as Practical Byzantine Fault Tolerance can be used. Because of the lack of a centralized authority, a distributed ledger allows the system to conduct transactions in a way that is both safe and nearlyimpossible to manipulate.

References

Anjana P S, Kumari S, Peri S, Rathor S, Somani A. *An efficient framework for optimistic concurrent execution of smart contracts.* In2019 27th Euromicro International Conference on Parallel, Distributed and Network-Based Processing (PDP) 2019 Feb 13 (pp. 83-92). IEEE.

Biswas, Kamanashis, Vallipuram Muthukkumarasamy, and Wee Lum Tan. *"Blockchain-based wine supply chain traceability system."* Future Technologies Conference (FTC) 2017. The Science and Information Organization, 2017.

Crosby M, Pattanayak P, Verma S, Kalyanaraman V. Blockchain technology: Beyond bitcoin. *Applied Innovation*. 2016 Jun; 2(6-10):71.

Dhinakaran, D, and P M Joe Prathap. *"Ensuring Privacy of Data and Mined Results of Data Possessor in Collaborative ARM."* Pervasive Computing and Social Networking. Springer, Singapore, 2022. 431-444.

Dias J P, Reis L, Ferreira H S, Martins Â. Blockchain for access control in e-health scenarios. *arXiv preprint arXiv*:1805.12267. 2018 May 31.

Georgakakos G and S Ioannidis. The 2019 edition of the IEEE 24th International Workshop on Computer-Aided Modeling and Design of Communication Links and Networks (CAMAD), pages 1–6, 2019.

Hasan, Haya R, and Khaled Salah. *"Proof of delivery of digital assets using blockchain and smart contracts."* IEEE Access 6 (2018): 65439-65448.

Joe Prathap, P M, and V Vasudevan. "Revised variable length interval batch rekeying with balanced key tree management for secure multicast communications." *IJCSNS* 8.4 (2008): 232.

Li, Jiaxing, Jigang Wu, and Long Chen. "Block-secure: Blockchain-based scheme for secure P2P cloud storage." *Information Sciences* 465 (2018): 219-231.

Menon, Varun G, and Joe Prathap P M. "Moving from topology-dependent to opportunistic routing protocols in dynamic wireless ad hoc networks: Challenges and future directions." Algorithms, methods, and applications in mobile computing and communications. *IGI Global*, 2019. 1-23.

Menon, Varun G, and P M Prathap. "Survey on latest energy based routing protocols for under waterwireless sensor networks." *International Journal of Computer Networks and Wireless Communications (IJCNWC)* 6.6 (2016): 4.

Naz M, Al-zahrani F A, Khalid R, Javaid N, Qamar A M, Afzal M K, Shafiq M. A secure data sharing platform using blockchain and interplanetary file system. *Sustainability*. 2019 Dec 10;11(24):7054.

Nizamuddin N, Hasan H, Salah K, Iqbal R. Blockchain-based framework for protecting author royalty of digital assets. *Arabian Journal for Science and Engineering*. 2019 Apr 1;44:3849-66.

Panescu, Adrian-Tudor, and Vasile Manta. "Smart contracts for research data rights management over the Ethereum blockchain network." *Science & Technology Libraries* 37.3 (2018): 235-245.

Park J S, Youn T Y, Kim H B, Rhee K H, Shin S U. Smart contract-based review system for an IoT data marketplace. *Sensors*. 2018 Oct 22;18(10):3577.

Praveen K V, Prathap P J, Dhanasekaran S, Punithavathi I H, Duraipandy P, Pustokhina I V, Pustokhin D A. Deep learning-based intelligent and sustainable smart healthcare application in cloud-centric IoT. *Computers, Materials & Continua*. 2021 Jan 1;66(2):1987-2003.

Premalatha, J, and P M Joe Prathap. *"A Survey on Underwater Wireless Sensor Networks: Progresses, Applications, and Challenges."* MATEC Web of Conferences. Vol. 57. EDP Sciences, 2016.

Selvi, M, and P M Joe Prathap. "WSN data aggregation of dynamic routing by QoS analysis." *Journal of Advanced Research in Dynamical and Control Systems* 9.18 (2017): 2900- 2908.

Showkat S, Qureshi S. Securing the Internet of Things Through Blockchain Approach: Security Architectures, Consensus Algorithms, Enabling Technologies, Open Issues, and Research Directions. *International Journal of Computing and Digital Systems*. 2023 Jan 29;13(1):97-129.

Shrestha, Ajay Kumar, and Julita Vassileva. *"Blockchain-based research data sharing framework for incentivizing the data owners."* International Conference on Blockchain. Springer, Cham, 2018.

Swetha, S, and P M JoePrathap. *"A Study on a Decentralized Network Secured Data Sharing using Blockchain."* 2022 1st International Conference on Computational Science and Technology (ICCST). IEEE,2022.

Xu Z, Liang W, Li K C, Xu J, Jin H. A blockchain-based roadside unit-assisted authentication and key agreement protocol for internet of vehicles. *Journal of Parallel and Distributed Computing*. 2021 Mar 1;149:29-39.

Zheng S, Pan L, Hu D, Li M, Fan Y. *A blockchain-based trading platform for big data.* In IEEE INFOCOM 2020-IEEE Conference on Computer Communications Workshops (INFOCOM WKSHPS) 2020 Jul 6 (pp. 991-996). IEEE.

Zhu L, Wu Y, Gai K, Choo K K. Controllable and trustworthy blockchain-based cloud data management. *Future Generation Computer Systems*. 2019 Feb 1;91:527-35.

Chapter 6

Blockchain-Powered Secured Health Care Data Transmission: An Overview

K. Kavitha[1], PhD
and J. Poongodi[2,*], ME, PhD
[1]Department of Electronics and Communication Engineering,
Kumaraguru College of Technology, Coimbatore, India
[2]Department of Computer Science and Business Systems,
KGISL Institute of Technology, Coimbatore, India

Abstract

Blockchain is a distributed, digital ledger that stores blocks of transactions, and all the blocks are digitally chained together on numerous computers in a way that prevents any record from being changed unilaterally altered because alteration reflects in subsequent blocks. The data in the blockchain is stored on the distributed network unlike traditional storage where the data is maintained in central storage; hence no one can change the data easily without the users' consensus. It provides a secure method of keeping patients' private and sensitive information utilizing a decentralized system of records, preventing a single attack. The cryptographic hash that connects each data block with newly updated information block records determines the value of a Blockchain hash; hence, integrity is good in the blockchain. In COVID-like pandemic situations, it's necessary to seek access to healthcare systems to meet rapid mass screening, rapid contact tracing, managing vaccines and drug supply, online consultations, and trading. It needs a secured multipurpose platform for coordinating personal and sensitive information. The emerging digital technologies IoT, 5G and AI were adopted effectively to meet the requirements during the unprecedented

* Corresponding Author's Email: poongodiapcse@gmail.com.

In: Blockchain Enabled Secure Big Data Computing …
Editors: L. Ashok Kumar, D. Karthika Renuka, V. Chandrasekar et al.
ISBN: 979-8-89113-243-6

lockdown. However, security and data liquidity are major open problems unresolved still. An emerging topic that has the potential to lay new groundwork for data sharing in healthcare is the trend towards patient-centred interoperability. Data liquidity in clinical, research and operational is mandated for patient-centred interoperability which needs novel, secured technology. Blockchain technology is one of the solutions suggested by researchers globally which enables secured data interoperability. A detailed survey of current research on secured data transfer using blockchain technology is given in this chapter.

The objective of the survey is to consolidate the needs of modern healthcare systems such as predictive medication, online consultation, ensuring vaccination during the pandemic, etc. Analyze the advantages and suitability of the evolving blockchain technology for secured healthcare data transmission. Summarizing the different proposed blockchain-based solutions for secured data transmission.

Keywords: blockchain, health care sector, security, distributed data storage, data liquidity, IoT, distributed ledger

Introduction

Advancement in digital technology influences and changes the human lifestyle all over the globe. The healthcare sector is also adapting the advantages of sensor technology, signal processing, high-speed data transmission, IoT etc. The advantages are secured data storage and transfer, better user experience, personalized medication, accuracy in diagnosis, managing pandemic situations etc., however, there are unsolved issues in secured data storage and personalizing the data sharing (Poongodi, J et al., 2022).

The domains of data storage, data sharing, and privacy protection have all given blockchains a lot of attention recently. Blockchain is irrevocable and transparent. Also, any data maintained on a blockchain must be encrypted (Gong. T et al., 2015). The most thorough record of a person's identification is their medical history, which is why it needs to be kept securely. Medical records' total integrity and security are guaranteed from the outset thanks to blockchain encryption, which prevents the information from being changed or erased. (Azaria, A et al., 2016) The use of health information technology is frequently cited as one of the solutions for increasing the efficiency and security of healthcare. With blockchain technology, transactions are recorded on a shared, decentralized ledger. Decentralization, verifiability, and immutability are qualities of blockchain that are crucial to the medical and healthcare sectors, particularly in the management of medical data. Medical information protection, medical data storage and exchange, medical data applications, forecast analysis, etc., are the primary topics of the current study in the blockchain for the healthcare industry. Data leaking is less likely thanks to blockchain technology.

Needs of Modern Healthcare Systems

The healthcare sector used a paper-based system, or a handwritten mechanism, to maintain medical records before the development of modern technology. Issues in the system were insecure and unorganized. Data duplication in all the hospitals where the patients approached results in data redundancy and poor privacy. In recent years, Electronic Health Records (EHR) systems were formulated by combining paper-based records and electronic records. The electronic form of the real-time patient record is called an HER and is made available to the patients and doctors wherever and whenever in a secure way.

Advantages of EHR

Improves patient care is achieved through, Coordinated and quick access to patient data, Increased decision support, clinical support, Timely reminder and quality reporting, Accuracy in billing and secured data transfer.

Increase patient participation: Healthcare service providers can easily manage appointments electronically. Through mail conversations, symptoms of the disease can be identified by the service providers before scheduling appointments with specialists.

Better patient coordination: Because of EHRs, each provider may access the same accurate and current patient data. With patients seeing several specialists, this is extremely crucial. Also, it is helpful in emergency care, which can lower medical errors and prevent pointless tests.

Superior Diagnostics & Outcomes: Improved care coordination can result in higher-quality treatment and better outcomes for patients. Medical errors can be lessened or even prevented with the use of electronic health records (EHRs), which can lead to better patient outcomes.

- ✓ Efficiency in operations and cost savings
- ✓ Decreased expenses for transcribing
- ✓ Enhanced coding automation and documentation capabilities

- ✓ Lower expenses for pulling charts, storing them, and re-filing
- ✓ Fewer medical errors because to improved patient data access and error prevention warnings
- ✓ Improved disease management and patient education will result in enhanced patient health and treatment quality.

Advantages and Suitability of the Evolving Blockchain Technology for Secured Healthcare Data Transmission

Blockchain

Blockchain technology is a time-stamped data model that serves as a distributed, tamper-proof ledger to register transactions that include an organized list of records connected by chains, or blocks, and is used to store data. It is a hash chain featuring independence, verifiability, and immutability properties. Blockchains are built by having system nodes "mine" blocks or add to the structure using a hash of the transactions that users have recorded on the blockchain (Xia Q et al., 2017).

How Security Is Achieved through Blockchain (Yue, X et al., 2016)

- The blockchain storage and sharing system is independent of outside parties, and no one entity has complete control over how transactions are handled.
- An unchangeable list of records is continuously increasing on a blockchain network. Because of this, block chain technology allows for the secure distribution of assets among clients who are not trusted.
- Blockchain technology provides safe cryptographic methods to recognise and verify users and systems, enabling scalable, distributed, and secure access control.
- The sender of valid transactions maintained in a blockchain digitally signs and timestamps the transaction, providing cryptographically indisputable proof of a record's provenance and its existence at a specific time.

Advantages

- For many IoT applications, block chain acts as a tamperproof mechanism. IoT and blockchain technology together may offer a cutting-edge solution that enables accessibility and data transparency without the use of a third party.

- Based on blockchain technology, the system offers data provenance, auditing, and control for shared data in cloud repositories among big data institutions.
- A group of people and organisations can instantly exchange the data that has been recorded on the blockchain.
- The ownership, storage security, privacy protection, tamper-proof, and practical interoperability are the key features of the blockchain.
- A blockchain has no single point of failure and is simple to verify.
- All transaction records are stored on the blockchain and made available to all network nodes. This permits upholding integrity, openness, and toughness.
- Improved Speed.
- Reduce the cost of production.

Various Blockchain and Comparison

Blockchain networks come in four basic categories:

1. Public blockchains [Trustless/ Permissionless]
   - Anybody with internet access can log on to a blockchain platform to become an authorised node because the public blockchain is open and permissionless.
2. Private blockchains
   - Closed network. Only specific people are allowed to validate the data or transaction In a private blockchain, a company manages the system's access rights.
3. Consortium blockchains
   - Collaboration among multiple organisational members on a decentralised network. A number of companies operate the consortium blockchain. It is only possible for system participants to access the blockchain..
4. Hybrid blockchains
   - Hybrid blockchains are those that are overseen to some extent by the public blockchain but are managed by a single enterprise.

The below table 1 indicates the comparison of the various blockchain technics.

Table 1. Comparison of types of block chain

Particulars	Public	Private	Consortium	Hybrid
Characteristics	• No central Authority • Open Source • Transparency	• Individual organization based • Allowed only limited users • Owned by private authority but no central authority	• Transactions can be started or received by member nodes, with the validator node being in charge of initiating, receiving, and validating them.	• Combination of private and public
Advantage	• Anyone with internet can participate • Trust among users • High Transparency	• Faster • Strong Privacy • Increased performance and scalability	• more secure, scalable, and efficient	• Reduce Computational cost • Secured
Disadvantage	• Less energy efficiency • Less Privacy • Less performance scalability	• Not decentralized requires permission • Frequent rule changes as per owners' need	• Lack of Transparency Lack of Transparency	• Transparency

Table 2. Summary of included articles

Model	Types of Block chain	Techniques	Advantage
Blockchain-based searchable attribute-based encryption for electronic health record sharing:	permissioned blockchains	Searchable Attribute-based encryption	Ensure data confidentiality and access control High retrieval Efficiency.
Medical Data Sharing Scheme Based on Attribute Cryptosystem and Blockchain Technology	consortium block chain	Attribute cryptosystem for data sharing	Confidentiality Unforgeability
Tamper-resistant mobile health using Blockchain technology	permissioned blockchains	Combining mHealth with blockchain technology	tamperproof system Accessibility and data transparency without third party.
Blockchain-based medical records secure storage and medical service framework	Permissioned blockchains	Combination of hash chain with blockchain technology	Patients control their personal medical data Privacy Protection Secure Storage
Blockchain Technology applied to Electronic Health Records	permission less blockchain	Privacy preserved the use of personal health data	Scalability, data standardization and cost of operation
Privacy-preserving framework for access control and interoperability of electronic health records using blockchain technology	permissioned blockchain	Ancile framework	Performance cost and gas limit, Access Control Security

Block Chain Literature Review in Medical Internet of Things

Several scientists, authors, and academics have provided reviews on a variety of IoT and blockchain use cases in healthcare to process different types of data. This review provides analysis of blockchain technology and its significance in the medical field.

1. *Blockchain-based searchable attribute-based encryption for electronic health record sharing: (Niu.S et al., 2020).* This model protects the privacy of electronic health records by storing the encrypted information and the keyword index on the hospital's server and in a permissioned block chain. During the authentication process, this technique assists in preventing rogue doctors from uploading fake information. Also, it helps doctors diagnose patients quickly and precisely and increases the effectiveness of hospital activities.
2. *Medical Data Sharing Scheme Based on Attribute Cryptosystem and Block chain Technology: (Yang. X et al., 2020).* This approach combines the benefits of blockchain technology and cloud storage. The attribute-based cryptosystem is used in this scheme to guarantee the secrecy of medical data stored in the cloud and to confirm the legitimacy of the medical data source. High performance in terms of computing overhead and security is provided by this strategy.
3. *Tamper-resistant mobile health using Blockchain technology (Ichikawa, D et al., 2017).* For mHealth, blockchain acts as a tamper-proof system. Blockchain technology and mHealth may be combined in a novel way to offer accessibility and data transparency without the need for a third party like a contract research organisation.
4. *Blockchain-based medical records secure storage and medical service framework (Chen, Y et al., 2019).* The blockchain is considered as a storage supply chain where every transaction may be immutably recorded, confirmed, and held accountable. It is suggested that medical data be stored, shared, and used utilising this storage system and service framework based on the blockchain. In the medical blockchain network, price-adjustable, convenient movement of valuable medical data is possible.
5. *Blockchain Technology applied to Electronic Health Records (Alkhushayni et al., 2019).* The overall cost and the amount of time required to share patient data among various healthcare organisations could both be greatly reduced as a result of this research. The patients

may benefit from having more control over their own data thanks to this technology. Access to patient data that is owned by other entities may be available to medical practitioners and institutions like hospitals. Blockchain technology can be utilised to manage EHR while enhancing operational efficiency through process simplification and transparency. The concept drastically cuts the time required for patient data to be shared among various healthcare companies while also lowering overall costs.

6. *Privacy-preserving framework for access control and interoperability of electronic health records using blockchain technology (Dagher GG et al., 2018) (Zhang A et al., 2018).* By leveraging cryptographic approaches, this technology focuses on preserving the confidentiality and privacy of patients while enabling access control. a blockchain-based platform that protects the privacy of patients' sensitive data while enabling secure, interoperable, and quick access to medical records by patients, providers, and third parties. The Ancile framework uses advanced cryptographic methods for additional security and makes use of smart contracts in an Ethereum-based blockchain for increased access control and data complexity. This approach presents a blockchain system that achieves a high level of decentralization by giving the patient ownership and final control over EHRs. The author prioritises security and access control, which is why they use encryption and authentication throughout the blockchain.

Based on the above survey summarize in table 2.

Conclusion

The block chain is one of the most important technologies that will significantly affect all economic sectors in the years to come. This technological advancement will also have a significant impact on the healthcare industry. Numerous healthcare solutions that only rely on Internet of Things technology are already available on the market. However, the fact that they use a centralized database is one of the issues with their system. As a result, the data that is accessible could not be safe, and the companies might also increase the price that they charge for storage. However, these problems are addressed by the utilization of blockchain technology, and the user is made the total owner of his data in this manner. Without his authorization, no one else can have access to the information.

The literature review made it absolutely evident that most methods have adequate levels of precision when it comes to blockchain technology, analyzing its unique characteristics, and unsolved problems, and discovering as well as examining potential integration strategies for IoT and blockchain. To solve this challenge, a distributed and decentralized ledger technology called blockchain is used. This tackles both the centralized problems that arise while managing the record system model as well as the scalability difficulties that are present in the system. Transparency will be made possible by this tactic, which will also stop invaders from destroying patient records as they previously accomplished. Since, the data stored on these blockchains, which are distributed throughout a number of fog zones, will only be accessible to approved members of the medical community.

Disclaimer

None.

References

Azaria, A. Ekblaw, T. Vieira, and A. Lippman, *"Medrec: Using blockchain for medical data access and permission management,"* in 2016 2nd International Conference on Open and Big Data (OBD), 2016, pp. 25–30.

Azaria, A., Ekblaw, A., Vieira, T., Lippman, A. *MedRec: Using Blockchain for medical data access and permission management.* In Proceedings of the 2016 2nd International Conference on Open and Big Data (OBD), Vienna, Austria, 22–24 August 2016; pp. 25–30.

Alkhushayni. S, Al-Zaleq. D, and Kengne N. L. G., *"Blockchain Technology applied to Electronic Health Records,"* in International Conference on Computer Applications in Industry and Engineering, ser. EPiC Series in Computing, vol. 63. EasyChair, 2019, pp. 34–42.

Chen, Y., Ding, S., Xu, Z., Zheng, H., Yang, S. Blockchain-based medical records secure storage and medical service framework. *J. Med. Syst*. 2019, 43, 5.

Dagher G. G., Mohler J., Milojkovic M., Marella P. B. Ancile: Privacy-preserving framework for access control and interoperability of electronic health records using blockchain technology. *Sustainable cities and society*. 2018;39:283–297.

Gong. T., Huang. H., Li. P., Zhang. K., and Jiang H., *"A Medical Healthcare System for Privacy Protection Based on IoT,"* in 2015 Seventh International Symposium on Parallel Architectures, Algorithms and Programming, 2015, pp. 217–222.

Ichikawa, D., Kashiyama, M., Ueno, T. Tamper-resistant mobile health using Blockchain technology. JMIR mHealth uHealth 2017, 5, e111.

Niu S., Chen. L., Wang. J., and Yu. F., *"Electronic Health Record Sharing Scheme With Searchable Attribute-Based Encryption on Blockchain,"* IEEE Access, vol. 8, pp. 7195–7204, 2020.

Poongodi, J., K. Kavitha, and S. Sathish. "Healthcare Internet of Things (HIoT) data security enhancement using blockchain technology." *Journal of Intelligent & Fuzzy Systems* Preprint (2022): 1-11.

Xia Q., Sifah E., Smahi A., Amofa S., Zhang X. BBDS: Blockchain-based data sharing for electronic medical records in cloud environments. *Information*. 2017;8(2):44.

Yue, X., Wang, H., Jin, D., Li, M., Jiang, W. Healthcare data gateways: Found healthcare intelligence on Blockchain with novel privacy risk control. *J. Med. Syst.* 2016, 40, 218–218:8.

Yang. X., Li. T., Pei. X., Wen.L., and Wang C., *"Medical Data Sharing Scheme Based on Attribute Cryptosystem and Blockchain Technology,"* IEEE Access, vol. 8, pp. 45 468–45 476, 2020.

Zhang A., Lin X. Towards secure and privacy-preserving data sharing in e-health systems via consortium blockchain. *J. Med. Syst.* 2018;42(8):140.

Chapter 7

The Integration of Blockchain with Internet-of-Things (BCOT) in Smart Cities

M. C. Jayaprasanna[1], M.E
and S. Sujatha[2], PhD

[1]Anjalai Ammal Mahalingam Engineering College, Kovilvenni, Thiruvarur Dist., Tamilnadu, India
[2]BIT Campus, Anna University, Tiruchirappalli, Tamilnadu, India

Abstract

A smart city could be a framework that makes use of data and communication technology to boost productivity and improve the lives of residents while dealing with the issues of expanding urbanization. IoT devices are connected to the network and share data because IoT technology has been a major pillar of smart cities. Block Chain technology might function as a decentralized digital ledger that frees users from the need for a central administrator and allows for highly reliable and transparent data sharing. Blockchain can raise the IoT's quality by offering a trusted sharing service where data is accurate and possibly traceable. Data can be traced back to a source at any time, and it doesn't change over time, boosting data security. Food traceability might be a crucial factor in determining food safety. Anywhere along the chain, there could be a leak of information that could result in fraud and slow down business operations. As a result, the combination of blockchain and IoT may result in the production of trustworthy and secure data. A new technology called blockchain promises to address the scalability, dependability, and security issues that the IoT paradigm raises. The counterfeit product that was moved from one location to another can be uniquely identified using the Block-Chain of Things

In: Blockchain Enabled Secure Big Data Computing …
Editors: L. Ashok Kumar, D. Karthika Renuka, V. Chandrasekar et al.
ISBN: 979-8-89113-243-6

(BCOT) method presented in this study. The product's Traceability Identification (Tid) can then be stored on a blockchain, which cannot be altered. Product message exchanges are handled as blockchain transactions that smart contracts verify. A Block-Chain of Things (BCOT) system can be used to verify who owns a food item. It could be applied to increase food safety in cities. As a result, it increases speed while lowering risk in food transportation.

Keywords: blockchain, internet of things, security, counterfeiting product, food traceability

I. Introduction

Blockchain is a cutting-edge database technology that prevents data from being deleted or modified without a consensus network. This technology will provide a secure, immutable ledger for tracing values such as payments, user accounts, and orders. Since blockchain supports decentralized nature, it enables all transactions to be viewed transparently. It can be done by connecting to a node or by using blockchain viewers that let users see real-time transactions.

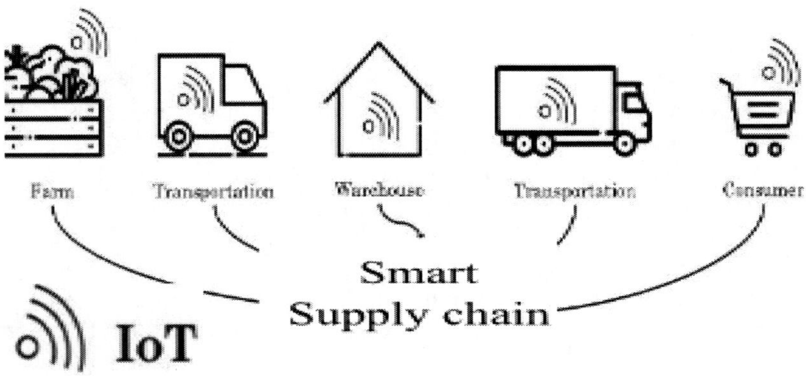

Figure 1. Internet of Things in Supply chain.

An interconnected system of machines, vehicles, and appliances known as the Internet of Things (IoT) allows for the collection and sharing of data without the need for human contact. IoT devices gather data and transfer it to a central data server, where it is analyzed and used to facilitate a variety of operations (Veena et al., 2015). There are numerous facets of our lives where the Internet of Things can be employed. Internet of Things devices are a useful tool to track and authenticate goods and shipments utilizing Internet technology in the supply chain. They can also keep an eye on how products are being stored along the entire supply chain (Niels et al., 2017), as seen in Figure 1. We can expect to have both IoT and Blockchain coming together more efficiently in the future.

II. Integration of IoT Has the Following Benefits

Blockchain-based IoT improves fault tolerance and eliminates the centralized server bottleneck. Blockchain network users have the ability to confirm the accuracy of the data they receive as well as the sender's identity. In order to increase security in BCOT, it can also enable privacy and security features like confidentiality and authentication (Antorweep et al., 2013).

Motivation

The development of wearable technology, the decline in price of embedded computers, advancements in storage technology, and expanded Internet and wireless access all contribute to the IoT's continued steady growth on a global scale. A more decentralized strategy is necessary to enable the long-term expansion of the IoT and to avoid failure, even while users are required to improve privacy and security within the centralized network topologies. Currently used centralized methods are managed by other entities. Users must have faith in these organizations to handle their IoT with various Block chain applications (Fran Casino et al., 2019).

Contribution

Dealers, clients, and raw material manufacturers are only a few of the numerous parties involved in the supply chain. The system is made more unstable and insecure by the engagement of several parties. With the use of BCOT in Figure 2, it is possible to ensure the ability to track shipments, make trustworthy payments, and other crucial characteristics. Blockchain is used to store, validate, and verify all of the sensors' sensitive data. BCOT is used by manufacturers and distributors of food goods to increase transparency by building a secure ledger that is only available to authorized users.

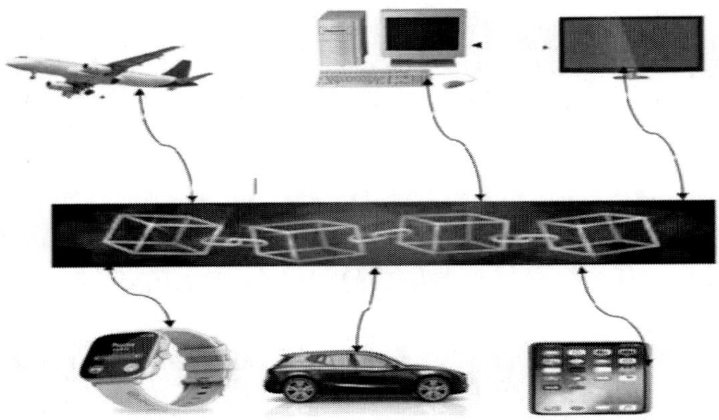

Figure 2. Secure Block Chain Internet of Things (BCOT).

III. Background

Overview

Without the use of a centralized third party, blockchain enables the exchange of transactions and information between untrustworthy parties. It is made up of time-stamped records of data that have been permanently, safely, and securely kept (Bhabendu K M et al., 2018). It is determined that private blockchain is the most appropriate type to be integrated with IoT and provide value to it based on the features of IoT networks and the classification of blockchain. The blockchain's structure is shown in Figure 3. Before being added to the chain, information contained in a block must first be verified.

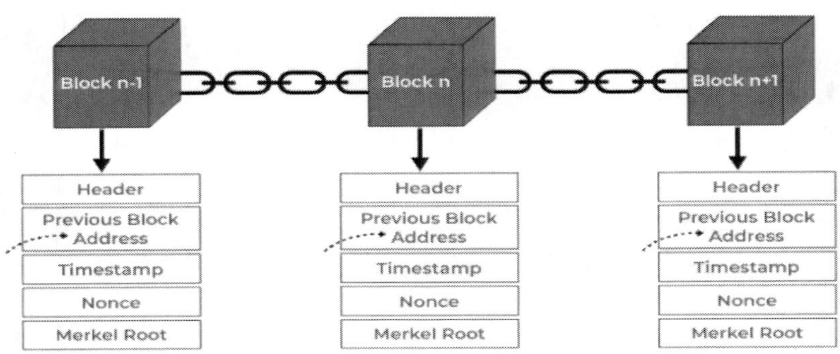

Figure 3. Block chain Architecture.

Features: Key Characteristics of Blockchain

Decentralization

Each transaction must be approved by the central trust agency (such as the central bank), which causes issues with cost and server performance.

The blockchain does not need a third party when operating in decentralized mode. Blockchain consensus methods are employed to preserve data stability in a decentralized network.

Persistence

Once a transaction is added to the blockchain network, it cannot be deleted or rolled back. Transactions that are invalid are not carried out.

Transparency

All transactions are carried out by the owner of the block connected to the address; the process is open, and no party to the transaction suffers a loss.

Distribution

Each node captures a copy of the data records, which are updated frequently.

Security

Security will be provided with the help of the Hash algorithm. The data cannot be changed or erased.

Comparison

The following Table 1 briefly addresses the differences between the centralized and the decentralized platforms.

Table 1. Comparison between centralized and decentralized system

Features	Centralized	Decentralized
Mode of Transaction	Centralization	Decentralization
Consuming Resource	High	Low
Data Storage	Centralized DB	Decentralized Ledger
Information Transparency	Low	High
Data Privacy	Low	High

Related Work

Various research on the security and traceability of logistics items using IoT in the supply chain exists. Traceability offers a product a digital identity and establishes a link with the user by giving them a fantastic view of the product's journey throughout its existence (Madumidha et al., 2019). IOT devices effectively reduce human intervention in the system. The traceability system is seen to be crucial to the food supply chain in order to prevent the risk of unmonitored products and the constant rise in food crises related to information about product safety and quality (Manu Bharadwaj, 2022).

IV System Design

Basic Model

Figure 4 shows the steps it takes to get a food product from the manufacturer to the consumer are represented by the extensive network known as the food supply chain. The procedures involve transferring raw resources, processing them into finished goods, transporting those goods, and giving them to the final consumer. Because of the intricate global food supply chain structure, product tampering and misrepresentation have dramatically increased. Because every member in the food supply chain keeps paper-based records, it is exceedingly challenging to identify the root of a problem. Manual record keeping leaves it open to supply chain changes. The food supply chain needs to promote transparency, accuracy in food traceability, and public access to source verification with sufficient information.

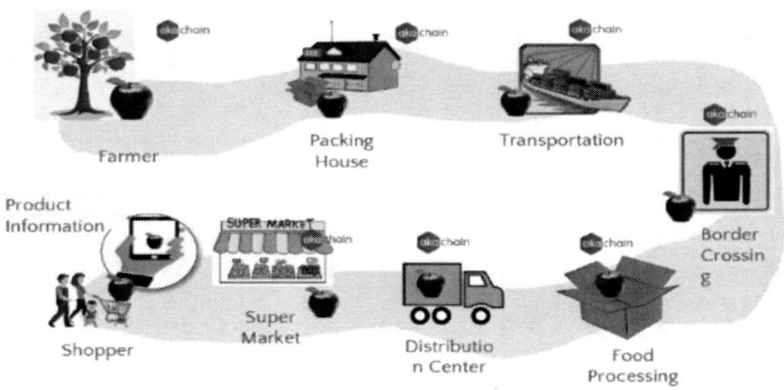

Figure 4. Food Distribution System.

Operation Flow

The suggested study introduces a Block-Chain of Things (BCOT) system that can identify a fake product even after it has been moved from one location to another. The product's Traceability Identification (Tid) can then be stored on a blockchain, which cannot be altered. Product message exchanges are handled as blockchain transactions that smart contracts ratify (Fabrizio et al., 2013 & Bhabendu et al., 2018). As shown in Figure 5, a BCOT system can be used to verify who owns a particular food item. It could be applied to increase food safety in cities. As a result, it increases speed while lowering risk in food transportation.

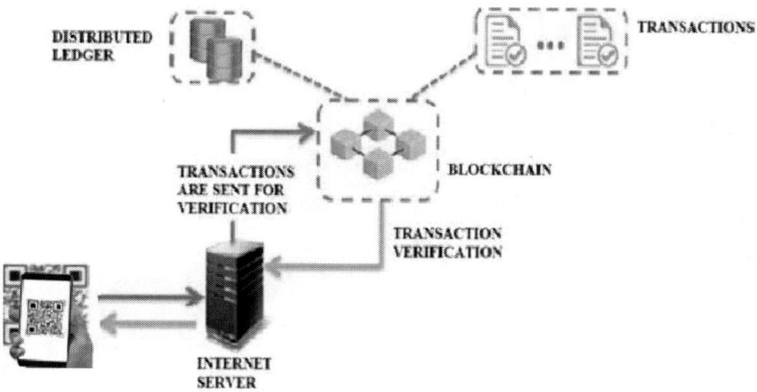

Figure 5. The proposed IoT-Blockchain integrated architecture.

V. Implementation

QR Code

In Figure 6, the QR (Quick Response) Code's extract of the product's unique identifier enables retrieval of the associated authentication parameter. It helps to manage and track data. QR code is then decoded and transferred to web browser for account creation.

Figure 6. Generation of QR code.

Account Creation

It helps users create (Figure 7) an account and connect with a blockchain that either stores them securely or is encrypted in their native browser. Since Wallet is a web wallet, we need to visit its official website to access the wallet. A password is used to confirm the ownership of the wallet and protect the information on it.

Blockchain Connection

The wallet lets users to access applications secretly and securely and allows programmes to communicate with the blockchain using a web wallet in Figure 8 and Figure 9. The transaction is sent to the Blockchain along with the created id. Everyone has a wallet on the blockchain platform, and to ensure the immutability of the data registered, the JSON generated is forwarded to a blockchain node.

Figure 7. Web Wallet Account Creation.

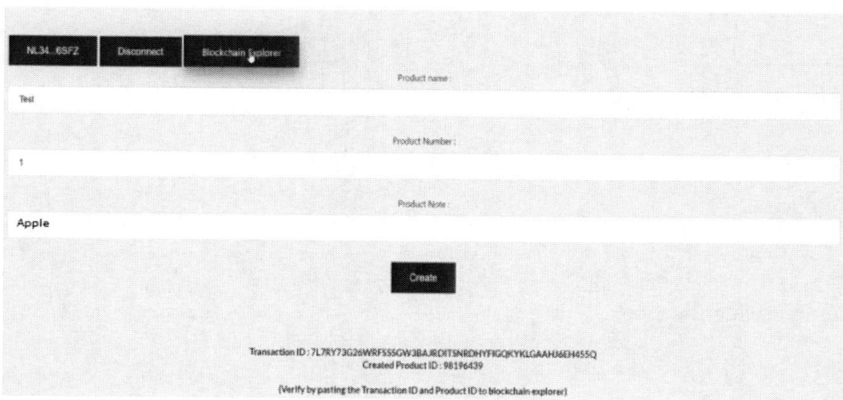

Figure 8. Transaction-id Creation.

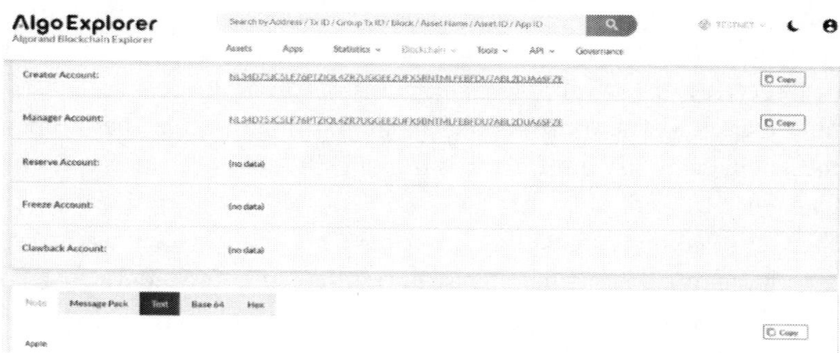

Figure 9. Connection with Block chain.

Verification

Finally, it will be verified (Figure 10) with the help of Algo Blockchain Explorer. Also, there was an asset created. The created ID will not be modified or used by a third party. Hence, security (Daniel M et al., 2018) is provided to enhance traceability and ensure the flow of products in the supply chain during transportation.

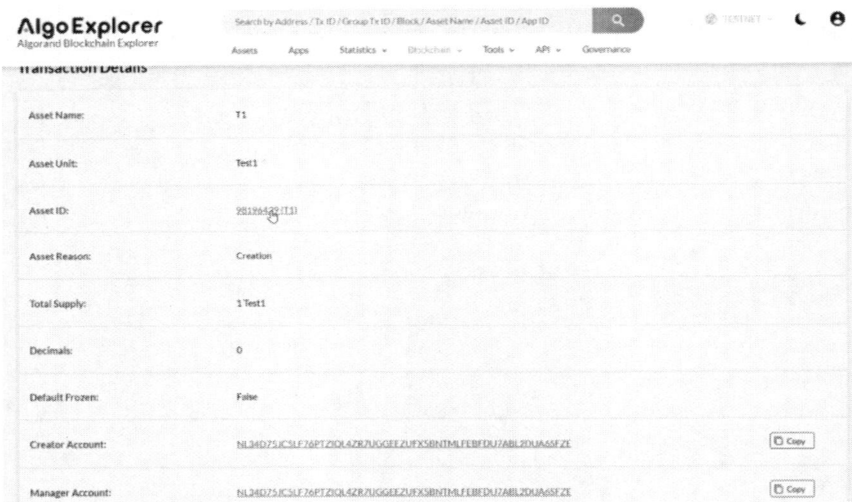

Figure 10. Verification.

Conclusion

Block-Chain of Things (BCOT) technology can recognize a counterfeit item even after it has been moved from one location to another. The product's Traceability Identification (Tid) can then be stored on a blockchain, which cannot be altered. The most popular method of information sharing between businesses in the food supply chain is through QR codes. Businesses only need to scan the QR codes on the packaging to upload the data to the shared blockchain system. To obtain a complete traceability history of the product back to the source, consumers can follow the same procedure. It could be applied to increase food safety in cities. As a result, it increases speed while lowering risk in food transportation.

References

Antorweep Chakravorty, Tomasz.W.Wlodarczyk, and Chunming Rong (2013), "Privacy-preserving data analytics for smart homes," in *Security and Privacy Workshops* (SPW), 23–27, IEEE.

Bhabendu Kumar Mohanta, Soumyashree S Panda and Debasish Jena (2018),"*An Overview of Smart Contract and Use cases in Blockchain Technology,"ICCCNT, IEEE-43488.*

Daniel Macrinici, Cristian Cartofeanu and Shang Gao (2018), "Smart Contract Applications within Blockchain Technology: A Systematic Mapping Study,"*Telematics and Informatics*, doi: https://doi.org/10.1016/j.tele.

Fabrizio Dabbene., Paolo Gay., & Cristina Tortia (2013)," Traceability issues in food supply chain management: A review,"*Bio systems engineering*, 120, 65-80.

Fran Casino, Thomas K. Dasaklis and Constantinos Patsakis (2019),"A systematic literature review of blockchain-based applications: Current status, classification, and open issues," *Telematics and Informatics,* vol. 36, pp. 55–81.

Manu Bharadwaj, 2022, https://tracextech.com/food-traceability-transforming-food-and-agriculture-industry.

Niels Hackius., & Moritz Petersen (2017),"*Blockchain in logistics and supply chain: trick or treat?," HICL*, 3-18.

S. Madumidha, Dr. P. Siva Ranjani, Ms. S. Sree Varsinee, and Ms. P.S. Sundari (2019)," Transparency and Traceability: In *Food Supply Chain System using Blockchain Technology with the Internet of Things*," ICOEI, ISBN: 978-1-5386-9439-8.

Veena Pureswaran, Sanjay Panikkar, Sumabala Nair and Paul Brody (2015), "Empowering the edge practical insights on a decentralized internet of things,"*IBM Institute for Business Value, Executive Report.*

Chapter 8

The Impact of Big Data in Smart Cities: Current Research Concepts and Future Perspectives

Gopika, G. S.[1]
Sreekrishna, M.[1]
Rajasree, R. S.[2]
Chithra, K.[3]
and Manikandan, M.[4]
[1]Department of CSE, Sathyabama Institute of Science and Technology, Chennai, India
[2]Department of AI & ML, New Horizon College of Engineering, Bangalore, India
[3]Department of Computer and Communication Engineering, NMAM Institute of Technology, Karnataka, India
[4]Department of Computational Intelligence, Faculty of Engineering and Technology, SRMIST, Chennai, India

Abstract

The rise of Internet of Things (IoT) techniques and the growth of big data have had a substantial effect on the viability of urban planning projects. The IoT allows for the connection of sensing, radio-frequency identifiers, and Bluetooth inside the real-world situation utilizing highly connected services, and big data offers the possibility for cities to gain useful insights from a significant quantity of data collected from multiple sources. Future smart cities face new and intriguing issues as a result of IoT and big data coming together in research areas that have not yet been fully revealed. In order for municipalities to realize the goal, concepts, and objectives of the uses for smart cities through achieving the key traits

In: Blockchain Enabled Secure Big Data Computing …
Editors: L. Ashok Kumar, D. Karthika Renuka, V. Chandrasekar et al.
ISBN: 979-8-89113-243-6

of a smart environment, these new difficulties mostly focus on issues related to business as well as technology. The technological devices and clever applications now in use in the world of smart cities are laid out in this article. Concentrating on how big data might radically alter urban populations at various levels, the ambitions of big data analytics to enable urban areas are examined.

Keywords: smart city, big data, internet of things, smart technologies, communication technologies

Introduction

In order for municipalities to realize the goal, concepts, and objectives of the use cases for smart cities through achieving the essential qualities of a smart environment, these new difficulties mostly focus on issues linked to business as well as technology. The technological devices and clever applications now in use in the world of smart cities are laid out in this article. Concentrating on how big data might radically alter urban populations at various levels, the ambitions of using big data analytics to promote urban areas are discussed. The usage of sensors can simplify routine tasks in everyday life to save labor and cut down on cooperation time. Large data volumes, which call for great agility and effectiveness, bring additional benefits to the table. Businesses used to gather, evaluate, and update data for potential decision-making purposes, but today's technology makes it much easier for businesses to identify ideas quickly (Alvear O et al. 2018). The biggest benefit in businesses today that has never been before and it's quicker processing. Dealing with massive data is crucial, and making conditions rapid, conditions. Results ought to serve as the main objective.

Apanaviciene R et al. (2020) gives the examination of deviations from the trends that can affect the results of business processes is undertaken, along with clustering as well as a better comprehension of the data driven by regulated industrial knowledge. Urban areas that can be developed to address technical problem problems are being considered as part of the creation of "smart cities" in a bid to boost the economy. The term "Big Data" refers to the vast amount of data, both organized and unorganized, that is controlled by using conventional databases. Big data uses both organised and unstructured data to depict the exponential expansion and accessibility of knowledge (Artyushina A et al. 2020). Big data systems will store information about smart city offerings in an effective manner thanks to programs for green infrastructure. This improved information is beneficial to the judgement in a number of smart public infrastructures. The various characteristics of big data include volume, diversity, velocity, unpredictability, and significance.

Perspectives of IoT and Big Data

There are numerous difficulties encountered while implementing smart cities using "big data & IoT" as the smart city movement grows. Austin LM et al. (2021) Suggested the largest problem is using standard data processors for handling because there are so many different sources and types. The problem posed by various surroundings and smart objects is information integration (Awais M et al. 2021). The situation becomes more complex as data is distributed, and the smart urban solution prevents the confidentiality of personal minds. The information has been collected by various organizations, and keeping the data is very difficult. With the integration of big data, data must be stored in a precise way without compromising quality of data (Figure 8.1). The enormous data size inside the big information context makes it difficult to secure the data. A safe metropolitan area is planned, and any data breaches must be prevented and adequately protected (Badidi E et al. 2020).

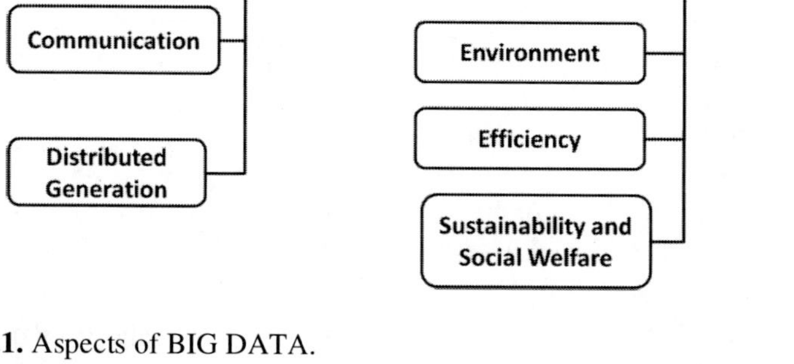

Figure 8.1. Aspects of BIG DATA.

Various Aspects of Smart Environments

Several elements that make together an intelligent town are shown in Figure as its individual components. It normally has five elements to intelligent city uses: gathering data, transmission and reception, storage, and fourth, the use of the information (Badii C et al. 2018). Sensory growth throughout many different areas has been greatly influenced by the collecting of data, which is application dependent (Badii C et al. 2020). Transferring the information from the information obtaining devices to the cloud for preservation and analysis makes up the second component of the process.

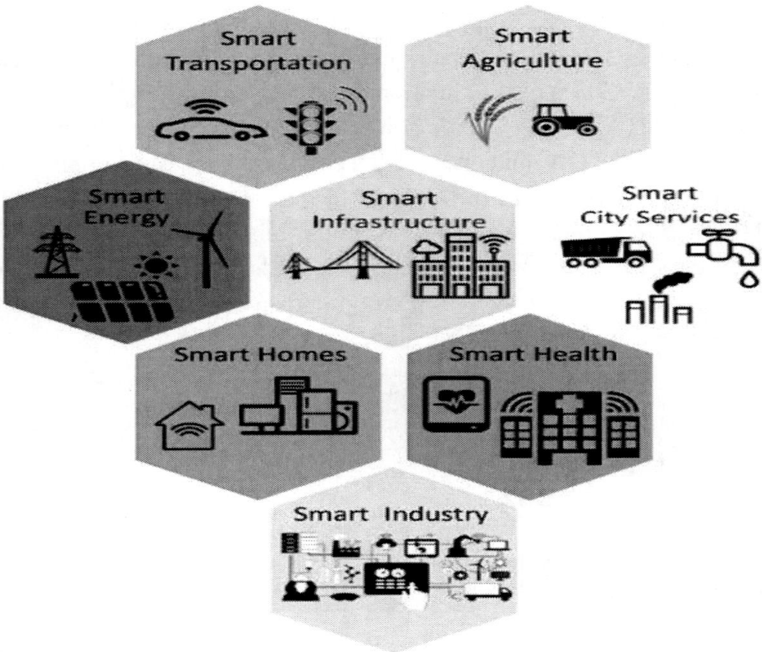

Figure 8.2. Constituents of Smart City.

Data analysis is the process of drawing conclusions and trends from obtained data to use in decision-making (Bauer M et al. 2021). In other circumstances, straightforward methods that involve simple making choices and aggregation would also be effective (Figure 8.2). The openness of the cloud enables the collection, processing, and instantaneous evaluation of varied information using statistical techniques, neural networks, and deep neural networks for more complicated decision-making (Beale M et al. 2021).

Smart Agriculture

Among the more crucial elements of the goals set forth in 2030 of the UN is the safety of food. Belli L et al. (2021) recommended the need to guarantee that agriculture is made viable and that scarce assets like water can be utilized successfully has been a top priority for locations all over the world due to the growing global population and escalating worldwide warming that results in unpredictable weather in food-producing regions of the world. Bhattacharya S et al. (2022) suggested the deployment of sensors in crops or farms to assess multiple variables to aid in making decisions while avoiding illnesses, insects, and other issues is known as "smart agriculture" (Bıyık C et al. 2021).

Smart Energy

The main generating source for most electrical systems, which is often a hydraulic or fossil fuel-based power plant, only provides energy in one direction (Burhan M et al. 2018). The energy production scheme used in such networks requires that the electricity generated by those sources outstrip the demand by a significant amount to ensure a constant supply of power because power generation is managed via feedback provided by substations, but because there isn't any data input from the consumer end (Cepeliauskaite G et al. 2021). Determining flaws in these systems and taking corrective action is likewise a time-consuming task.

Smart Health

ICT is used in the context of "smart health" to increase access to and the caliber of medical care. This subject has attracted the attention of scholars and healthcare professionals alike because of the growing population and rising healthcare expenses. The overcrowded health services of today are unable to meet the growing demand of the public. With telemedicine solutions and improved diagnosis support for doctors using AI, smart health strives to guarantee that health is accessible to as numerous individuals as feasible.

Smart Home

The smart residence is a key element of the Smart City concept since it serves as the focal point of city residents' lives (Chang F et al. 2020). Sensing devices that are positioned all over an individual's residence and offer data regarding the home and its occupants are used in "smart homes." These sensors may include devices which monitor behavior among users, such as movement trackers, environmental sensors, and strength/energy usage.

Smart Industry

Businesses all over the world are actively working to become more effective and profitable while lowering costs (Coardoş D et al. 2019). The concept of an interconnected factory that has all of its middleman functionaries fully integrated and cooperating with one another is at the heart of the sector 4.0 paradigm (Demirel D et al. 2021). The global web of things has made this feasible. The sector has benefited from the use of IoT in production and manufacturing processes, as well as digital physical systems combining workers and machines, including improved speed and innovation, efficiency of manufacturing schemes (resources as well as procedures), better product quality, and increased safety for factory workers. However, working with a variety of diverse equipment and gadgets poses a unique set of difficulties, necessitating cyber-physical systems with flexibility in configuration, accessibility, and quick deployment for application in internet of things (IoT) apps for smart businesses (Dogru N et al. 2018).

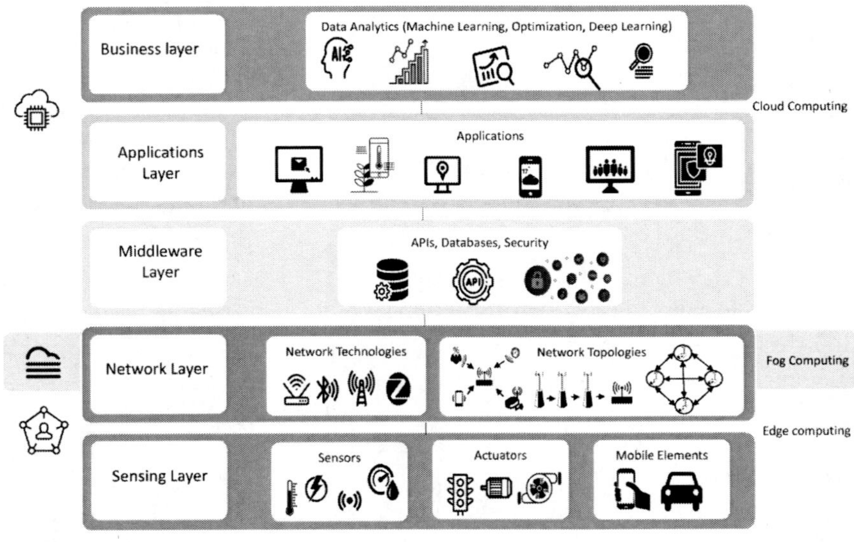

Figure 8.3. Architectural Framework for Smart City.

Smart Infrastructure

The built environment of a city is crucial to its standard of life, thus municipalities must develop new bridges, roads, and structures for residents' use, as well as carry out upkeep to ensure continuous use (Ejaz W et al. 2019). By employing acceleration sensors and intelligent substances to track the structural wellness of buildings and bridges, intelligent construction helps cities make sure that everything is in good condition and usable. These monitors' data collection enables predictive repair of these crucial components, preserving municipal operations as usual (El Fawal AH et al. 2017).

Smart City IoT Architectures

By making use of services offered by the cloud, the Internet of Things integrates the functions of data sensing, transfer/reception, processing, and storage [43]. Based on science and technology, the typical IoT design has five levels, each of which operates on data from the layer before. Additionally, it displays the three various IoT system designs that exist (Fadhil JA et al. 2020). The ability to sense layer, also known as the perception layer, is made up of controllers that may interact with real objects, including RFID readers for detecting RFID tags, and also sensors that can gather data about physical quantities relevant to any application (Ganesan M et al. 2019). The Middleware layer receives the data read by the layer of sensing by using the networking layer and wireless network technologies including Wi-Fi, Bluetooth, Zigbee, and cellular internet, between others (Figure 8.3). The software layer leverages the data via multiple APIs and databases management tools to offer users with services, while its middleware level offers an umbrella interface for the sensor layer's devices (Garmaroodi MSS et al. 2020).

Features of Big Data and Smart City

The cities with smart technology continuously produce a vast volume of data from many applications, such as medical care, energy management, traffic management, environment surveillance, etc. Processing information is a significant difficulty because the rate of data production varies for various sensors (depending on the application requirements) (Golbabaei F et al. 2020). For instance, temperature devices might produce data every hour, whereas GPS devices might produce data every few seconds. Given that the data that smart cities produce comes from a variety of sources, data quality is also crucial (Figure 8.4). The source of the information should be reliable, data should be gathered from a variety of people, and the amount of sampling of the information should be raised in order to assure the integrity of the data (Huang DC et al. 2018).

Computation in Infrastructure

This is a guide to the many computing frameworks that are typically employed for massive data sets originating from smart towns. Informatics facilities have the ability to deal with data in batches or in genuine/near-real time, based on the data demands. For instance, ignite is frequently used for processing in real time whereas Hadoop is often implemented for processing batches (Huseien GF et al. 2021).

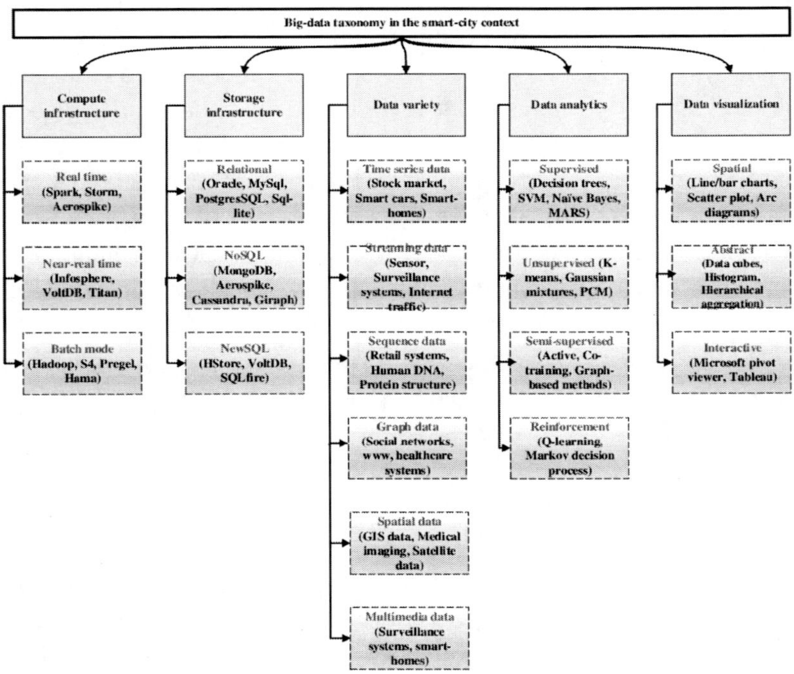

Figure 8.4. Features of BIGDATA.

Storage

The information gathered from the smart city initiatives is very diverse, covering everything from text to media (Jin J et al. 2014). Since a large portion of sensor data is naturally unorganised, additional databases must be used in contrast to the standard relation-based database design. As a result, the second vertical that chooses the kind of storage required depending on the massive amounts of data type is store architecture (Kang Y-S et al. 2016).

Variety

Details like websites, the internet, body part networks, etc. are well fitted to be treated as a graph data structure and is therefore identified to as vertex data (Kaur P et al. 2019). Information gathered from resources like satellite imagery, mapping systems, or contents from health imaging is referred to as data pertaining to space (Kézai PK et al. 2020). Images, footage, and audio all feature in data that is multimedia. Each of the information categories discussed here has distinct qualities and is examined using different methods for data mining.

Analytics

Li Y et al. (2018) suggested to gather knowledge from the massive data produced by intelligent towns for making forecasts, seeing patterns, uncovering concealed data, or making judgements, a wide range of algorithms based on machine learning are employed. A suitable analytical strategy must be chosen based on the needs. Unsupervised approaches are typically used for grouping and source signal separation, whereas algorithms with supervision are employed for categorization, prediction, and regress (Lombardi M et al. 2021). On huge data sets without labels, where conventional algorithms with supervision cannot be used, pick up truck-supervised techniques are used. The structural similarities across labelled and raw data are effectively used by semi-supervised methods of learning to expand the functional translation across huge data sets (Lopes SF et al. 2020).

Visualization

One of the most important features of big data in an intelligent urban infrastructure is that it provides a visual context for the data, allowing the many people involved to comprehend its relevance (Mekki K et al. 2019). Through the use of geospatial display schemes, data elements can be mapped to specific locations on a coordinate system, allowing for a more straightforward depiction of a complicated data collection. Such a method can be seen in line charts, bar charts, scatter diagrams, etc. Large-scale data are consolidated using abstract display approaches before being presented to display units. Some examples include hierarchical aggregation, statistical binning, and data cubes (Mogro AE et al. 2021). The third type of representation, known as interactive presentation, includes methods that enable real-time user engagement and representation.

Sensing Methodologies

Mohapatra S et al. (2018) gives the core of innovations in smart cities is detecting. Sensors give the information and data that are used to produce breakthroughs for intelligent cities. Numerous sensors are employed as part of these efforts due to the greatly varying characteristics of these projects & their various parts (Moscholidou I et al. 2020). The main IoT element in urban smart city structures, sensors enable communication among the intelligent city network and the people who live there and enable the creation of new services. A factor to keep in mind is that most of these sensors have numerous uses, as was said (Nesti G et al. 2020). Additionally, every application calls for the employment of numerous distinct types of sensors and the measurement of various physical values (Figure 8.5).

Ambient Sensors

Detectors that measure tangible values representing the conditions of the environment, such as humidity, Fahrenheit, brightness, and tension, are referred to as environmental sensors (Nkonyana T et al. 2019). Ambient sensors are employed in a wide range of applications associated with smart cities, including intelligent homes where they are utilised to control the level of comfort, as well as for clever city amenities.

Challenge	Mitigation/Research Direction
Security and Privacy	Encryption New authentication mechanisms New standards to anonymize data Prevent data leakage
Smart Sensors	Interoperability: Open Standards Reliability and Robustness: Decentralized and distributed architectures and decision making Power and Memory: Energy harvesting, Low power sensors, New database storage systems
Networking	Low power networks, Network schemes that ensure fluent mobility and routing
Big data analytics	New algorithms which work with different natured data, Develop scalable and explainable AI

Figure 8.5. Sensing Methodologies for various fields.

Bio Sensors

Biosensors are used to assess many aspects of an organism's health (Oztemel E et al. 2020). In smart neighborhoods, biological sensors are employed to monitor people for medical reasons. These sensors comprise oximetry, blood sugar, electroencephalography (EEG), electromyography (EMG), electrocardiography (ECG), resistance measurement on the skin, respiration, breathing detectors, and many.

Chemical

Sensors for chemicals are employed for determining the chemical makings of materials. Examples include smoke detectors, pH and other water-related sensors, and gas sensors that can monitor/detect the levels of carbon monoxide (CO), atmospheric carbon dioxide (CO2), & other gases for surveillance of air quality (Popa D et al. 2019).

Electric Sensors

Electric sensors are frequently used in smart electricity systems and smart homes to track the energy usage of consumers and equipment (Qiu C et al. 2020). These sensors allow for accurate measurement of electricity. In measuring current and voltage, devices include electricity transformers and voltmeter sensors.

PIR Sensors

Presence sensors signal the arrival of people or things (Queralta JP et al. 2021). Bates switching can be utilized on doorways and windows for security, passively Infrared (PIR) sensors, which are widely common and used for recognising human motion, and inductive loop sensors, which employ induction of electricity, may be utilized to detect activity in transportation systems (Quadar N et al.2021). Object distance is also measured using ultrasonic sensors. This kind also includes capacitive sensors, which may be employed to measure location.

Communication Technologies for Smart City

RFID

The device that reads the signal and the tag are two completely different types of devices used in RFID connection. When a tag approaches the reader, which is often powered, a transmission of information occurs after permission as a tag draws power to the reader's battery (Ramírez-Moreno MA et al. 2021). These tags are referred to as inactive tags, although there additionally exist active keywords that have independent power.

Near Field Communication

In opposition to RFIDs, NFC requires electricity and allows data transmission and receiving in each direction. In contrast to RFID, whose is unable to be employed for measurements or control tasks, this makes it possible to employ NFC to manage and setup devices (Ray PP et al. 2021). While NFC and RFID both use resonances, NFC is only used over extremely tiny distances. NFC technology is widely employed in intelligent residences and in projects including mobile payments.

Bluetooth

The format of the protocol is made for low-bandwidth, short-range interactions in a setup where objects can quickly enter or quit the internet

(Salhaoui M et al. 2019). Due to the existence of a controlling device at the heart of the communication mechanism, Bluetooth naturally supports the star topology (Shahrour I et al. 2021). It is capable of processing 2 Mbps of data at its highest possible rate while operating across the 2.4 megahertz ISM band. Due to its direct connection to cell phones without requiring a middleman hub device, the Bluetooth protocol has become widely employed in smart homes (Singh D et al. 2017).

Cellular

3G, 4G, and 5G wireless connections are referred to as mobile technology. All three of these technologies were the most widely used IoT enablers, besides Bluetooth and Zigbee (Ahmed S et al. 2014). In comparison to other guidelines, mobile signaling has a high transfer rate and facilitates more content-rich programmes. They are favored for numerous tasks where energy is not a concern due to the long range that they offer.

Current Problems and Future Perspectives

In addition to the benefits of big data analytics outlined above, there are additionally a number of open-ended queries (Sousa PHF et al. 2019). To provide new scholars in the area with future recommendations, these kinds of problems are being discussed.

Data Integration

The data assimilation from various businesses, various settings, and a large range of sensor devices make smart-cities conceivable. Even inside a company, data integration presents significant difficulties, particularly in the IT industry (Alhussein M et al. 2014). Development of open standards throughout the information technology and communications sectors may therefore aid in lowering the technological obstacles, but it is most difficult to overcome the societal and administrative ones (Syed AS et al. 2021). Thus, adequate attention should be paid to the creation of standards that will direct the creation of future smart cities. A city that is smart has recently incorporated a number of technologies, lowering the technological obstacles to handling the data.

Security and Privacy

Cybersecurity becomes a crucial concern as smart cities interconnect various gadgets to the World Wide Web (Tian S et al. 2019). The current Mirai spyware, which infected connected devices and recruited them into a botnet, causing internet service to be interrupted for thousands of people, demonstrates the dangers associated with this smart paradigm. The main issues that security experts are concerned about are risks related to shoddy pairing and discovering protocols that may reveal device knowledge, inadequate authorization, shoddy encryption when communicating that can reveal private information, and vulnerabilities in the equipment/sensors that could enable remote spying (Zimmermann L et al. 2018).

Data Analytics

A smart city's success is largely dependent on the capability of data analysis. To make sense of the vast amounts of data that a smart city collects, new information-mining algorithms and visualization techniques are needed. Real-time analytics are significantly more important for their better operation than the conventional "store and process later" scenario. Thus, the difficulties are posed by stringent time constraints that can impair the efficiency of a smart city, in addition to the magnitude and variability of the data. The robustness, productivity, and efficacy of the current methods used for data mining should also be kept when the amount of data increases (Tian S et al. 2021).

Conclusion

When examining speculations in every variation, technological support is always necessary. There is always quick access to numerous community services, including local agencies, security forces, free libraries, and educational institutions. In the present article, a system built on the Internet of Things is described that will benefit both the government and the populace.

The public benefits in addition to the executive agencies from effective super city design, excellent safety, and an atmosphere that is healthy. The Hadoop ecosystem, which sits at the top of the stack, processes large amounts of data that arrive quickly. By using the datasets from the current intelligent networks, the effectiveness of the system is assessed. The privacy and safety of IoT with smart neighborhoods is a significant field for research in terms of cryptography methods, authenticating protocols, anonymization of information methods, and other approaches to prevent unauthorized use of the IoT network. As previously indicated, technologies like blockchain might be employed to implement access monitoring and management, secured device discovery, spoofing protection, as well as information loss prevention while maintaining the use of all aspects of encryption. The potential uses of intelligent towns where big data analytics might have an important effect were also explored, as well as a potential company model and architectural with the purpose of controlling big data for intelligent cities. Additionally, many case studies were looked at. In order to give fresh researchers in the field research directions, a number of open research problems were finally presented. Big data can be useful for gathering crucial information and for aiding in decision-making. Big data research in the context of smart cities is still in its infancy, but the issues that have been raised and need to be resolved make it a useful area of study.

References

Ahmed S. (2014). Performance analysis of Mobile WiMAX Technology. In: *Proceedings of the international conference on computing for sustainable global development* (INDIACom), New Delhi, India; March 5-7, p. 959-61.

Alhussein, M., Muhammad, G., Hossain, M. S., and Amin, S. U. (2018). Cognitive IoT-cloud integration for smart healthcare: case study for epileptic seizure detection and monitoring. *Mob Netw Appl.*, 23(6), 1624-35. doi: 10.1007/s11036-018-1113-0.

Alvear, O., Calafate, C. T., Cano, J. C., and Manzoni, P. (2018). Crowdsensing in smart cities: overview, platforms, and environment sensing issues. *Sensors (Basel).*, 18(2), 460. doi: 10.3390/s18020460, PMID 29401711.

Apanaviciene, R., Urbonas, R., and Fokaides, P. A. (2020). Smart building integration into a smart city: comparative study of real estate development. Sustainability., 12(22), 9376. doi: 10.3390/su12229376.

Artyushina, A. (2020). Is civic data governance the key to democratic smart cities? The role of the urban data trust in Sidewalk Toronto. *Telemat Inform.*, 55, 101456. doi: 10.1016/j.tele.2020.101456.

Austin, L. M., and Lie, D. (2021). Data trusts and the governance of smart environments: lessons from the failure of Sidewalk Labs. *Surveill Soc.*, 19(2), 255-61. doi: 10.24908/ss.v19i2.14409.

Awais, M., Raza, M., Singh, N., Bashir, K., Manzoor, U., Islam, S. U. et al. (2021). LSTM based Emotion Detection using Physiological Signals: IoT framework for Healthcare and Distance Learning in COVID-19. *IEEE Internet Things J.*, 8(23), 16863-71. doi: 10.1109/JIOT.2020.3044031, PMID 35582634.

Badidi, E., Mahrez, Z., and Sabir, E. (2020). Fog computing for smart cities' big data management and analytics: a review. *Future Internet.*, 12(11), 190, doi: 10.3390/fi12110190.

Badii, C., Bellini, P., Cenni, D., Marazzini, M., Nesi, P., Pantaleo, G., et al. (2018). Snap4City: A scalable IOT/IOE platform for developing smart city applications. *Proceedings of the 2018. IEEE SmartWorld, ubiquitous intelligence & computing, advanced & trusted computing, scalable, computing & communications, cloud & big data computing, Internet of people and smart city innovations*. Guangzhou, China, 8–12 October, doi: 10.1109/SmartWorld.2018.00353.

Badii, C., Bellini, P., Cenni, D., Mitolo, N., Nesi, P., Pantaleo, G., et al. (2020). Industry 4.0 Synoptics controlled by IoT applications in Node-RED. In: *Proceedings of the IEEE international conferences on internet of things (iThings)*, Rhodes, Greece; November 2-6.

Bauer, M., Sanchez, L., and Song, J. S. (2021). IoT-enabled smart cities: evolution and outlook. *Sensors (Basel).*, 21(13), 4511. doi: 10.3390/s21134511, PMID 34209436.

Beale, M., Uchiyama, H., and Clifton, J. C. (2021). IoT evolution: what's next? *IEEE Wirel Commun.*, 28(5), 5-7. doi: 10.1109/MWC.2021.9615126.

Belli, L., Cilfone, A., Davoli, L., Ferrari, G., Adorni, P., Di Nocera, F., et al. (2020). IoT-enabled smart sustainable cities: challenges and approaches. *Smart Cities.*, 3(3), 1039-71. doi: 10.3390/smartcities3030052.

Bhattacharya, S., Somayaji, S. R. K., Gadekallu, T. R., Alazab, M., and Maddikunta, P. K. R. (2022). A review on deep learning for future smart cities. *Internet Technol Lett.*, 5(1), e187. doi: 10.1002/itl2.187.

Bıyık, C., Abareshi, A., Paz, A., Ruiz, R. A., Battarra, R., Rogers, C. D. F. et al. (2021). Smart mobility adoption: a review of the literature. *JOItmC.*, 7(2), 146. doi: 10.3390/joitmc7020146.

Burhan, M., Rehman, R. A., Khan, B., and Kim, B. S. (2018). IoT elements, layered architectures and security issues: A comprehensive survey. *Sensors (Basel).*, 18(9), 2796. doi: 10.3390/s18092796, PMID 30149582.

Cepeliauskaite, G., Keppner, B., Simkute, Z., Stasiskiene, Z., Leuser, L., Kalnina, I., et al. (2021). Smartmobility services for climate mitigation in urban areas: case studies of Baltic countries and Germany. *Sustainability.*, 13(8), 4127. doi: 10.3390/su13084127.

Chang, F., and Das, D. (2020). Smart nation Singapore: developing policies for a citizen-oriented smart city initiative. In: *Developing national urban policies: ways forward to green and smart cities*. Berlin, Heidelberg, Germany: Springer, p. 425-40.

Coardoș, D., Tîrziu, E., and Gheorghe-moisii, M. (2019). A general framework based on IoT technology for smart governance 2019. IE (informatics in economy) international conference. Bucharest, Romania, May 30-31, doi: 10.12948/ie2019.06.03.

Collini, E., Nesi, P., and Pantaleo, G. (2021). Deep learning for short-term prediction of available bikes on bike-sharing stations. IEEE Access., 9, 124337-47. doi: 10.1109/ACCESS.2021.3110794.

Demirel, D. (2021). How the smart governance model shapes cities? Cases from Europe. *J Enterpr Commun.*

Diez, A., Khoa, N. L. D., Makki Alamdari, M., Wang, Y., Chen, F., and Runcie, P. (2016). A clustering approach for structural health monitoring on bridges. *J Civ Struct Health Monit.*, 6(3), 429-45. doi: 10.1007/s13349-016-0160-0.

Dobre, C., and Xhafa, F. (2014). Intelligent services for big data science. *Future Gener Comput Syst.*, 37, 267-81. doi: 10.1016/j.future.2013.07.014.

Dogru, N., and Subasi, A. (2018). Traffic accident detection using random forest classifier. In: *Proceedings of the 2018 15th Learning and Technology Conference*, L and T 2018, Jeddah, Saudi Arabia; February 25-26, p. 40-5. doi: 10.1109/LT.2018.8368509.

Ejaz, W., and Anpalagan, A. (2019). Internet of things for smart cities: technologies, big data and security. Berlin, Heidelberg, Germany: Springer, doi: CrossRef.

El Fawal, A. H., Mansour, A., Najem, M., and Le Roy, F. (2017). Le Jeune, D. LTE-M adaptive eNodeB for emergency scenarios. In: *Proceedings of the international conference on information and communication technology convergence (ICTC),* Jeju, Korea; October 18-20, p. 536-41.

Fadhil, J. A., Omar, O. A., and Sarhan, Q. I. (2020). A survey on the applications of smart home systems. In: *Proceedings of the 2020 international conference on computer science and software engineering (CSASE)*, Duhok, Iraq; April 16-18, 168-73. doi: 10.1109/CSASE48920.2020.9142103.

Ganesan, M., and Sivakumar, N. (2019). IoT based heart disease prediction and diagnosis model for healthcare using machine learning models. In: *Proceedings of the 2019 IEEE international conference on system, computation, automation and networking. Pondicherry*, India: ICSCAN; March 29-30, p. 1-5. doi: 10.1109/ICSCAN.2019.8878850.

Garmaroodi, M. S. S., Farivar, F., Haghighi, M. S., Shoorehdeli, M. A., and Jolfaei, A. (2020). Detection of anomalies in industrial IoT systems by data mining: study of CHRIST Osmotron water purification system. *IEEE Internet Things J.*, 8(13), 10280-7. doi: 10.1109/JIOT.2020.3034311.

Golbabaei, F., Yigitcanlar, T., and Bunker, J. (2021). The role of shared autonomous vehicle systems in delivering smart urban mobility: A systematic review of the literature. *Int J Sustain Transp.*, 15(10), 731-48. doi: 0.1080/15568318.2020.1798571.

Guo, J., Xie, X., Bie, R., and Sun, L. (2014). Structural health monitoring by using a sparse coding-based deep learning algorithm with wireless sensor networks. *Pers Ubiquit Comput.*, 18(8), 1977-87. doi: 10.1007/s00779-014-0800-5.

Huang, D. C., Lin, C. F., Chen, C. Y., and Sze, J. R. (2018). The Internet technology for defect detection system with deep learning method in smart factory. In: *Proceedings of the 2018 4th international conference on information management*. Oxford, UK: ICIM; May 25-27, p. 98-102. doi: 10.1109/INFOMAN.2018.8392817.

Huseien, G. F., and Shah, K. W. (2021). Potential applications of 5G network technology for climate change control: A scoping review of Singapore. *Sustainability.*, 13(17), 9720. doi: 10.3390/su13179720.

Jin, J., Gubbi, J., Marusic, S., and Palaniswami, M. (2014). An information framework for creating a smart city through internet of things. *IEEE Internet Things J.*, 1(2), 112-21. doi: 10.1109/JIOT.2013.2296516.

Kang, Y. S., Park, I. H., Rhee, J., and Lee, Y. H. (2016). MongoDB-based repository design for IoT-generated RFID/sensor big data. *IEEE Sens J.*, 16(2), 485-97. doi: 10.1109/JSEN.2015.2483499.

Kaur, P., Kumar, R., and Kumar, M. (2019). A healthcare monitoring system using random forest and internet of things (IoT). *Multimed Tools Appl.*, 78(14), 19905-16. doi: 10.1007/s11042-019-7327-8.

Kézai, P. K., Fischer, S., and Lados, M. (2020). Smart economy and startup enterprises in the Visegrád Countries—A comparative analysis based on the Crunchbase Database. *Smart Cities.*, 3(4), 1477-94. doi: 10.3390/smartcities3040070.

Li, Y., Cheng, X., Cao, Y., Wang, D., and Yang, L. (2018). Smart choice for the smart grid: narrowband Internet of Things (NB-IoT). *IEEE Internet Things J.*, 5(3), 1505-15. doi: 10.1109/JIOT.2017.2781251.

Lombardi, M., Pascale, F., and Santaniello, D. (2021). Internet of Things: A general overview between architectures, protocols and applications. *Information.*, 12(2), 87. doi: 10.3390/info12020087.

Lopes, S. F., Pereira, R. M., Lopes, S. O., Coutinho, M., Malheiro, A., Fonte, V. (2020). Yet a smarter irrigation system. Lect. Notes inst. comput. *Sci Soc Inform Telecommun Eng LNICST.*, 323, 337-46.

Mekki, K., Bajic, E., Chaxel, F., and Meyer, F. (2019). A comparative study of LPWAN technologies for large-scale IoT deployment. *ICT Express.*, 5(1), 1-7. doi: 10.1016/j.icte.2017.12.005.

Mogro, A. E., and Huertas, J. I. (2021). Assessment of the effect of using air conditioning on the vehicle's real fuel consumption. *Int J Interact Des Manuf.*, 15(2-3), 271-85. doi: 10.1007/s12008-021-00750-8.

Mohapatra, S., Patra, P. K., Mohanty, S., and Pati, B. (2018). Smart health care system using data mining. In: *Proceedings of the 2018 international conference on information technology. Bhubaneswar*, India: ICIT; December 19-21, p. 44-9. doi: 10.1109/ICIT.2018.00021.

Moscholidou, I., and Pangbourne, K. (2020). A preliminary assessment of regulatory efforts to steer smart mobility in London and Seattle. *Transp Policy.*, 98, 170-7. doi: 10.1016/j.tranpol.2019.10.015.

Nesti, G. (2020). Defining and assessing the transformational nature of smart city governance: insights from four European cases. *Int Rev Admin Sci.*, 86(1), 20-37. doi: 10.1177/0020852318757063.

Nkonyana, T., Sun, Y., Twala, B., and Dogo, E. (2019). Performance evaluation of data mining techniques in steel manufacturing industry. *Procedia Manuf.*, 35, 623-8. doi: 10.1016/j.promfg.2019.06.004.

Oztemel, E., and Gursev, S. Literature review of Industry 4.0 and related technologies. *J Intell Manuf.*, 31(1), 127-82. doi: 10.1007/s10845-018-1433-8.

Popa, D., Pop, F., Serbanescu, C., and Castiglione, A. (2019). Deep learning model for home automation and energy reduction in a smart home environment platform. *Neural Comput Appl.*, 31(5), 1317-37. doi: 10.1007/s00521-018-3724-6.

Qiu, C., Wu, F., Lee, C., and Yuce, M. R. (2020). Self-powered control interface based on Gray code with hybrid triboelectric and photovoltaics energy harvesting for IoT smart home and access control applications. *Nano Energy.*, 70, 104456. doi: 10.1016/j.nanoen.2020.104456.

Quadar, N., Chehri, A., Jeon, G., and Ahmad, A. (2021). Smart water distribution system based on IoT networks, a critical review. In: *Human centred intelligent systems*. Berlin, Heidelberg, Germany: Springer, p. 293-303. doi: 10.1007/978-981-15-5784-2_24.

Queralta, J. P., Gia, T. N., Tenhunen, H., Westerlund, T., and Edge, A. I. (2019). LoRa-based health monitoring: fall detection system with fog computing and LSTM recurrent neural networks. In:. In: *Proceedings of the 2019 42nd international conference on telecommunications and signal processing (TSP)*, Budapest, Hungary; July 1-3, p. 601-4.

Ramírez-Moreno, M. A., Keshtkar, S., Padilla-Reyes, D. A., Ramos-López, E., García-Martinez, M., Hernández-Luna, M. C., et al. (2021). Sensors for sustainable smart cities: a review. *Appl Sci.*, 11(17), 8198. doi: 10.3390/app11178198.

Ray, P. P. (2017). A survey on visual programming languages in Internet of Things [sci program]. *Scientific Programming.*, 2017, 1-6. doi: 10.1155/2017/1231430.

Salhaoui, M., Guerrero-González, A., Arioua, M., Ortiz, F. J., El Oualkadi, A., and Torregrosa, C. L. (2019). Smart industrial iot monitoring and control system based on UAV and cloud computing applied to a concrete plant. *Sensors (Basel).*, 19(15), 3316. doi: 10.3390/s19153316, PMID 31357720.

Shahrour, I., and Xie, X. (2021). Role of Internet of things (IoT) and crowdsourcing in smart city projects. *Smart Cities.*, 4(4), 1276-92. doi: 10.3390/smartcities4040068.

Singh, D., Merdivan, E., Psychoula, I., Kropf, J., Hanke, S., Geist, M., et al. (2017). Human activity recognition using recurrent neural networks. Lect. *Notes comput. Sci. (Including Subser. Lect. Notes Artif. Intell. Lect. Notes Bioinform.*), 10410 LNCS:267-74.

Sousa, P. H. F., Nascimento, N. M. M., Almeida,. J S., Rebouças Filho, P. P., and Albuquerque, V. H. C. (2019). Intelligent Incipient Fault Detection in Wind Turbines based on Industrial IoT Environment. *J Artif Intell Syst.*, 1, 1.

Syed, A. S., Sierra-Sosa, D., Kumar, A., and Elmaghraby, A. (2021). IoT in smart cities: A survey of technologies, practices and challenges. *Smart Cities.*, 4(2), 429-75. doi: 10.3390/smartcities4020024.

Tian, S., Yang, W., Le Grange, J. M. L., Wang, P., Huang, W., and Ye, Z. (2019). Smart healthcare: making medical care more intelligent. *Health.*, 3(3), 62-5. doi: 10.1016/j.glohj.2019.07.001.

Wang, J., Meng, C., Wang, C. T., Liu, C. H., Chang, Y., Li, C., et al. (2021). A fully self-powered, ultra-stable cholesteric smart window triggered by instantaneous mechanical stimuli. *Nano Energy.*, 85, 105976. doi: 10.1016/j.nanoen.2021.105976.

Zimmermann, L., Weigel, R., and Fischer, G. (2018). Fusion of nonintrusive environmental sensors for occupancy detection in smart homes. *IEEE Internet Things J.*, 5(4), 2343-52. doi: 10.1109/JIOT.2017.2752134.

Chapter 9

Internet of Things for Smart Cities: Overview and Challenges

N. Rajathi[1,*], PhD
V. Vanitha[1], PhD
and P. Parameswari[2], PhD
[1]Department of Information Technology, Kumaraguru College of Technology, Coimbatore, India
[2]MCA Department, Kumaraguru College of Technology, Coimbatore, India

Abstract

The Internet of Things (IoT) is a new technology that allows a vast network of objects to communicate with one another for creating innovative and valuable services. It refers to a diverse range of technologies, hardware, and software stacks. IoT is essentially a system where the Internet is connected to the physical world via a multitude of sensors. The technology penetration in India is low at the national level, a viable IoT architecture based on recent technological advancements is needed to provide long-term cost-effective solutions. Smart cities are an essential facet for offering effective solutions for any country's economic growth. The government has to provide a range of services to its citizens, and the IoT can be a huge help in this effort. While having regulated access to data will enable a range of digital services, it will be feasible to interact with a great number of homogenous and heterogeneous systems in a transparent and seamless manner.

Information and Communication Technologies (ICT) have been used because of the government's and the private sector's Smart City initiatives to find long-term, cost-effective solutions to problems with crime control, traffic congestion, waste management, unemployment,

[*] Corresponding Author's Email: rajathi.n.it@kct.ac.in.

In: Blockchain Enabled Secure Big Data Computing …
Editors: L. Ashok Kumar, D. Karthika Renuka, V. Chandrasekar et al.
ISBN: 979-8-89113-243-6

and other urban issues. The goal is to increase the city's liveability to previously unseen levels by maximizing resource utilization while lowering operational cost. The intension of this chapter is to examine the role of the IoT in the development of smart cities, as well as the key benefits and challenges of IoT-based smart cities.

Keywords: IoT, smart city, challenges

Introduction

A Smart City is a collaborative initiative aimed at enhancing the well-being of its residents and visitors in a sustainable manner. This involves the efficient management of resources, promoting education on sustainability, fostering harmony between people and the environment, and improving infrastructure and innovation in areas such as health, education, safety and others. In order to tackle emerging challenges like climate change and rapid urbanization, transparent Information and Communication Technology is utilized to monitor, analyze and optimize social aspects such as citizen involvement, equity, governance, urban features like green space, mixed land usage and biodiversity as well as physical systems like resources, waste and transportation.

Need for Smart City

The trend of migration from rural to urban areas is on the rise worldwide. It is projected that by the year 2050, 70% of the global population will reside in urban centers, and India is not immune to this phenomenon (https://www.worldometers.info/world-population/world-population-projections).

During the past 25 years, India's rapid modernization efforts have primarily targeted cities that have existed for a long time. In search of better earnings and lifestyles, people have been flocking to cities, as opportunities in rural areas remain limited. This has led to the unplanned expansion of areas adjacent to existing major cities.

Although a large number of suburban clusters have emerged in India to meet people's living needs, planned development and expansion of Indian cities have not received the attention they deserve. Despite the fact that amenities and urban infrastructure are constantly overwhelmed by the influx of people, India has managed to make do with incremental development and temporary solutions for cities. However, as more and more people aspire to live in cities in the future, this approach will no longer be sustainable. According to a report by India's NITI AYOG, Indian cities currently account for approximately 62-63% of India's GDP, a figure expected to increase to 75% by 2021.

The rapid expansion of cities has resulted in overpopulation, leading to a shortage of natural resources. Social and economic imbalances among citizens contribute to the challenges facing communities. Megacities are struggling to keep pace with population growth, and as such, smart cities are essential for creating more livable and sustainable urban environments. The evolution of technology, including the Internet of Things, Artificial Intelligence, and Big Data, can be leveraged to develop smart cities that address various societal issues in a collaborative manner.

Smart City Overview

The smart city, also known as an "intelligent city," a "digital city," a "eco-city," or a "sustainable city," is an urban development that uses next-generation information technologies, such as IoT, AI, and big data technologies, as well as cloud computing, to completely perceive, effectively integrate, and efficiently process various data and information within a city.

A smart city employs intelligent service management to coordinate various systems across the city. Data collected from citizens and mechanical devices is processed and analyzed to monitor and manage traffic and transportation systems, power plants, water supply networks, waste disposal, and other systems. This approach creates a secure space for businesses to grow, while also offering city residents a more convenient and high-quality living environment. The sustainable development of the urban economy, society, and environment is consequently supported.

The government must make significant investments to build a smart city. Nonetheless, it is one of the best lifestyle changes that can be made if done deliberately. Smart cities are intended to make the best use of available space and resources while also distributing benefits in the most efficient and effective way possible. It also seeks to improve connectivity at various levels among citizens and between the administration and the population. Public infrastructure, such as schools, roads, and hospitals, is being upgraded. The system can eliminate several redundancies in the current system, saving time and money. Because technology is rapidly advancing, it is clear that people must adapt their lifestyles to meet modern-day demands.

The smart cities vision is to create green urban environments and effective city systems that promote stress-free and enjoyable city living. Ideal city infrastructure enables people to work more easily, connect with nature, and relax their minds, resulting in improved quality of life for citizens. As a result, citizens become more aware and self-organize when their life-comfort and mental state improve.

Smart cities are designed to be environmentally friendly, with devices that can monitor air purity and other environmental and health factors. Investment in a smart city should include the retention of a conscientious workforce to monitor and adjust the system. To be successful, a smart city must also meet the social and psychological needs of its population. The goal of a smart city is to improve the quality of urban services while lowering their costs, with smart management, lifestyle, mobility, housing, and economy being key features. Smart cities aim to balance technological innovation with economic, social, and environmental challenges to enhance quality of life and foster sustainable development. Ultimately, the theme that unifies smart cities is how to live better together while protecting our environment.

Characteristics of Smart City

The key characteristics of a smart city were explored in detail (Zihan et al., 2021). Smart cities are known to possess four main features. Firstly, they utilize comprehensive sensing by integrating smart sensors into urban public facilities, enabling the core system of urban operation to be monitored. Secondly, they promote full integration by connecting the Internet of Things (IoT) and the internet system and integrating data into the city's core system to create a smart infrastructure. Thirdly, they stimulate innovation by encouraging the government, businesses, and individuals to develop new technologies and businesses for smart infrastructure, promoting development. Finally, they adopt collaborative operation, allowing critical systems and participants in the city to work together efficiently and harmoniously in line with smart infrastructure, leading to optimized smart city operation.

Benefits of Smart City

Smart cities can offer many benefits, below are some of the significant benefits that smart cities can offer:

1. Environmental impact: The primary aim of smart and sustainable cities is to reduce carbon dioxide emissions. They achieve this by improving energy efficiency and storage, waste management, and traffic conditions, among other measures.
2. Enhanced management of energy and water: Smart cities focus on efficient use of resources, such as energy and water. This is achieved through the implementation of smart grids and water management systems, which ensure the availability of energy and tap water quality throughout the city.
3. Transportation: Smart cities strive to optimize mobility and transportation, ensuring efficient and clean transportation of people, services, and goods. They leverage technologies such as real-time updates to alleviate traffic congestion and improve transportation systems.
4. Security: Safety and security are paramount in all cities. Smart cities have the potential to enhance citizen security by leveraging technologies such as CCTV cameras equipped with facial recognition software, motion and smoke detectors, and fire alarms to monitor and alert authorities of potential threats.

Smart Cities and Internet of Things

The concept of smart cities heavily relies on the internet of things, which is the technology that has made pervasive digitization possible. IoT involves the connection of devices to the internet, allowing them to transmit data to the cloud and receive instructions. This technology enables data collection and analysis, providing useful information for decision-making and policy formulation. In the context of smart cities, IoT enables sensors to collect and transmit data about the city's state to a central cloud. The data is then processed to extract patterns and aid in decision-making.

IoT technology can be leveraged to enhance the efficiency of cities in multiple areas, including traffic management, air pollution control, waste management, smart building design, and disaster preparedness. The implementation of IoT-enabled smart city solutions aims to create cities of the future that support economic development, improve infrastructure and the environment, and digitally optimize public assets. By leveraging IoT, cities can achieve more efficient and sustainable use of resources, enhanced citizen safety and comfort, and improved quality of life for residents. These solutions can enable cities to better address the challenges they face, such as urbanization, population growth, and climate change, and pave the way for a more connected and sustainable future.

The remainder of this chapter is organized as follows: the literature review is covered in section 2. IoT applications for smart cities are addressed in section 3 while IoT challenges for smart cities are discussed in section 4. Section 5 gives the research directions, while Section 6 provides the final remarks.

Literature Survey

According to (Samih, 2019), there are several ongoing smart city initiatives, and information and communication technology (ICT) plays a vital role in transforming traditional cities into smart cities. The authors highlighted the relationship between the Internet of Things (IoT) and smart city functions, emphasizing the importance of IoT in collecting and analyzing data for decision-making purposes. They also presented a futuristic scenario based on IoT technology that could optimize waste management processes in smart cities. Overall, the authors emphasize the potential of ICT and IoT in transforming cities into smarter and more efficient urban environments.

According to (Syed et al., 2021), Smart Cities rely heavily on IoT technology to provide a range of services. The authors offer a detailed analysis of the various domains of Smart Cities, including their architectures and the challenges involved in deploying these applications. They also delve into the sensing and networking technologies used in Smart City applications, as well as the use of AI in these environments. The authors further explored the various deployment types for these applications, based on the technologies and architectures presented, providing an overview of the current research scenario in IoT-based Smart Cities. Finally, the authors discussed the security and privacy concerns associated with IoT-based Smart Cities and presented a SWOT analysis. Overall, the article offers a comprehensive overview of Smart Cities, including their benefits and challenges, and highlights the crucial role of IoT technology in their implementation.

The study described in (Dizon et al, 2022) proposes an IoT-based solution for streetlights that can effectively reduce energy consumption. The researchers tested different lighting schemes and found that they resulted in varying degrees of energy savings. While the adaptive approach to lighting remains inconclusive, the study did show a clear reduction in power consumption based on the relationship between traffic flow and energy expenditure. The proposed solution could be implemented in other cities to help reduce energy consumption. The ultimate goal of the system is to create a smart city with energy-efficient street lighting and smart waste management systems. The system will also monitor air pollution by uploading the measured parameters to the cloud, and analyze the stored data to provide better services to citizens.

The authors (Manjunath et al., 2018) proposed an Internet of Things (IoT)-based system for monitoring and disposing of garbage, which utilizes ultrasonic and MQ4 sensors. The system uses an ultrasonic sensor to measure the amount of rubbish in the bins, while the MQ4 sensor measures the level of odour in the biodegradable rubbish bin. Once the threshold level is reached in the non-biodegradable smart bin, the information is sent to the municipal corporation through a garbage disposal application. Similarly, if the threshold level is exceeded in the biodegradable bin, the wastes are dumped into the underlying chamber via a sliding lid. This proposed system provides a smart solution to manage waste disposal in a more efficient and sustainable way.

Health is a state of complete physical, mental, and social well-being, and not just the absence of illness. The desire for a better life is closely linked to health, but the global health crisis has created challenges due to issues like poor health services, disparities between rural and urban areas, and a lack of healthcare professionals during critical times. Smart healthcare platforms utilize devices such as wearables, the Internet of Things, and mobile internet to facilitate the sharing of health information and connect individuals, resources, and organizations (Chowdhury et al., 2018). Intelligent medical treatment involves a range of actors, including physicians, nurses, hospitals, and research organizations, and encompasses disease prevention, identification, assessment, healthcare management, patient decision-making, and medical research.

Islam et al., (2020), presented a smart healthcare system that is capable of monitoring essential vital signs of patients, including heart rate, body temperature, and environmental factors such as humidity, CO and CO2 gas levels in the hospital room. The system demonstrated an accuracy rate greater than 95% in all cases. Although the tests were conducted outside of the hospital, authorized medical staff can access and track the data in real-time. This system could also assist healthcare workers during times of epidemics or crises as the raw medical data can be swiftly analysed.

The authors (Islam et al., 2020) proposed an intercity bus management system that uses a GPS tracker to monitor bus location in real time and a QR code to store ticketing information. Three distinct web-based applications were created for passengers, drivers, and bus company administrators.

In (Taufik et al., 2021), an Optimized Weight Elman Neural Network (OWENN) algorithm was proposed for traffic forecasting. The study also suggested a way to improve the performance of Intel 80,286 Microprocessor in traffic management. The OWENN algorithm was shown to be efficient in predictingtraffic, which can be useful in managing and improving traffic flow in urban areas. The proposed enhancement for the microprocessor can help in better utilization of the available resources and improving the overall performance of the traffic management system.

An interval-based scheme for detecting nodes exhibiting Sybil behavior is proposed (Neelakandan et al., 2021). The framework can detect Sybil behavior whenever nodes perform Co-operative Blackmailing attacks. The detection scheme is edge-based, with the edge node examining the connected IoT devices to identify Sybil behavior. Upon identification, suspected nodes are closely monitored by the edge node to confirm their behavior. This approach can prevent normal nodes in the network from being reduced by Sybil attacks through Co-operative Blackmailing. Simulation results demonstrate the effectiveness of the proposed method.

The authors highlighted the current issues faced by collaborative smart cities, such as low quality of service and lengthy service response times (Simpson et al., 2021). To mitigate these issues, edge computing has been introduced to address the problems faced by city terminals. However, security and privacy concerns in collaborative services are significant, and it is essential to ensure a secure edge computing environment. To address these concerns, the authors proposed an offloading method that can be used in smart cities to improve privacy, promote edge utility, and increase offloading efficiency. By combining edge computing with information entropy, the proposed method aims to strike a balance between collaborative service and privacy protection.

IoT Applications for Smart Cities

IoT applications in smart cities are used autonomously, without human intervention(Shakya et al., 2021). For a variety of purposes, many IoT devices are connected to one another and communicate.When it comes to analysing, managing, and controlling the IoT resources in the smart city, resource consumption and other devices are unquestionably playing a hugely important role. A typical smart city application has four sections: data gathering, transmission, storage, and analysis (Du, 2021).

The volume of data gathered from numerous appliances and devices is greater. Across numerous platforms and systems, big data provides management and analytical capabilities for enormous amounts of data. Big Data uses a range of storage technologies, including cloud computing and fog computing, which offer storage and fast access, to save huge amounts of data in addition to processing it quickly (Alam, 2021). Data analysis draws coclusions and patterns to aid in decision making .The various entities of smart city ecosystem is depicted in Figure 1 and shows how data is gathered from edge smart sensors and transformed for processing before being sent to the cloud.

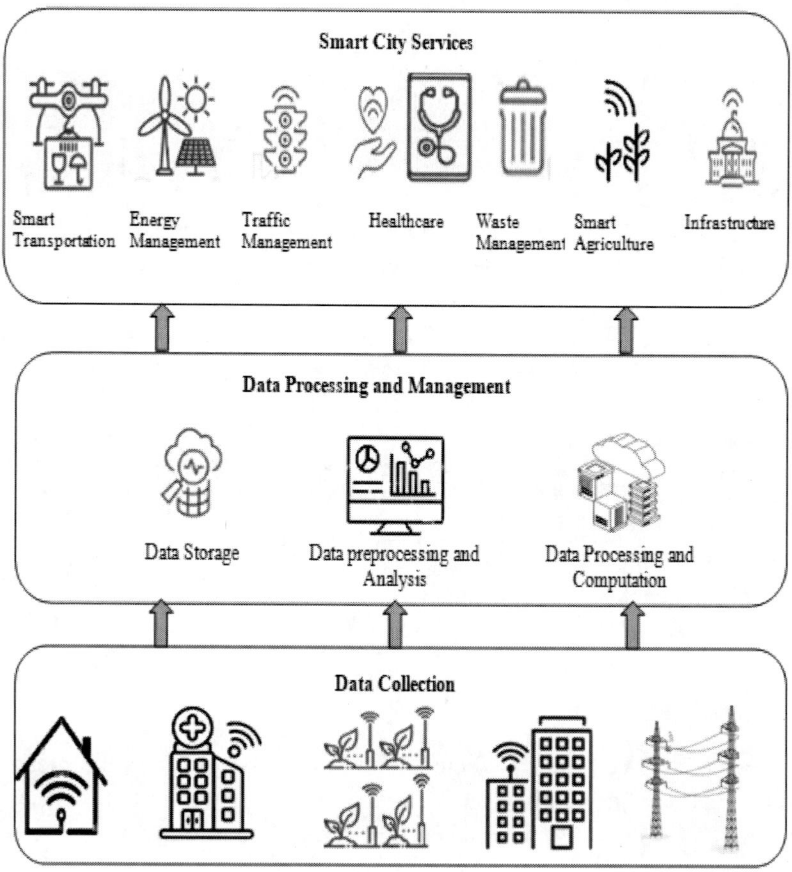

Figure 1. Block diagram of smart City System Components.

Smart city services encompass the activities that sustain a city's population, these involves air quality monitoring (Dutta et al., 2017), water quality (Chowdury et al., 2019), water usage and leakage(Rojek and Studzinski, 2019), fault detection in products (Wan et al., 2018), Health monitoring (Trencher et al., 2019; Alhussein et al., 2017; Kaur et al., 2019), plant disease detection (Saha, et al., 2018; Araby et al., 2019), smart parking (Al-Turjman Malekloo, 2019), structural health monitoring of a bridge(Guo et al., 2014; Dang et al., 2020; Gui et al., 2017), etc. The capacity of smart technologies has recently been developed throughout the city by a variety of sectors. These are presented in Figure 2.

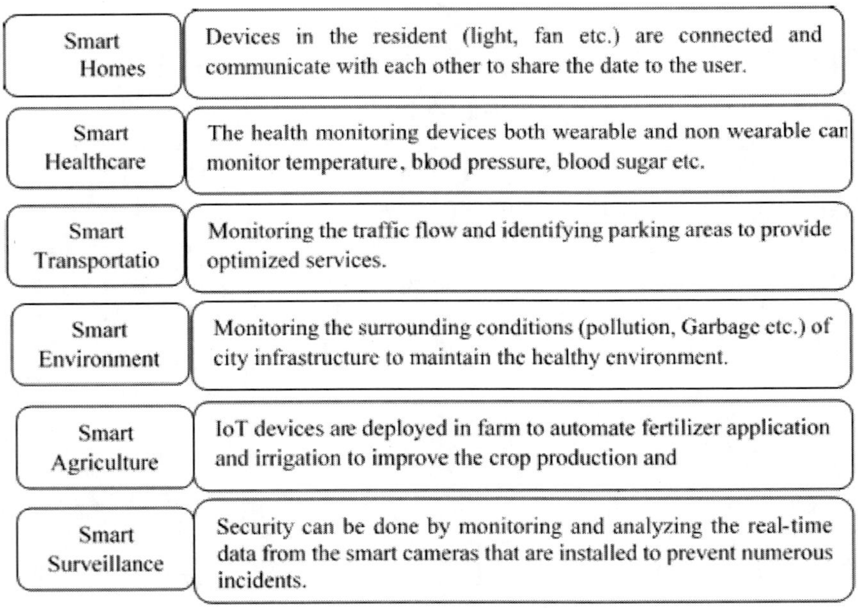

Figure 2. Different Sectors of Smart City.

IoT Challenges for Smart Cities

Diverse IoT devices are connected with each other and communicate for different tasks. Cities must manage enormous amounts of data, adopt IoT across all city functions, and protect critical infrastructure from an increasing number of threats if they are to become smart cities.

Based on the literatures, a number of issues in the digital eco system of smart cities need to be resolved in order for their services to continue to be useful to all stakeholders concerned. Some of the challenges faced are presented in the following Figure 3.

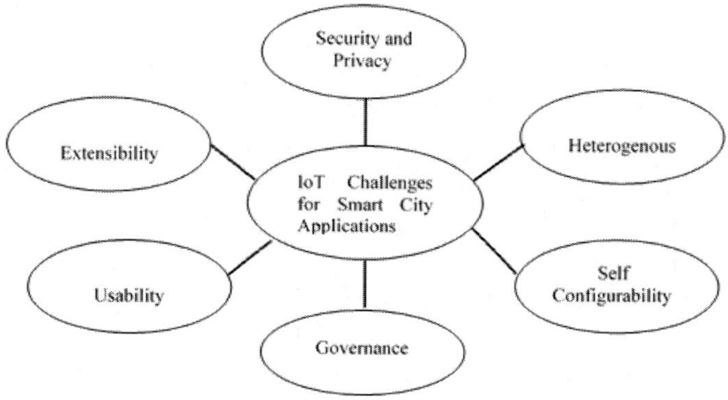

Figure 3. Various challenges for smart city applications.

- Security and privacy: Which protects against malicious attacks on IoT applications and devices? Heterogeneity: This is related to the use of a large variety of devices and protocols.
- Self-configurability: It is required in IoT systems to establish automatic configuration in the case of any physical change.
- Governance: IoT governance is a set of rules and guidelines for the technical architecture, data management, and information security of the IoT.
- Usability: Utilizing cutting-edge human-computer interface tools to make it easier for non-technical individuals to use IoT apps.
- Extensibility: This ensures that IoT systems may easily be extended to incorporate new functionalities, configurations, and technologies

Research Direction in Creating Sustainable IoT Based Smart City Ecosystem

The IoT growth, on the other hand, has dramatically generated a variety of research directions for the growth of a sustainable smart city. Below are some of the research directions.

- *Efficiency in Energy:* Energy efficient IoT gadgets and sensors will be created to assist smart city entities in using lesser energy, lowering carbon emissions and enhancing sustainability.
- *Data Handling*: Proper management of data is critical for the success of smart city ecosystems. Developing efficient data storage, processing, and analysis tools can help optimize data utilization, enhance service delivery, and promote sustainable development.
- *Privacy and Security:* Because of the volume of data collected and processed in smart cities, privacy and security are crucial challenges. It is essential to investigate methods for protecting sensitive data and personal information.
- *Compatibility and Standardization:* To enable seamless communication and data sharing, IoT devices and systems utilized in smart cities must be compatible and standardised. A more sustainable and interconnected ecosystem can be developed by creating standards and protocols for smart city systems and components.
- *Scalability:* Scalability assurance becomes a crucial research focus as smart city ecosystems expand. Scalable IoT systems and devices can ensure that smart cities can grow and manage to serve their residents' demands.
- *Sustainability Metrics:* Developing appropriate metrics and tools for measuring the sustainability of smart city ecosystems is crucial. Researching ways to measure and assess the social, environmental, and economic impacts of smart city projects can help ensure their long-term sustainability.
- *Resilience Metrics:* It is essential to develop suitable metrics and instruments for assessing the sustainability of smart city ecosystems. Initiatives related to smart cities can be made more long term sustainable by looking ways to assess their effects on the society, economy and environment.
- *Participation by Citizens:* Smart city projects must involve citizens in their creation and execution. It can be beneficial to investigate methods of including citizens and giving them access to information and decision-making processes to build more resilient, responsive, and egalitarian smart city ecosystems.

Overall creating a sustainable IoT based smart city eco system requires interdisciplinary research, celebration between stakeholders.

Conclusion

IoT research has transformed everyday needs, primarily in smart cities. IoT applications are used in the smart city without the involvement of humans. IoT devices are interconnected and communicate with one another for a variety of functions. These devices generate enormous volumes of data that are processed, managed, and stored in the cloud. In order to satisfy the demands of the expanding population, it is important to take into account a variety of smart city needs. The IoT innovations, on the other hand, have significantly shaped several research directions for the smart city. In light of this, a number of issues and potential directions for future research are covered in this chapter.

Disclaimer

None

References

Alam, T., 2021. Cloud-based IoT applications and their roles in smart cities. *Smart Cities*, 4(3), pp.1196-1219.

Alhussein, M., 2017. Monitoring Parkinson's disease in smart cities. IEEE Access, 5, pp.19835-19841.

Al-Turjman, F. and Malekloo, A., 2019. Smart parking in IoT-enabled cities: A survey. *Sustainable Cities and Society*, 49, p.101608.

Araby, A. A., Abd Elhameed, M. M., Magdy, N. M., Abdelaal, N., Abd Allah, Y. T., Darweesh, M. S., Fahim, M. A. and Mostafa, H., 2019, May. Smart iot monitoring system for agriculture with predictive analysis. In 2019 8th International Conference on *Modern Circuits and Systems Technologies* (MOCAST) (pp. 1-4). IEEE.

Chowdhury P, Sen R, Ray D,. Roy P and Sarkar S, "Garbage Monitoring and Disposal System for Smart City Using Iot," 2018 Second International Conference on Green Computing and Internet of Things (ICGCIoT), Bangalore, India, 2018, pp. 455-460.

Chowdury, M. S. U., Emran, T. B., Ghosh, S., Pathak, A., Alam, M. M., Absar, N., Andersson, K. and Hossain, M. S., 2019. IoT based real-time river water quality monitoring system. *Procedia Computer Science*, 155, pp.161-168.

Dang, H. V., Tran-Ngoc, H., Nguyen, T. V., Bui-Tien, T., De Roeck, G. and Nguyen, H. X., 2020. Data-driven structural health monitoring using feature fusion and hybrid deep learning. *IEEE Transactions on Automation Science and Engineering*, 18(4), pp.2087-2103.

Dizon, E. and Pranggono, B., 2022. Smart streetlights in Smart City: a case study of Sheffield. Journal of *Ambient Intelligence and Humanized Computing*, pp.1-16.

Du, J., 2021, March. Application analysis of IoT technology in smart cities. In 2021 *2nd International Conference on E-Commerce and Internet Technology* (ECIT) (pp. 264-269). IEEE.

Dutta, J., Chowdhury, C., Roy, S., Middya, A. I. and Gazi, F., 2017, January. Towards smart city: sensing air quality in city based on opportunistic crowd-sensing. In Proceedings of the *18th international conference on distributed computing and networking* (pp. 1-6).

Gui, G., Pan, H., Lin, Z., Li, Y. and Yuan, Z., 2017. Data-driven support vector machine with optimization techniques for structural health monitoring and damage detection. KSCE *Journal of Civil Engineering*, 21, pp.523-534.

Guo, J., Xie, X., Bie, R. and Sun, L., 2014. Structural health monitoring by using a sparse coding-based deep learning algorithm with wireless sensor networks. *Personal and Ubiquitous Computing,* 18, pp.1977-1987.

Islam, M. M., Rahaman, A. and Islam, M. R., 2020. Development of smart healthcare monitoring system in IoT environment. SN computer science, 1, pp.1-11.

Kaur, P., Kumar, R. and Kumar, M., 2019. A healthcare monitoring system using random forest and internet of things (IoT). *Multimedia Tools and Applications,* 78, pp.19905-19916.

Manjunath, P., Soman, R. and Shah, P. G., 2018, August. IoT and block chain driven intelligent transportation system. In 2018 *Second International Conference on Green Computing and Internet of Things* (ICGCIoT) (pp. 290-293). IEEE.

Neelakandan, S.B.M.A.T.S.D.V.B.B.I., Berlin, M. A., Tripathi, S., Devi, V. B., Bhardwaj, I. and Arulkumar, N., 2021. *IoT-based traffic prediction and traffic signal control system for smart city. Soft Computing,* 25(18), pp.12241-12248.

Rojek, I. and Studzinski, J., 2019. Detection and localization of water leaks in water nets supported by an ICT system with artificial intelligence methods as a way forward for smart cities. *Sustainability*, 11(2), p.518.

Saha, A. K., Saha, J., Ray, R., Sircar, S., Dutta, S., Chattopadhyay, S. P. and Saha, H. N., 2018, January. IOT-based drone for improvement of crop quality in agricultural field. In 2018 IEEE 8th *Annual Computing and Communication Workshop and Conference* (CCWC) (pp. 612-615). IEEE.

Samih, H., 2019. Smart cities and internet of things. Journal of Information Technology *Case and Application Research,* 21(1), pp.3-12.

Shakya, S., 2021. Collaboration of smart city services with appropriate resource management and privacy protection. *Journal of Ubiquitous Computing and Communication Technologies* (UCCT), 3(01), pp.43-51

Simpson, S. V. and Nagarajan, G., 2021. An edge based trustworthy environment establishment for internet of things: an approach for smart cities. *Wireless Networks*, pp.1-17.

Syed, A. S., Sierra-Sosa, D., Kumar, A. and Elmaghraby, A., 2021. IoT in smart cities: A survey of technologies, practices and challenges. *Smart Cities,* 4(2), pp.429-475.

Taufik, M., Hudiono, H., Rakhmania, A. E., Perdana, R. H. Y. and Sari, A. S., 2021. An Internet of Things Based Intercity Bus Management System for Smart City. *International Journal of Computing and Digital Systems*, 10.

Trencher, G. and Karvonen, A., 2019. Stretching "smart:" Advancing health and well-being through the smart city agenda. *Local Environment*, 24(7), pp.610-627.

Wan, J., Yang, J., Wang, Z. and Hua, Q., 2018. Artificial intelligence for cloud-assisted smart factory. *IEEE Access*, 6, pp.55419-55430.

Worldometers. World Population Forecast—Worldometers. 2019. Available online: https://www.worldometers.info/world-population/world-population-projections/.

Chapter 10

IoT Enabled Smart Transportation System

Premkumar Duraisamy[1,*]
Yuvaraj Natarajan[1]
Kathiresan K.[1]
and Niranjani V.[2]
[1]Department of Computer Science and Engineering, KPR Institute of Engineering and Technology, Coimbatore, Tamilnadu, India
[2]Department of Computer Science and Engineering, Sri Eshwar College of Engineering, Coimbatore, Tamilnadu, India

Abstract

The IoT-enabled intelligent transportation system is one of the emerging areas for researchers. This chapter discusses the general structure of STS (Smart Transportation System), Urban Transport System, Smart Traffic Management and Parallel Processing System. Smart Transport Systems combine sensors, intelligent traffic control, smart roads, and innovative cyber-physical security components. The STS should be dynamic because every vehicle needs to interact with other cars so that we can achieve high throughput with enhanced security. The Internet of Things (IoT) has significantly changed how we live and work. One of the areas where IoT is making a substantial impact is transportation. IoT-enabled intelligent transport systems can optimize traffic flow, reduce congestion, and improve safety. This chapter proposes an architecture for IoT-enabled intelligent transport systems that includes various components such as sensors, data analytics, communication networks, and control systems. The proposed architecture enables the collection and analysis of

[*] Corresponding Author's Email: technicalprem@gmail.com.

In: Blockchain Enabled Secure Big Data Computing …
Editors: L. Ashok Kumar, D. Karthika Renuka, V. Chandrasekar et al.
ISBN: 979-8-89113-243-6

data in real time, which can be used to optimize transport operations and enhance safety. We also discuss the potential benefits of IoT-enabled intelligent transport systems and the challenges that must be addressed to realize these benefits. The transportation industry has seen a significant increase in the demand for efficient and sustainable systems that can accommodate the growing volume of goods and people. The proposed system is based on a distributed architecture that incorporates sensors, actuators, and IoT devices to enable real-time monitoring and control of transportation operations. The chapter describes the system architecture, the key components, and their functionality, and discusses the potential benefits and challenges of implementing the proposed system.

Keywords: IoT, smart transport systems, sensors, data analytics, communication networks, control systems

Introduction

Transportation is a critical aspect of our daily lives, and its systems face several challenges, including increasing congestion, safety concerns, and environmental issues. The IoT technology is a promising solution to tackle these problems by creating intelligent transport systems capable of optimizing traffic flow, reducing congestion, and enhancing safety (Boreiko et al, 2017). This chapter proposes an architecture for IoT-enabled intelligent transport systems comprising various components that collect and analyze data in real-time. Table 10.1 shows the comparison of different IoT technologies used in different review work.

Table 10.1. Literature survey

Author(s)	Title	IoT Technologies Used	Application/Use Case
D. Wu, et al.	A Survey of the Internet of Things in Smart Transportation	Wi-Fi, Bluetooth, DSRC, Cellular, Satellite	Smart parking, intelligent traffic management, public transport management
M. Hasan, et al.	IoT-Based Intelligent Transportation Systems: A Comprehensive Review	Wi-Fi, Bluetooth, Cellular, DSRC	Traffic flow management, accident detection and response, driver behavior monitoring
S. K. Jha, et al.	Internet of Things Based Intelligent Transport System: A Review	Wi-Fi, Cellular, DSRC, Bluetooth, Zigbee	Traffic signal control, vehicle tracking, parking management

R. M. Shakeri, et al.	A Comprehensive Review of Internet of Things (IoT)-Based Solutions.	Wi-Fi, Cellular, DSRC, Zigbee, Bluetooth, RFID	Traffic congestion management, emergency response, public transport management

Challenges Facing Transportation Systems

Transportation systems face increasing congestion, safety concerns, and environmental issues. Congestion wastes time and money and contributes to air pollution and greenhouse gas emissions. According to a Texas A&M Transportation Institute report, traffic congestion cost US drivers nearly $88 billion in 2019. According to the World Health Organization, road traffic accidents cause over 1.3 million deaths each year globally (Boreiko et al., 2016). Transportation systems also contribute significantly to environmental issues such as air pollution and greenhouse gas emissions.

Potentiel Benefits of IoT-Enabled Smart Transport System

IoT-enabled intelligent transport systems have the potential to address many of the challenges facing transportation systems today. Intelligent transport systems can optimize traffic flow, reduce congestion, and improve safety by enabling real-time data collection and analysis. They can also reduce environmental impact by optimizing transport operations and lowering emissions.

Proposed Architecture

The proposed architecture for IoT-enabled intelligent transport systems consists of several components, as shown in Figure 10.1 These components include sensors, data analytics, communication networks, and control systems.

Sensors

Sensors are a crucial component of IoT-enabled intelligent transport systems. They can collect a wide range of data, such as traffic flow, weather conditions, vehicle speed, and occupancy. This data is used to optimize transport operations and improve safety (Brega et al., 2022). For instance, sensors can detect congestion and adjust traffic signals to optimize traffic flow.

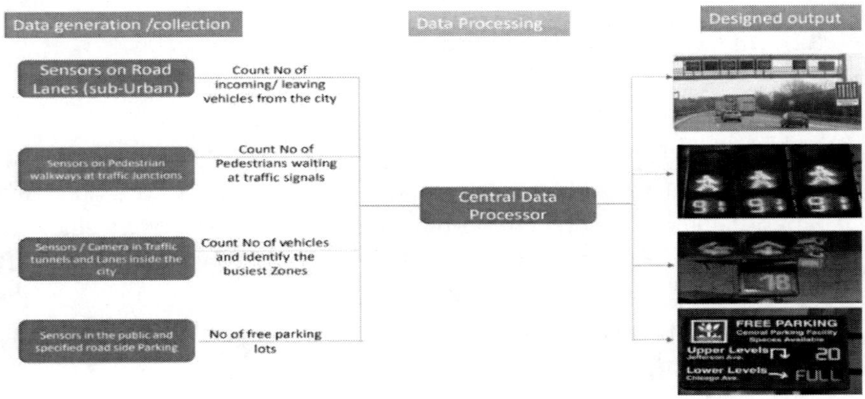

Figure 10.1. Architecture for IoT Enabled Smart Transport Systems.

There are several types of sensors are shown in Figure 10.2 and it can be used in IoT-enabled smart transport systems.

Traffic Sensors

These sensors are installed on roads, highways, and intersections to measure traffic flow, vehicle speed, and occupancy. This data can be used to optimize traffic signal timing and reduce congestion.

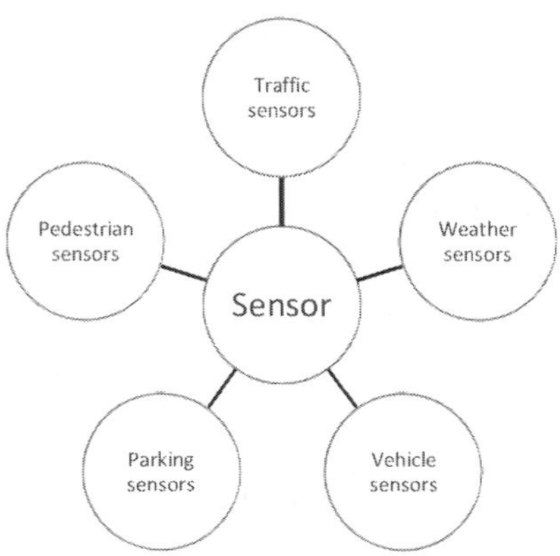

Figure 10.2. Types of sensors.

Parking Sensors

These sensors are used to detect available parking spaces and transmit this information to drivers in real-time. This can help reduce congestion and improve the parking experience for drivers.

Pedestrian Sensors

These sensors are used to detect pedestrian activity and adjust traffic signal timing accordingly. This can improve safety for pedestrians and reduce congestion at intersections.

Data Analytics

Data analytics is another critical component of IoT-enabled intelligent transport systems. It involves the processing and analysis of data collected from various sensors (Bhatia et al., 2023). This data is used to identify patterns and insights that can be used to optimize transport operations and improve safety. For example, data analytics can predict traffic congestion and adjust traffic signals accordingly. Data analytics is another crucial component of IoT-enabled intelligent transport systems. The massive amounts of data generated by the various sensors in the transport system must be collected, stored, and analyzed to provide insights that can be used to optimize operations, reduce costs, and improve safety.

Data analytics can help transport operators understand traffic patterns, identify congestion areas, and optimise vehicle routes. It can also be used to identify potential safety hazards and proactively address them before they become a problem.

Some several techniques and tools can be used for data analytics in IoT-enabled intelligent transport systems, including:

Machine Learning

Large volumes of data may be analyzed using machine learning algorithms to find trends and anomalies. These algorithms are able to anticipate traffic patterns, spot aberrant driving behavior, and spot possible dangers.

Data Visualization

Tools for data visualization can be used to show complicated data in an easy-to-understand manner. Transport professionals may use this to visualize traffic flow patterns, detect congested regions quicker, and make more educated choices.

Predictive Analytics

On the basis of past data, predictive analytics techniques may be used to predict future events. This may be applied to better route planning, traffic pattern prediction, and the detection of possible vehicle maintenance concerns.

Real-Time Analytics

Data may be analyzed in real-time using real-time analytics systems, which can also provide alerts when certain criteria are reached. For instance, if a vehicle exhibits unusual behavior, the transport operator can be notified and instructed to take the necessary action.

Overall, data analytics is a crucial component of IoT-enabled intelligent transport systems as it provides insights that can be used to optimize operations, reduce costs, and improve safety. By leveraging the power of machine learning, data visualization, predictive analytics, and real-time analytics, transport operators can make informed decisions and improve the overall efficiency and safety of the transport system (Duraisamy et al., 2023).

Communication Networks

IoT-enabled innovative transport systems rely heavily on communication networks to enable devices and sensors to communicate with each other and the central management system. These systems can include vehicles, traffic lights, road sensors, and other components, all of which need to communicate in real time for the system to function effectively (Duraisamy et al., 2020).

There are several communication network technologies that can be used in IoT-enabled smart transport systems, including:

(a) Cellular networks use existing cellular infrastructure to transmit data between devices and the central management system. They are reliable and widely available but may have limited bandwidth in certain areas.
(b) Wi-Fi networks transport data between devices and the central management system using wireless access points. Although they have a small range and can only be accessible in certain locations, they are affordable and offer a high bandwidth.
(c) **Dedicated short-range communications (DSRC):** This method enables communication between neighboring devices by using a real frequency spectrum. It has a small range but works well for vehicle-to-vehicle and vehicle-to-infrastructure communication.
(d) **Bluetooth:** This technology is commonly used for short-range communication between devices, such as vehicles and smartphones. It is cost-effective and widely available but has a limited range.
(e) Satellite networks use satellites to transmit data between devices and the central management system. They provide global coverage but are expensive and have limited bandwidth.

The choice of communication network technology depends on the specific requirements of the intelligent transport system and the available infrastructure in the area. In some cases, combining different technologies may provide optimal coverage and performance.

Control Systems

Control systems are responsible for controlling various aspects of transport operations, such as traffic signals, public transport schedules, and toll booths. They use data collected from sensors and data analytics to optimize transport operations and improve safety. Control systems can be automated or manually operated, depending on the requirements of the transport system (Fitriani et al., 2021).

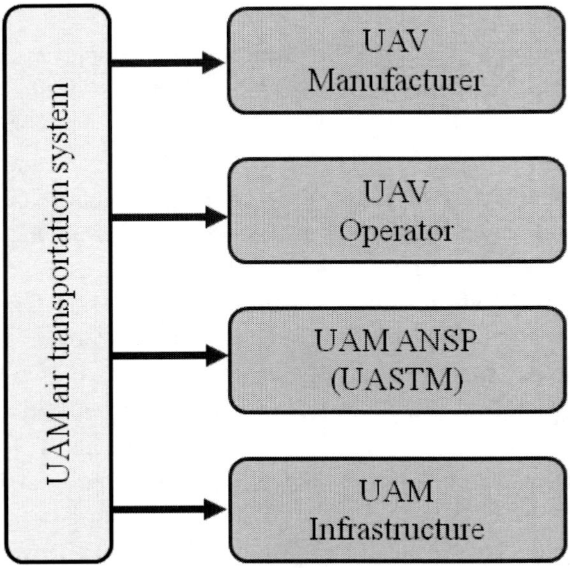

Figure 10.3. UAM Air Transportation System.

Section A: Urban Transport System

Urban Air Mobility (UAM)

The UAM idea of using vertical take-off and landing aircraft (VTOL) is gaining increasing interest in industry and research. Helicopter-based taxi services exist and continue to live in major cities worldwide, but air mobility with VTOL vehicles is not new [9]. However, technological advances have only recently enabled the development of the next generation of so-called personal aviation vehicles (PAVs) or electric VTOL vehicles (eVTOLs). Figure 10.3 shows the UAM general system components.

UAM air transport is believed to have less environmental impact than conventional aircraft due to eliminating greenhouse gases, reducing air pollution, and reducing noise (Galland et al., 2011). As UAM is currently in the planning and testing stages and related procedures and standards are being developed by legal bodies, aviation authorities, and other bodies, the certification and ownership of UAM operators are controversial (Gomides et al., 2023). Increase, Regarding regulations, permits, operations, infrastructure, maintenance and certification procedures, UAM can represent an entirely new aviation industry as an innovative mode of air transport. UAM air travel has less environmental impact than traditional modes of transportation due to eliminating greenhouse gases and reducing air pollution and noise.

In addition, since UAM is a new branch with cutting-edge applications and high-tech design concepts, responsible aviation authorities must ensure operational quality and flight safety by applicable regulations.

Next-generation VTOLs, based on improved electric and distributed propulsion and increased battery energy density, produce less noise and emissions than traditional helicopters, making urban air transport economically viable, and it can be made socially acceptable (Gopi et al., 2023).

The depicts the UTM architecture, which presents a layered/hierarchical framework for managing urban air traffic, with organizations including UAS operators, supplemental data service providers, and UAS service suppliers collaborating with the UASs to make operational decisions (Horazdovsky et al., 2023). For UAM operations, it is envisioned that the various industry players—including UAS operators, USSs, and SDSPs—will coordinate with one another. The Flight Information Management System is a vital component that guarantees the overall safety of the airspace used by conventional air vehicles like airplanes, helicopters, and UAS agents (FIMS). A central data exchange gateway for ATM, UTM, public service, and regulatory bodies are called FIMS. The use is authorized for functionality, quality of service, and other criteria connected to the FIMS in the UTM architecture (Isawi et al., 2023). FIMS UTM functions include registration and licensing, ensuring fair use of airspace, sharing information on aircrew restrictions and notices (NOTAMs), identification and authentication of data/service providers, and general airspace conditions and status. Surveillance and the possibility of conflict resolution of flight plans from different USS serving the same airspace, most importantly. However, UTM's effectiveness depends on the deployment paradigm for these tasks (Janos et al., 2018).

For instance, considering tactical deconfliction in a centralized model may improve control and become a significant bottleneck as operation volume increases. On the other hand, the need for standardization can make a federated model complex (Kazi et al., 2018).

Note that the UTM architecture incorporates the idea of vehicle-to-vehicle (V2V) communication capabilities, allowing the UAS to provide relevant information (e.g., four-dimensional proximity to the UAS). as an operator). Recognize the situation and respond appropriately. NASA and others have studied the importance of maintaining air separation. Vehicle technology is designed to anticipate and resolve conflicts based on common intent in ADS-B and similar data (Morri et al., 2018).

Through the work described in this chapter, it is hoped to improve our ability to communicate information to enhance our collective situational awareness and reason by using that information to learn more about how the situation will develop. When combined with an appropriate onboard AI decision engine, such information will allow the UASs to put corrective measures into place either passively through proactive action or through inter-UAS conversations. Certain circumstances call for direct UAS-UAS discussion with little USS participation for approval since USS-USS negotiation may be slow, unsafe, or useless in acquiring vehicle utilities. Because these USS are planned to be highly automated bots, latencies are anticipated to be significantly lower than they would be for human-centric operations. Nevertheless, most risks, such as communication network congestion, latency, brownout, etc. are reduced by the UAS level of decision-making (Makarova et al., 2017).

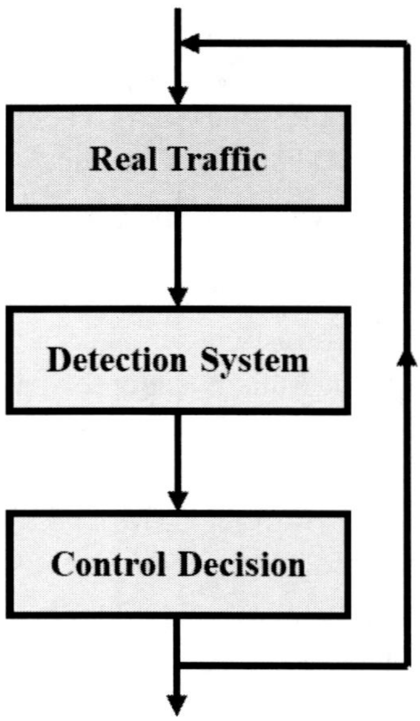

Figure 10.4. Intelligent transportation system (ITS).

Section B: Smart Traffic Management System

Using IoT technology, the Smart Traffic Management System (STMS) seeks to increase the effectiveness and security of transportation networks. To improve traffic patterns and avoid accidents, this technology offers real-time data on traffic flow, road conditions, and weather. Modern cities cannot function without STMS, and the use of STMS systems is spreading around the globe. The latest IoT technologies utilised in STMS and their applications will be thoroughly covered in this chapter (Nikhilesh et al., 2020).

A technology known as the Internet of Things (IoT) enables actual things to be connected to the internet and transmit and receive data. There have been previous applications of IoT-based systems for solving everyday problems, such as in building smart homes and supporting disabilities, as well as improving safety in natural resource extraction and monitoring the environment. Urban areas have complex interdependent subsystems, including the traffic system, which plays a significant role in the world economy. However, with increasing populations, vehicles on the roads also increase, resulting in daily traffic congestion and delaying emergency services such as ambulances. This delay can impact the life-saving capabilities of emergency services, as 65-70% of deaths could have been prevented if emergency teams had responded more quickly. Despite government and town planners' efforts to add more roads and flyovers to ease traffic congestion, the results have not been promising. Predicting travel times, fuel consumption, and managing panic situations become challenging with traffic congestion. However, managing traffic data is essential to overcome traffic congestion and make dynamic decisions to drive traffic on roads. In recent years, research efforts have focused on integrating wireless sensors and different devices on the IoT platform to make real-time traffic monitoring, analytics, and dynamic congestion management a reality. This can also lead to innovative traffic management systems like the proposed smart barricade prototype (Pecar et al., 2017). Figure 10.4 shows the working cycle of intelligent transportation system.

Section C: IoT Technologies in Smart Traffic Management System

Internet of Things (IoT)

In order to gather and send real-time data on traffic, weather, and road conditions, numerous sensors and devices must be integrated into STMS, which is made possible by IoT. By giving STMS precise information on the number of cars on the road and their speed, IoT enables STMS to improve traffic patterns and ease congestion.

Smart traffic management systems (STMS) based on the Internet of Things (IoT) gather real-time data on traffic conditions from a network of connected devices and sensors, then utilise this data to improve traffic flow and lessen congestion.

Consider a scenario in which a city wishes to improve the junction traffic lights to lessen rush-hour congestion. At junctions, the city installs cameras and sensors linked to an IoT platform. These sensors gather data on traffic volume, speed, and vehicle kinds in real-time. The data is sent to the IoT platform, where it is processed by machine learning algorithms in real-time to forecast traffic trends.

The IoT platform may modify traffic lights in real-time to enhance traffic flow and lessen congestion based on the traffic data. For instance, traffic lights can be changed to give the route with heavy traffic priority if another road has little traffic. This can lessen traffic congestion and wait times.

An IoT-based STMS can utilise dynamic message signs, variable speed restrictions, and lane control signals in addition to traffic signals to improve traffic flow. For instance, emotive message signs can warn drivers in real time about traffic conditions and recommend detours. Likewise, variable speed limits allow for the adjustment of speed restrictions in response to current traffic circumstances (Rathod et al., 2016).

In order to gather and send real-time data on traffic, weather, and road conditions, numerous sensors and devices must be integrated into STMS, which is made possible by IoT. By giving STMS precise information on the number of cars on the road and their speed, IoT enables STMS to improve traffic patterns and ease congestion. Smart traffic management systems (STMS) based on the Internet of Things (IoT) gather real-time data on traffic conditions from a network of connected devices and sensors, then utilise this data to improve traffic flow and lessen congestion.

Table 10.2. Issues of the smart traffic

Issue	Details
Importance	Intelligent Transportation Systems (ITS) monitor traffic data
Methods	Image processing, RFID, and induction loops
Drawbacks	These methods have certain shortcomings
Solution	Proposes a density-based traffic management system
Approaches	Precision traffic monitoring, Efficient modelling and control
Stochastic reasoning	Using techniques involving stochastic reasoning to reduce traffic congestion

Let's say a city wishes to improve the junction traffic lights to lessen rush-hour congestion. At junctions, the city installs sensors and cameras that are connected to an IoT platform. Real-time information on traffic volume, speed, and vehicle kinds is gathered by these sensors. The IoT platform receives the data, which is then processed in real-time by machine learning algorithms to forecast traffic trends. The IoT platform may modify traffic lights in real-time to enhance traffic flow and lessen congestion based on the traffic data. For instance, traffic lights can be changed to give the route with heavy traffic priority if another road has little traffic. This may shorten wait times and enhance traffic (Rawdah et al., 2016).

An IoT-based STMS can utilise dynamic message signs, variable speed restrictions, and lane control signals in addition to traffic signals to improve traffic flow. For instance, emotive message signs can warn drivers in real time about traffic conditions and recommend detours. Moreover, speed restrictions can be changed dependent on the flow of traffic in the present. Overall, incorporating IoT technology into STMS may greatly enhance commuters' driving experiences by reducing congestion, improving traffic flow, and overall. Table 10.2 shows the issues related to smart traffic system.

A traffic management system that operates intelligently without human intervention is essential. By utilising advanced technology such as Chirp-based Spread Spectrum and LoRaWAN, a system can offer faster and more transparent access to destinations. The current traffic management system needs to be improved, resulting in delays and inconvenience for commuters. Introducing an intelligent IoT system that can operate the traffic lanes is crucial (Stankov et al., 2021).

For instance, a city in India experienced a common problem where ambulances had to wait long intervals for clearance in the lane, leading to delays in medical emergencies. The existing traffic management system needed the necessary feasibility for ambulances and other vehicles. This required a radical change, and an intelligent traffic monitoring system was introduced. The plan worked by installing AD HOC sensor nodes that gathered real-time information on traffic density and transmitted it to the traffic officials on site. Based on this data, the traffic light management system was enhanced to give high priority and a green signal to essential vehicles such as an ambulance, fire engine trucks, or police patrols. The method utilized IoT and an intelligent mechanism operated without human intervention (Sliacky et al., 2015).

This system brought several advantages, including significantly reducing annoyance and inconvenience for commuters. The system also minimized the number of accidents caused by traffic congestion by identifying and avoiding such scenarios. The intelligent traffic management system was remotely managed and provided clear and faster access to all vehicles' destinations.

The case study highlights the significance of an intelligent traffic management system employing advanced technology and IoT to operate traffic lanes. The system can provide faster and more transparent access to destinations for all vehicles, reduce annoyance and inconvenience, and, most importantly, provide a higher priority to essential vehicles such as ambulances during medical emergencies.

Wireless Sensor Networks (WSNs)

WSNs are essential in STMS as they allow the collection of real-time data on traffic flow, weather, and road conditions. These sensors can be installed on roads, traffic signals, and intersections to collect data and transmit it to a central control system.

Wireless Sensor Networks (WSNs) consist of many small, low-power, and low-cost sensors distributed throughout a physical environment, such as a road network. These sensors are equipped with various sensors, such as temperature, humidity, and motion, to detect the physical parameters of the environment. They are also equipped with wireless communication modules, allowing them to communicate with each other and a central control system.

In the context of STMS, WSNs are used to collect data on traffic flow, weather, and road conditions. These sensors are typically installed at strategic locations throughout the road network, such as on roads, traffic signals, and intersections. They can collect data on the number of vehicles on the road, speed, and congestion at a given location. They can also collect data on weather conditions impacting traffic flow, such as temperature, humidity, and precipitation.

The operation of WSNs in STMS typically involves the following steps:

Deployment

The sensors are installed at strategic locations throughout the road network. The deployment strategy depends on the specific requirements of the STMS application and the physical layout of the road network.

Data Collection

The sensors collect data on traffic flow, weather, and road conditions. This data is typically collected at regular intervals, such as every few minutes, and is transmitted to a central control system.

Data Transmission

The sensors communicate the gathered data to a centralised control system using wireless connection. Several wireless communication protocols, such as Wi-Fi, Zigbee, or cellular, can be used for this work.

Data Processing

The central control system processes the gathered data to extract pertinent data, such as the quantity of moving cars, their speeds, and the degree of congestion at a certain site. Using this information will improve traffic patterns and ease congestion.

Control Actions

The central control system can take various steps based on the processed data, such as adjusting traffic signals, redirecting traffic, or providing real-time information to drivers.

Example

Consider an example of a WSN deployed on a busy highway. The WSN consists of sensors installed at regular intervals along the route. These sensors are equipped with motion sensors and Wi-Fi communication modules. Batteries power them and can operate for several months without recharging.

The sensors gather information on traffic flow, speed, and congestion at each place. This information is sent to a central control system, which analyses it and gives drivers real-time information. To reroute traffic to alternate routes, the control system could, for instance, identify significant congestion at a certain area. Drivers may also receive real-time information from the system, such as the projected time of arrival at their location and other routes to avoid traffic (Sridharan et al., 2017).

WSNs are:

- An essential component of STMS
- Providing real-time data on traffic flow
- Weather
- Road conditions

These sensors enable STMS to optimize traffic patterns and reduce congestion by providing accurate data on the number of vehicles on the road and their speed.

Radio Frequency Identification (RFID)

RFID is a cutting-edge technology that makes it possible to identify drivers and cars, easing traffic jams and increasing the effectiveness of transportation systems as a whole. Vehicles that have RFID tags affixed may interact with traffic lights and get real-time data on traffic trends.

In order to automatically recognize and track tags affixed to things, including cars, RFID (Radio Frequency Identification) technology employs electromagnetic fields. A tiny chip and an antenna that may be fastened to an automobile are what make up RFID tags.

In an STMS, RFID technology can monitor traffic flow and improve efficiency. For example, RFID readers can be installed at traffic signals, and RFID tags can be installed in vehicles. As a car approaches a traffic signal, the RFID reader sends a signal to the RFID tag on the car, which responds with the vehicle's identification information.

When the central control system receives this information, it may modify the timing of the traffic signals to ease congestion and enhance traffic flow. The system can improve traffic patterns in real time by using RFID technology to track the number of cars on the road and their speeds. The use of electronic toll collection (ETC) systems is one illustration of the application of RFID technology in STMS. ETC systems eliminate the need for manual toll collection and enhance traffic flow by using RFID tags implanted in cars to automatically collect tolls as the vehicle passes through toll booths.

Another example is the use of RFID technology in parking management systems. RFID tags can be used to track the entry and exit of vehicles from parking lots, allowing the system to optimize parking space usage and reduce congestion in and around parking facilities.

Overall, RFID technology is a valuable tool for STMS, allowing for the collection and analysis of real-time data on traffic patterns and vehicle movements. This information can be used to optimize traffic flow, reduce congestion, and improve overall transportation efficiency (Tagiltseva et al., 2021).

Global Positioning System (GPS)

GPS is a popular technology in STMS, enabling real-time data collection on vehicle location, speed, and direction. GPS data is used to optimize traffic patterns and provide real-time information to drivers on road conditions and traffic flow.

Design Specifications

The Internet of Things (IoT) is defined by linked devices, such as processing units and actuators, that interact with one another to accomplish a shared objective. However, experts have discovered that approximately 85% of networks are unlinked, preventing them from exchanging data with one another or the cloud. Nonetheless, the Internet of Things idea seeks to allow connectivity and a collaborative effort to connect online objects to gather and analyse data. The Internet of Things architecture aims to give devices autonomy while enabling them to contact a website for data. Furthermore, the Internet of Things is a system of active involvement that has spread worldwide.

Most traffic-predicting models concentrate on metropolitan traffic management and major roads, with few studies on concentration roads and restricted campuses. Researchers proposed an IoT-based system for collecting, processing and transmitting traffic data in real-time to handle this problem.

The main aim is to improve transportation by disseminating traffic information about gridlock and odd traffic events through wayside texting devices. The gadget also gathers traffic statistics from the state government, giving people early alerts during peak hours to save time.

Another option is an IoT-based traffic management system, in which infrared sensors monitor car density and flow algorithms handle traffic light timing. The system uses a Wi-Fi subsystem to send data to a distant network port. During an emergency, the system gives drivers precedence or provides a lengthy signal time to cross the street. Additionally, cars that run red lights can be identified and fined using a traffic wallet Smartphone software.

For more than a century, traffic signals have been used as signalling instruments to control traffic flows. However, inadequate facilities cannot keep up with the increasing vehicles and road users. Several steps have been suggested to reduce traffic, including constructing other motorways and bypassing roads, flyovers, and restoration, while others use traffic control models such as RFID or image processing. The area traffic management system watches traffic within a particular region and reacts dynamically to variations to avoid congestion. Figure 10.5 depicts "Keep Going Hong Kong," an initiative utilizing the area traffic management system for clever transportation. Overall, traffic is an essential aspect of any community, and fast population development has created many problems, including traffic congestion, time waste, and criminal activity. More extended traffic jams cause air pollution, which harms the environment and business productivity (Vakula et al., 2017).

The operational flow chart for signaling control of the proposed system can be seen in Figure 10.6. This system aims to decrease delay time through consideration of the arrival time of vehicles.

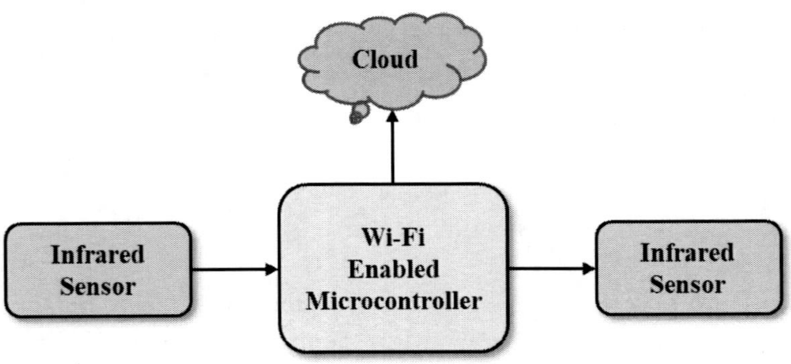

Figure 10.5. Block Diagram of Traffic Management.

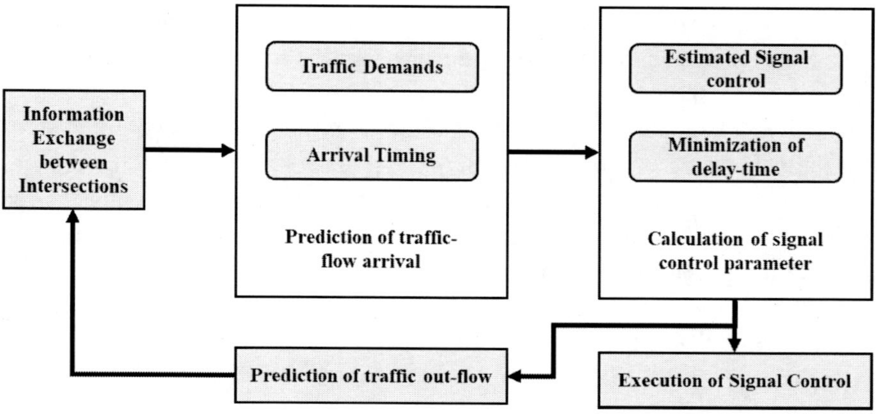

Figure 10.6. Flow chart for signal control operation.

Prototype Implementation

The outlined system uses essential electrical components such as LEDs for showing passenger traffic signals, infrared sensors to determine the number of vehicles, an RF emitter and receiver to illustrate the operation of priority vehicles, and a microprocessor (8051). To ascertain traffic intensity, the system uses infrared communication, with the microcontroller getting IR impulses from an infrared radiation sensor and reacting correctly based on traffic numbers. When there is more activity on one side, the microcontroller sends green signals to the side, with the fewest IR sensors getting calls. The system contains LEDs to demonstrate traffic control light operation, and its purpose is to send a signal to the green LED in a high-traffic direction. In a 4x2 lane configuration, there are two sensors for each lane, enabling the microcontroller to decide the path and orientation based on traffic volume. This system also claims to deliver acceptable outcomes regarding emergency car clearance (Yang et al., 2014).

On the other hand, traditional traffic management methods use road traffic sign lights with pre-set time delays for each way, creating unnecessary blockages in one channel while the other lanes are vacant. The author proposes using numerous sensors instead of conventional traffic control devices to provide high-priority traffic clearing for paramedics. The sensors identify moving cars and transmit the information to the microcontroller, which regulates the timing circuit. When traffic volume is high, the LEDs illuminate longer than usual; when traffic volume is low, the LEDs glow shorter. In the four-lane dense traffic system illustrated in Figure 10.7, a priority car in one lane needs precise and convenient sensor operation for appropriate clearance. In such cases, the proprietor of the fire engine truck can use the siren, which transmits RF signals that cause the monitors on the road to give green signs for the vehicle. Alternatively, the Blynk app creates the signal and identifies the primary vehicle's position. Each lane has a set period for turning the red LED indication on and off. The microcontroller determines the intellectual functioning of the system based on the cadence in each channel.

Applications of STMS

Traffic Management and Control

Urban traffic flow is managed and controlled using STMS, which lowers congestion and boosts the overall effectiveness of transportation networks. In order to help drivers and authorities, STMS can recognise and address traffic congestion.

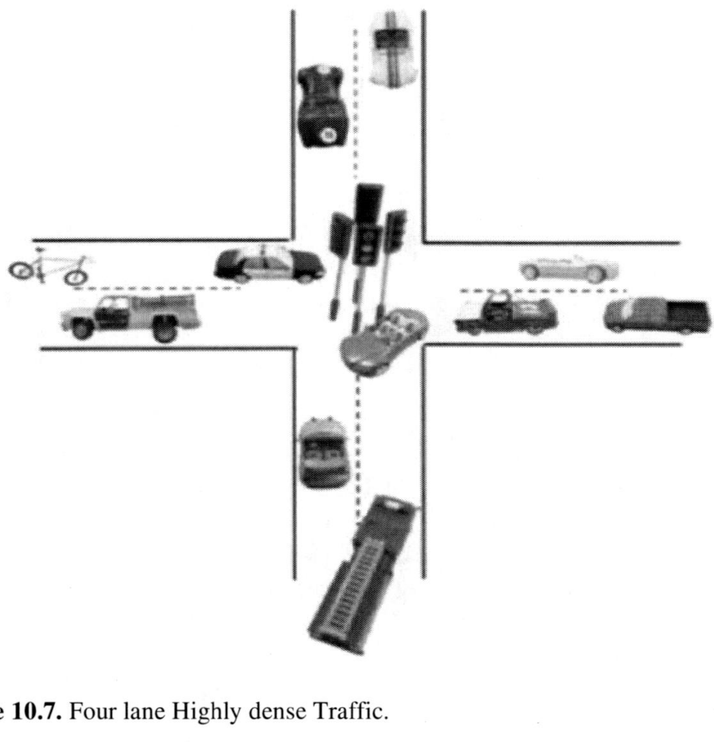

Figure 10.7. Four lane Highly dense Traffic.

Road Safety

STMS also improves road safety by detecting and responding to accidents and dangerous road conditions. STMS can detect and react to hazards such as flooding, ice, or debris, reducing the risk of accidents.

Public Transportation

Moreover, STMS is used to manage and enhance public transportation networks, enhancing their general effectiveness and lowering traffic congestion. Authorities may adjust routes and timetables using real-time data from STMS on the position and speed of shared vehicles. Environmental Impact

STMS can also reduce the environmental impact of transportation systems by lowering congestion and optimizing traffic patterns. This can reduce emissions and improve air quality, making urban areas healthier and more liveable.

Section D: Parallel Transport System Using IoT

The proper operation of transportation is crucial for both economic development and social well-being in today's society. There is a rising demand for transportation systems that can deliver efficient, dependable, and sustainable services due to the number of products and people that need to be carried. To ensure the timely and secure delivery of products and people, transportation operations must be effectively managed due to their complexity. By providing real-time monitoring and management of transportation operations, the Internet of Things (IoT) has emerged as a potent technology that can revolutionize the transportation sector. In this research, we suggest an IoT-based parallel transport system to boost the effectiveness and performance of transportation networks (Gomides et al., 2023).

System Architecture

A distributed network of sensors, actuators, and Internet of Things (IoT) gadgets make up the proposed parallel transport system, which enables real-time monitoring and management of transportation operations. The system design relies on a parallel processing approach in which many transportation routes are analysed concurrently to cut down on delivery times and transportation expenses. IoT devices are employed to gather information on the transport activities, and the cloud computing component gives the system access to storage and processing power. The IoT devices' data is processed by the data analytics component, which then offers insights into the transport operations. The data analytics component of the decision support system uses insights to inform choices. The user is given a visual representation of the transport activities via the user interface component (Janos et al., 2018).

The sensing layer is responsible for collecting data from various sources, such as GPS sensors, temperature sensors, and humidity sensors, to monitor the transportation operations. The data collected is transmitted to the processing layer, where it is analyzed and processed to generate insights that can be used to optimize the transportation routes. Figure 10.8 uses machine learning algorithms and data analytics techniques to analyze the data and generate insights. The actuation layer is responsible for controlling the transportation operations, such as adjusting the speed of the vehicle or changing the route, based on the insights generated by the processing layer (Garvanov et al., 2021).

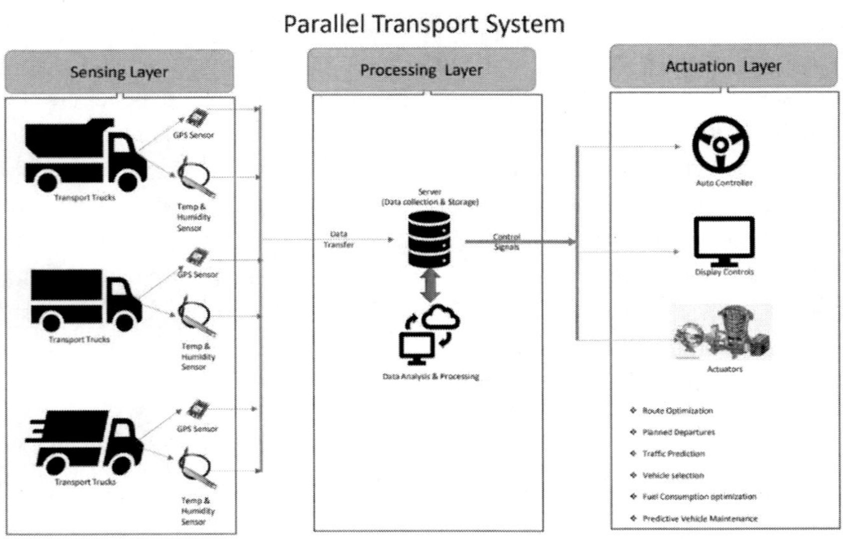

Figure 10.8. System Architecture of Parallel Transport system.

Key Components and Functionality

The sensing layer consists of various sensors that collect data about the transportation operations. These sensors include GPS sensors that track the location of the vehicle, temperature sensors that monitor the temperature inside the vehicle, and humidity sensors that monitor the humidity levels. The data collected by these sensors is transmitted to the processing layer, where it is analyzed to generate insights that can be used to optimize the transportation routes.

The processing layer uses machine learning algorithms and data analytics techniques to analyze the data collected by the sensing layer. The algorithms used include clustering, classification, and regression algorithms, which are used to analyze the data and generate insights. The insights generated by the processing layer are used to optimize the transportation routes by adjusting the speed of the vehicle, changing the route, or optimizing the delivery schedule.

The actuation layer is responsible for controlling the transportation operations based on the insights generated by the processing layer. This layer uses actuators, such as motors and valves, to adjust the speed of the vehicle, change the route, or optimize the delivery schedule. The actuation layer is designed to be responsive and flexible, so it can quickly adapt to changing conditions and optimize the transportation operations accordingly.

Benefits of the System: The parallel transport system using IoT has several potential benefits, including increased efficiency, reduced costs, improved safety, and reduced environmental impact. The system can optimize transport operations in real-time, which can reduce delivery times and improve the overall efficiency of the transport system. The system can also reduce costs by minimizing fuel consumption and optimizing the use of transport resources. The system can improve safety by providing real-time monitoring of transport operations and identifying potential safety hazards. Finally, the system can reduce environmental impact by minimizing the carbon footprint of transport operations.

Conclusion

IoT-enabled intelligent transport systems have the potential to transform the transportation sector by optimizing traffic flow, reducing congestion, and improving safety. The proposed architecture for IoT-enabled intelligent transport systems consists of several components, including sensors, data analytics, and communication networks. There are several use cases for IoT-enabled smart transport systems, such as Intelligent traffic management, Real-time vehicle tracking, Predictive maintenance, Smart Parking, Autonomous vehicles, and Public transport management, where the IoT- enabled innovative transport systems are used to optimize operations. The proposed framework for a parallel transport system using IoT can provide real-time monitoring and optimization of transport operations. The system can optimize transport operations in real-time, which can increase efficiency, reduce costs, improve safety, and reduce environmental impact. However, the implementation of such a system poses several challenges, including data privacy and security, interoperability, complexity, and cost. Further research is needed to address these challenges and to develop effective parallel transport systems using IoT for different contexts and applications.

References

Boreiko, O., and Teslyuk, V. (2017). "Model of a controller for registering passenger flow of public transport for the "smart" city system," *2017 14th International Conference The Experience of Designing and Application of CAD Systems in Microelectronics (CADSM)*, Lviv, Ukraine, pp. 207-209, doi: 10.1109/CADSM.2017.7916116.

Boreiko, O., and Teslyuk, V. (2016). "Structural model of passenger counting and public transport tracking system of smart city," *2016 XII International Conference on Perspective Technologies and Methods in MEMS Design (MEMSTECH)*, Lviv, Ukraine, pp. 124-126, doi: 10.1109/MEMSTECH.2016.7507533.

Brega, A. V., and Erokhina, O. V. (2022). "Smart City Transport Technologies: Infrastructure of the Future," *2022 Systems of Signals Generating and Processing in the Field of on Board Communications, Moscow*, Russian Federation, pp. 1-5, doi: 10.1109/IEEECONF53456.2022.9744352.

Bhatia, S., Gautam, D., Kumar, S., and Verma, S. (2023). "Automatic Seat Identification System in Smart Transport using IoT and Image Processing," *2023 3rd International Conference on Intelligent Communication and Computational Techniques (ICCT)*, Jaipur, India, pp. 1-6, doi: 10.1109/ICCT56969.2023.10075664.

Duraisamy P, M. Duraisamy, M. Periyanayaki and Y. Natarajan, "Predicting Disaster Tweets using Enhanced BERT Model," 2023 7th International Conference on Intelligent Computing and Control Systems (ICICCS), Madurai, India, 2023, pp. 1745-1749, doi: 10.1109/ICICCS56967.2023.10142660.

Duraisamy P, Y. Natarajan, V. Niranjani and K. Parvathy, "Optimized Detection of Continuous Gravitational-Wave Signals using Convolutional Neural Network," 2023 3rd International conference on Artificial Intelligence and Signal Processing (AISP), VIJAYAWADA, India, 2023, pp. 1-5, doi: 10.1109/AISP57993.2023.10134809.

Fitriani, L., Dianti, S., Kurniadi, D., Mulyani, A., and Setiawan, R. (2021). "Mapping-Based Using Geographic Information Systems for Smart Transportation," *2021 International Conference on ICT for Smart Society (ICISS)*, Bandung, Indonesia, pp. 1-5, doi: 10.1109/ICISS53185.2021.9533202.

Garvanov, I., and Garvanova, M. (2021). "New Approach for Smart Cities Transport Development Based on the Internet of Things Concept," *2021 17th Conference on Electrical Machines, Drives and Power Systems (ELMA)*, Sofia, Bulgaria, pp. 1-6, doi: 10.1109/ELMA52514.2021.9503084.

Galland, S., Lamotte, O., and Gaud, N. (2011). "MetroB: Evaluation and simulation of public transport system," *IET International Conference on Smart and Sustainable City (ICSSC 2011)*, Shanghai, pp. 1-5, doi: 10.1049/cp.2011.0278.

Gomides, T. S., De Grande, R. E., Pereira, R. S., Meneguette, R. I., de Souza, F. S. H., and Guidoni, D. L. (March 2023). "An Urban Traffic Management System based on Vehicle Cooperation," in *IEEE Latin America Transactions*, vol. 21, no. 3, pp. 441-449, doi: 10.1109/TLA.2023.10068848.

Gopi, B., Premalatha, J., Kalaivani, R., and Ravikumar, D. (2023). "Cloud based Landslide Detection and Alerting Nearby People by using IoT Technology," *2023 Third International Conference on Artificial Intelligence and Smart Energy (ICAIS)*, Coimbatore, India, pp. 1287-1292, doi: 10.1109/ICAIS56108.2023.10073773.

Horazdovsky, P., Kozhevnikov, S., and Svítek, M. (2019). "Dynamic Public Transport in Smart City using Multi-agent system," *2019 Smart City Symposium Prague (SCSP)*, Prague, Czech Republic, pp. 1-5, doi: 10.1109/SCSP.2019.8805716.

Isawi, O. A. A., Jaafari, K. A. A., and Sumaiti, A. S. A. (2023). "Impact of Integrating Renewable Energy Systems on the Smart Grid-Transportation Nexus Operation under Electric Vehicle cyber-attacks," *2023 IEEE PES Conference on Innovative Smart Grid Technologies - Middle East (ISGT Middle East)*, Abu Dhabi, United Arab Emirates, pp. 1-6, doi: 10.1109/ISGTMiddleEast56437.2023.10078591.

Janos, V., and Kříž, M. (2018). "Smart urban transport," *2018 Smart City Symposium Prague (SCSP)*, Czech Republic, pp. 1-5, doi: 10.1109/SCSP.2018.8402664.

Kazi, S., Bagasrawala, M., Shaikh, F., and Sayyed, A. (2018). "Smart E-Ticketing System for Public Transport Bus," *2018 International Conference on Smart City and Emerging Technology (ICSCET)*, Mumbai, India, pp. 1-7, doi: 10.1109/ICSCET.2018.8537302.

Morri, N., Hadouaj, S., and Ben Said, L. (2020). "Towards an Intelligent control system for public transport traffic efficiency KPIs optimization," *2020 Global Congress on Electrical Engineering (GC-ElecEng)*, Valencia, Spain, pp. 33-37, doi: 10.23919/GC-ElecEng48342.2020.9286268.

Makarova, I., Shubenkova, K., Mavrin, V., Boyko, A., and Katunin, A. (2017). "Development of sustainable transport in smart cities," *2017 IEEE 3rd International Forum on Research and Technologies for Society and Industry (RTSI)*, Modena, Italy, pp 1-6, doi: 10.1109/RTSI.2017.8065922.

Nikhilesh, K. S., Raaghavendra, Y. H., Madhu Soothanan, P. J., and Resmi, R. (2020). "Low-cost IoT based weather monitoring system for smart community," *2020 Fourth International Conference on Inventive Systems and Control (ICISC)*, Coimbatore, India, pp. 482-486, doi: 10.1109/ICISC47916.2020.9171156.

Pecar, M., and Papa, G. (2017). "Transportation problems and their potential solutions in smart cities," *2017 International Conference on Smart Systems and Technologies (SST)*, Osijek, Croatia, pp. 195-199, doi: 10.1109/SST.2017.8188694.

Rathod, R., and Khot, S. T. (2016). "Smart assistance for public transport system," *2016 International Conference on Inventive Computation Technologies (ICICT)*, Coimbatore, India, pp. 1-5, doi: 10.1109/INVENTIVE.2016.7830206.

Rawdah, R., and Ali, S. S. (2021). "Proposing a Real-Time Ticket Monitoring System for Public Transport in NSW, Australia," *2021 2nd International Conference on Robotics, Electrical and Signal Processing Techniques (ICREST)*, DHAKA, Bangladesh, pp. 221-224, doi: 10.1109/ICREST51555.2021.9331127.

Stankov, I., and Stefanova-Stoyanova, V. (2021). "Urban Intelligent Transport Management Systems," *2021 29th National Conference with International Participation (TELECOM), Sofia, Bulgaria*, pp. 113-116, doi:10.1109/TELECOM53156.2021.9659791.

Sliacky, M., Jerabek, M., Borka, J., and Skuhra, R. (2015). "Process analysis of Transport Ticketing and Information Systems for Czech public transport," *2015 Smart Cities Symposium Prague (SCSP), Prague, Czech Republic*, pp. 1-3, doi: 10.1109/SCSP.2015.7181572.

Sridharan, S., Prasad, R. V., and Srinarayan, S. (2017). "Designing a smart transport system application for South Indian traffic scenarios — A modern approach towards digitalizing the transport systems," *2017 2nd International Conference on Communication and Electronics Systems (ICCES)*, Coimbatore, India, pp. 1-6, doi: 10.1109/CESYS.2017.8321124.

Tagiltseva, J. A., Kuzina, E. L., Drozdov, N. A., Vasilenko, M. A., Kuzina, M. A., and Korenyakina, N. N. (2021). "The International Cargo Transportations in South Russian Transport and Logistics System," *2021 International Conference on Quality Management, Transport and Information Security, Information Technologies (IT&QM&IS)*, Yaroslavl, Russian Federation, pp. 248-251, doi: 10.1109/ITQMIS53292.2021.9642872.

Vakula, D., and Raviteja, B. (2017). "Smart public transport for smart cities," *2017 International Conference on Intelligent Sustainable Systems (ICISS)*, Palladam, India, pp. 805-810, doi: 10.1109/ISS1.2017.8389288.

Yang, X., and Zhang, L. (2014). "A dynamic method to monitor public transport based on smart card and GPS Data," *17th International IEEE Conference on Intelligent Transportation Systems (ITSC)*, Qingdao, China, pp. 924-929, doi: 10.1109/ITSC.2014.6957807.

Chapter 11

Internet of Things Enabled Electric Vehicles in Smart Cities

V. Ebenezer[1], PhD
M. Robinson Joel[2], PhD
E. Bijolin Edwin[1], PhD
M. Roshni Thanka[1], PhD
S. Stewart Kirubakaran[1], PhD
P. Joy Kiruba[3]
and M. Joe Marshell[4], PhD

[1]Division of Computer Science and Engineering, Karunya Institute of Technology and Sciences, Coimbatore, Tamilnadu, India
[2]Department of Computer Science and Engineering, Kings Engineering College, Chennai, Tamilnadu, India
[3]Department of Computer Science and Engineering, PSNA College of Engineering and Technology, Dindigul, Tamilnadu, India
[4]Department of Electronics and Instrumentation Engineering, SRM Valliammai Engineering College, Chengalpattu, Tamilnadu, India

Abstract

The Internet of Things is establishing itself as a brand-new wireless technology platform in every industry. It has exhibited its scope and development mostly in the area of electric vehicles. A sensor, a smart gadget that is integrated into the body of an electric car that is both effective and affordable, is used to bring back electric automobiles. Additionally, this aids in environmental protection and resolving all problems with current vehicles. In this article, it is discussed how electric

In: Blockchain Enabled Secure Big Data Computing …
Editors: L. Ashok Kumar, D. Karthika Renuka, V. Chandrasekar et al.
ISBN: 979-8-89113-243-6

vehicles need to be equipped with IoT-based technology that can track their battery life. An online monitoring system has been employed to continuously (day-by-day) monitor all vehicles because these electric vehicles are implemented with the internet. The online outcomes are then displayed visually. The result of the proposed work is evaluated through visual analysis and with other performance parameters.

Keywords: Internet of Things, wireless technologies, electric vehicles, sensor

Introduction

There have been more developments in car manufacturing in recent days, with all firms making advancements in vehicle production and working toward a smart vehicle environment. As a result, obsolete engines have been phased out in favour of newer models that pose far less environmental risks. As a result, electric vehicles that emit far less pollution have been developed, along with a number of features that are quite similar to those found in current automobiles. Electric cars (EVs) will be ubiquitous in upcoming near future, necessitating efficient and intelligent charge scheduling strategies. Smart grids were outfitted with instruments and sensors that help in the monitoring and management of the responsibility scheduling process. As a result, combining EVs with latest technology smart grids in a bigger IoT entity can help accelerate the widespread adoption of EV cars in the transition to a green and intelligent transportation age.

Traditional automobiles rely on fossil fuel energy for the majority of their power. Fossil fuels are currently regarded as the world's principal energy supplies. Fuels are at the heart of all modern technical achievements in agriculture and transportation. Fuels are becoming more important in the field industry. The biggest sources of fuel consumption like gasoline, petrol, diesel requires global transportation. Furthermore, these fuels were considered as non-renewable energy supplies that last in the run out in the not-too-distant place. Due to the release of carbon dioxide (CO2), burning fuels is the most significant source of Green House Gas (GHG). Furthermore, it has a detrimental influence on global warming, which is affecting the entire planet's environment. The key benefit of using gasoline in cars is that they are often less classy to buy or maintain. These forms of gasoline cars are also less expensive to maintain. Also, it's far easier to find parts and mechanics for gasoline-powered cars than it is for electric vehicles. When compared to electric cars, gasoline vehicles also tend to perform better in terms of speed and acceleration.

The speed at which an electrical vehicle can operate is more comparable to that of a diesel or petrol engine automobile. Their range with a gasoline automobile and an electric car is comparable. Furthermore, the efficiency of electric vehicles is comparable to that of gasoline and diesel automobiles. Furthermore, these electric cars feature a battery supply, which is beneficial for long-distance travel and even for novices. Electric vehicles are growing in popularity among both individuals and companies looking to reduce their carbon footprint. Electric automobiles run on a battery pack that is electrically charged. This can be charged using a regular wall outlet, but it is often charged using a dedicated charging station. It is compulsory to power the motor and move the wheels. Simply plugging the vehicle's charging point into an outlet allows the battery to be powered by the vehicle. Because no fuels are used to directly power the car, they are the most environmentally friendly option. In terms of emissions, they also have no effect on the environment. An electric vehicle is a vehicle that is powered by more than one electric motor. It can be run on a battery or by a collector system that pulls power from the outside of the car.

The main benefit of an electric car over ones fueled by petrol or diesel is that they are more environmentally friendly. Your expenses for petrol are eliminated because electric vehicles only run on electricity, which means they don't consume any fossil fuels. The use of public charging stations is, of course, subject to tariffs, but refuelling can be facilitated by the use of a fuel card. In addition to being much quieter than cars with petrol or diesel engines, electric vehicles may also be significantly more economical, especially when moving at a moderate speed through crowded cities.

A smart city is a technologically advanced metropolitan region that employs a variety of electronic technologies. Voice activation methods and sensors to gather particular data from all sources are the ways. Asset management is aided by the information gleaned from such data. They can be utlisized efficiently to manage resources and services. These data are utilised to enhance city operations, particularly in the smart city. This covers information gathered from all citizens and gadgets. These devices also gather information from buildings and assets, which is then analysed. They're also scrutinised in order to keep track of and manage traffic and transportation networks. They're also utilised to keep an eye on the data and utilities of power facilities. The devices were also employed to collect data on a regular basis by the water supply networks and the waste management system. Crime detection and information systems proposed by Bellavista, 2013 were also employed in the construction of smart cities. Devices are also utilised to collect data from schools and the finest libraries for the available book collection. Similarly, gadgets are used to collect data in hospitals and other outreach program. Smart cities are described as such in terms of both how their governments use technology and how they monitor it. The data collected is utilised to study and plan the developments that govern the city that has been prposed by A Bui Zanella, 2014.

Cassandras, 2016 proposed that to increase the efficiency of city operations and the services while also connecting residents, the concept of a "smart city" take part statistical information and communication technology (ICT) with a variety of physical devices connected to the Internet of things (IoT) network. During this stage of monitoring, a surveillance and monitoring device has been inserted into relevant elements of electric cars to replace the previous technology of the Global Positioning System (GPS). An intelligent gadget known as wireless sensors has performed the same function as GPS. Even with electric automobiles, sensor technologies will be more useful than GPS gadgets. When wireless sensors are deployed, all users will find it easier to maintain their vehicles. Even the introduction of electric cars will make the environment more user-friendly, and all locations can be transformed into smart cities and smart villages. Furthermore, because these electric cars are equipped with a sensor system that can monitor all traffic circumstances, they will be extremely useful in congested areas. This device might also be used to identify individuals in the event of a minor cardiac arrest, preventing numerous accidents.

Environmentalists continue to warn that hazardous emissions from gasoline and diesel automobiles are damaging the environment. In many metropolitan locations, air pollution generated by cars is particularly dangerous. Vehicles that run on alternate fuels are required in this climate. Although electric cars have yet to make a substantial effect in India, their popularity is expected to skyrocket in the coming years. In this case, we've produced a list of benefits and drawbacks for people considering purchasing an electric vehicle. Electric automobiles use a lot less energy than cars that operate on gasoline or diesel. For example, a petrol automobile that must go 40 kilometres per day and has a mileage of 13 kilometres per litre might cost up to one lakh rupees each year. To fully charge an electric automobile, it will require ten units. It reaches a maximum distance of 85 kilometres. A 40-kilometer electric car costs over 5 units of power. If a unit costs Rs 5, it will only cost Rs 9,000 per year in Indian currency. Electric vehicles have far reduced fuel expenses. These calculations are made in Indian currency.

Electric automobiles do not create smoke, making them ideal for urban environments. These vehicles will be completely ecologically friendly. The workplace is suitable for commuting and conducting business in the city. The small shape of the Mahindra E2O automobile for sale in India makes it easy to manage and park. Electric vehicles, like gasoline and diesel vehicles, are built differently. There is no risk of fire since the electricity will be stopped immediately in the event of an accident. Also, the same crash test requirements apply to gasoline vehicles. Electric vehicles do not require routine maintenance. The amount of maintenance required will be low. Petrol and diesel autos have extremely high maintenance expenses.

It is likely to incur increased maintenance expenditures after two or three years. Electric automobiles, on the other hand, have substantially cheaper maintenance expenses. Petrol and diesel cars, without a doubt, provide a livelier and more noiseless ride than automobiles. However, this might be harmful at times. We've listed a number of drawbacks as to why this is the case. Used for the country's economic growth. That instance, cutting imports of gasoline and diesel will lessen the need to pay in dollars. As a result, the strengthening of the Indian rupee will be aided. The pattern of reliance on Gulf nations for gasoline and diesel will also change.

Recharging the battery in electric automobiles allows it to go short distances. Customers are apprehensive because of this. Furthermore, it is not possible to use it for long-distance travel in existing Indian electric automobiles. There is also a scarcity of lodging. There aren't enough charging stations to keep electric cars' batteries charged. Extensive electricity loading stations, similar to gas stations, would undoubtedly be beneficial. Pickup, speed, and other factors differ between petrol and diesel vehicles. This is something that automotive enthusiasts find perplexing. Many people consider it a waste of time to charge the battery for too long. After a certain amount of use, the batteries in these electric automobiles must be replaced, just like car tyres. It's also rather costly. Even if the cost of maintenance is cheap, the expense of replacing this battery in a few years will be a significant hardship for customers. These electric vehicles cannot be used in places where there are frequent power outages or shortages. Travellers benefit from noise and vibration reduction, but it can occasionally lead to an accident without the driver even realising the car is approaching.

Literature Review

The smart city idea has recently been enabled in numerous application areas, with the goal of improving living quality while lowering expenses. For some cases, the urban IoT term is used to define the smart city notion that has been proposed by Farmanbar, 2019. The primary barrier in politics is how decision-making is distributed among numerous interests. Consolidating the strategic planning and management of smart city features under a single, specialised city department is one way to overcome this roadblock by institutionalising the entire decision-making and implementation process formulated by M Helmy et al., 2018. Emerging smart grids have demonstrated the capacity to combine non-renewable and sustainable energy sources, easing environmental concerns and lowering operational costs in light of the imminent shortage of conventional fossil fuels. Plans for Sujata Joshi et al., 2016 work examines the evolution of nano grids, renewable energy-based applications, energy-consumption technologies, and technology-based water grids in urban cities to clarify the role of urban grids in launching smart technology-based cities. A seurvey of the key European urban city initiatives has also been conducted. These will provide academics with a broader perspective on smart-grid system operation, monitoring, control, and audit.

By using the efficient and scalable ability of the alternating orientation technique of multiplier, an enhanced decentralised charging strategy is also presented to coordinate the charging of large-scale EVs in numerous residential buildings (ADMM). Extensive research has shown by the team Yao L et al., 2015 that the suggested approach beats two previous decentralised systems by providing greater final mean state-of-charge (SOC) levels to linked EVs and resulting in a lower utility cost. Furthermore, the suggested pricing method has been shown to be superior to the centralised approach in terms of consuming less communication overhead and providing greater resilience in protecting user privacy. R Sousa et al., 2018 proposed a smart recharging system that enables real-time regulation for the electricity-based vehicle's where battery charging and current depends on demand at the dwelling, which is detected via a domestic wireless sensor network. To get the vehicle's information, an Android mobile application was created. In order to communicate with the wireless sensor nodes of an intra-vehicular wireless sensor network (IVWSN), this programme employs the BLE protocol. Additionally, a real-time notification system was put in place to alert consumers to certain events, such low and full rechargeable batteries. The essential components of the suggested IoT system are supported by experimental results.

The goal of Savari et al., 2020 and his team was to develop a better electric vehicle charging system by using the benefits of the Internet of Things (IoT) technology. Using a range of sensors and broadcasting methods, the IoT paradigm provides current facilities with a real-time interactive picture of the physical environment. A real-time server-based forecasting application is proposed in this research paper. The first and most important item is to give schedule management in order to prevent waiting times. The second goal is to give real-time charging station recommendations for EVs that are both affordable and quick to charge. Furthermore, the proposed system eliminates third-party intrusion and safeguards EV user privacy as well as sophisticated information flow between the user and the charging station. The charging station may be used by end-users according to their needs.

M Asaad et al., 2017 proposes a real-time Battery Monitoring System (BMS) that uses the coulomb method of counting for SoC estimate and the messaging-based Message Queue Telemetry Transport (MQTT) protocol for interaction. The suggested BMS is built utilising suitable sensor technologies, a central CPU, interfacing devices, and the Node-RED environment on a hardware platform. To enable smart charging, an optimization approach focused at maximising trade income for electric Vehicles aggregators is proposed. The notion of one-to-one care towards the performance of the vehicle using various IoT techniques were proposed by D Mills et al., 20223, allowing for direct monitoring of IoT products. Two essential elements make up the recommended Internet of Things-based battery monitoring system: a piece of monitoring equipment, and user interaction. According to test results, the system is able to recognise declining battery performance and notify the user for additional action.

Peris Ortiz et al., 2022 offered a study paper that looks at the many components and dimensions of a smart city. In order to get deeper understanding of the notion of smart cities, a framework has been developed to cover the gap in the literature regarding the growth of smart cities and its activities. Based on a thorough analysis of the literature in the fields of culture, organisation, socioeconomics, law, technology, and sustainability (SMELTS), we identified six key pillars for building the framework. The research illustrates how these factors might contribute to the success of the smart city concept. The suggested framework has been utilised to determine numerous research goals and its practical consequences have been traced. A smart city may be thought of as a system that uses modern communication technology to provide value-added services for exceedingly intricate cities and their management. The paradigm may be used in a variety of fields, including industry 4.0, smart appliances, and medical assistance. In the same way, it's used in mobile healthcare and city-based aged support. In emerging cities, power systems and intelligent power management are also important. It's also used in the automobile industry, traffic system management, and a variety of other fields by Philips, 2014.

An urban setting that offers a fresh perspective on environmental monitoring and energy distribution to the city's most vulnerable residents. A Smart City can also refer to transportation, commercial initiatives, and a variety of other socio-economical activities proposed by Patrice Seuwou et al., 2020. A "Smart City" is defined as the integration of multiple heterogeneous dispersed devices into a shared environment supported by software. Smart Cities, according to the study, have a number of common characteristics, including being sensible many times, connectable at all times, constantly available, and omnipresent. They're also social, shareable across devices, and visible or augmented.

In order to manage a city's resources in a way that will improve quality of life, foster a more sustainable environment, and boost economic value and efficiency, a number of ICT solutions may be combined. Future city design will need to include Internet of Things (IoT) concepts, which have exploded in popularity in recent years. [15] A smart city is one that is implanted, linked, and smart. The following features, although not limited to the ones above, can be used to define and quantify a smart city. The first is Smart Electricity, and the second is Intelligent Data, which is used to gather information. Intelligent Transportation creates a traffic-free environment, while Smart Infrastructure allows cities to live more intelligently. As, Malmö Stad, 2013 postulated that the information transmitted were additionally aided by devices connected and know that having IoT is at the heart of all of these aspects, contributing to better network management, efficient production, and power distribution. Zanella and their team propose an architecture that makes use of the Internet of Things (IoT) as a technology solution capable of meeting the needs of all complicated application scenarios in emerging cities. According to the study, including a wide range of devices such as sensors and actuators for monitoring and regulating cities might result in a myriad of digital services.

Wireless sensor networks are a domain that has a significant influence on linking things over a long distance. This link is critical for smart city design and implementation. WSNs, on the other hand, have been popular in academic circles rather than industry for more than a decade. As a result, it should be mentioned that combining diverse discoveries from academic knowledge with industry standards might help to better offer possible insights to the smart city paradigm. Another great strategy was to have multiple standardisation bodies codify the architecture, protocols, and capabilities of the IoT. IEEE, HART, IETF, and ISA, for example, have focused the usage of IoT for supporting smart city services. Another technology that may be used in Smart City architectures is Wireless Multi-hop Infrastructures (WMIs). WMIs might provide uniform and quickly deployed wireless routers to allow catastrophe resilient network access for smart cities, as smart cities are heavily reliant on communications technologies. The technology also adds to the Smart City paradigm's long-term viability by ensuring high-resilience components suggested by R Teng et al., 2016.

In Malmö, Sweden's Hyllie smart networks, energy usage is controlled through the use of smartphones and smart metres. The particular initiative was successful in providing new energy solutions since it accepted user participation in energy usage as well as renewable energy options. The customers are able to manage difficulties related to energy generation, electric vehicle charging, and home heating thanks to the integrated infrastructure system. Users are urged to monitor and be conscious of their energy use. Furthermore, the infrastructure informs the user of the most appropriate period for excessive power usage. The project highlights include potential transportation methods proposed by ecmapping, 2018 in addition to smart grids. The implementation of a lighting control system has reduced the amount of energy used and the amount of money spent on lighting. Users can adjust the lighting levels using the system's smartphone software. The system serves as a conduit for anonymous data collection on temperature, humidity, and room occupancy through integrated sensors, assisting management in making decisions on utilisation and adjustments. The technology enables intelligent decision-making while maintaining high levels of energy and operational efficiency suggested by I Vilajosana et al., 2013.

As Zanella et al., 2014 proposed that by the end of the century, fully autonomous vehicles will be available to the general public, but there are still impediments to mainstream use. During this time, however, the promise for connected automobiles is exciting, inter, and high-growth sector in the IoT's growth, allowing for improved road safety, smart traffic management, advanced navigation aid, passenger amusement, and much more.

IoT in Electric Vehicles

The Internet of Things (IoT) emerges as a significant solution that aids organisations in staying ahead of the competition in this Big Data era where actual data and increased connectivity are crucial in many industries. Every aspect of human existence will be impacted by the spread of "smart gadgets" and digitalization initiatives, which will also spur commercial innovation and help firms stay competitive and address global issues. Because the Internet of Things creates massive volumes of data, it is a major driver for big data analytics systems and applications that are designed to process and analyse large amounts of data from various connected or embedded sensors. Electrical Vehicles (EVs) are becoming more common all around the world as worries about carbon dioxide emissions, energy conservation, and unsustainable fossil fuels grow. Smart charging stations are becoming more necessary as the industry expands. The majority of electric vehicle charging stations in use today are dispersed and have complicated structures, making it challenging for on-site staff to operate and maintain them. EV charging stations, for instance, must always be accessible because consumers charge their cars through an app; nevertheless, if a station unexpectedly goes offline, it must be rectified as soon as feasible. Additionally, there is a problem with data transfer because crucial data is typically provided late.

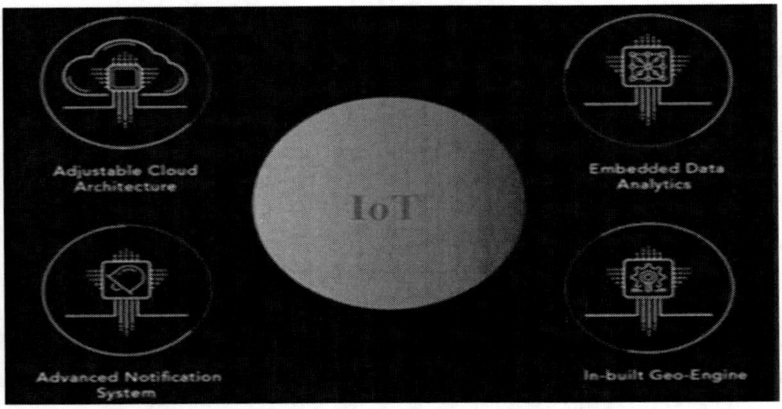

Figure 11.1. IoT enable Architecture for data collection.

As shown in Figure 11.1, the architecture for collecting data compromise four different components. It's also feasible to create a whole IoT framework for complex electric vehicle charging stations, controls, consumption, and administration. IoT connects and maintains various engineering devices via a single network. Users may add private, public, or hybrid cloud storage as needed thanks to the platform's adaptive cloud architecture. Advanced data analytics systems analyse data from a variety of sources before presenting it in a user-friendly style. The IoT platform alerts the appropriate people immediately if there are any problems or changes. Furthermore, a built-in geo-engine makes monitoring and controlling distributed devices a breeze.

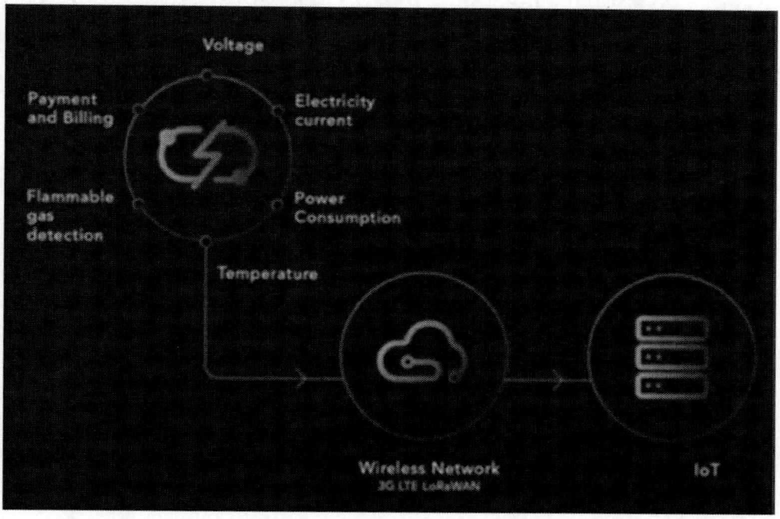

Figure 11.2. IoT for EV charging station monitoring and management.

In addition, the multi-level help desk enables consumers to seek advanced support from a service business if a charging network operator is unable to remedy the issue. To make things simpler for drivers, the integrated data system may analyse their data and find patterns in their behaviour in order to provide them with extra services and better assistance. As in Figure 11.2, EV charging stations are becoming more efficient and convenient as a result of IoT technology, not just for drivers but also for service staff. Using IoT, electric vehicle charging stations may be easily monitored and operated from a single platform. The Internet of Things connects electric vehicles with charging stations, making it easier for them to find one. They're enabling data from drivers and charger manufacturers. Local service firms and station owners can also benefit from the Internet of Things by working more effectively.

Results and Discussion

Figure 11.3 depicts a typical Battery Management System structure with major battery parameter measurements. Current, voltage, and temperature are examples of such phenomena. The process of making essential calculations/estimations in order to extract relevant information about an energy storage system, such as the State of Health (SoH), State of Charge (SoC), and operational temperature range. Controlling steps are done based on these computed characteristics to preserve the battery's longevity and safety against potential risks. As a result, the primary goal of monitoring is to assess various variables, log occurrences, produce alerts, record use profiles, and show this information to the user both locally and remotely.

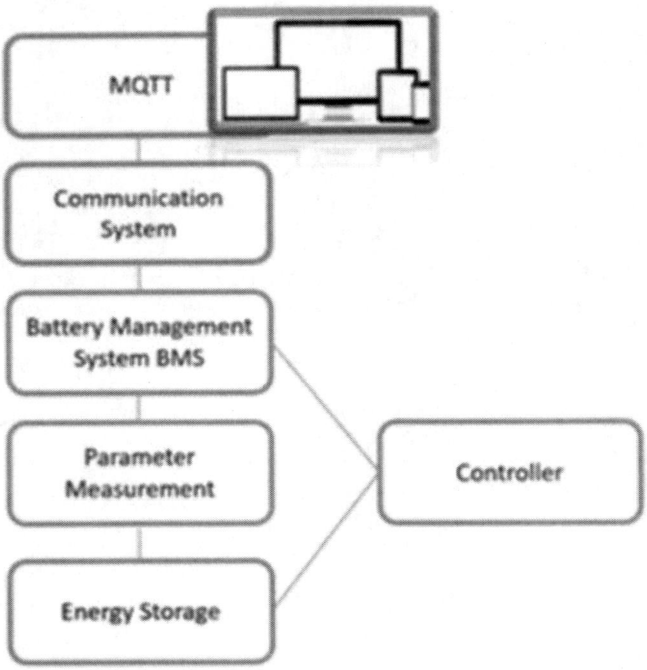

Figure 11.3. Battery System Framework.

The proper usage of a battery necessitates an understanding of its state of charge (SoC). The correct evaluation of the battery's remaining capacity is required for the formulation of an appropriate control plan. In practice, a battery is defined as the ratio of its existing dimensions (Q(n,t)) to its nominal dimensions (Qn). The minimal capacity is specified by the battery manufacturer and indicates the maximum amount of charge that may be held in the battery. THe SoC can be calculated as follows in Equation 1:

$$SoC(n,t) = \frac{Q(n,t)}{Qn} \quad (1)$$

A practical design of battery charging, liquidation, and eventually sleep practises can increase the battery-operated pack's overall longevity. For instance, if the SoC climbs above 10%, the discharge should be allowed, and if it falls below 10%, the discharge should be stopped. The battery charging must stop when the SoC reaches 95%. Cyclic full charge/discharge increases the battery pack's lifespan. The process needs to be kept below the recommended temperature to prevent the battery from exploding.

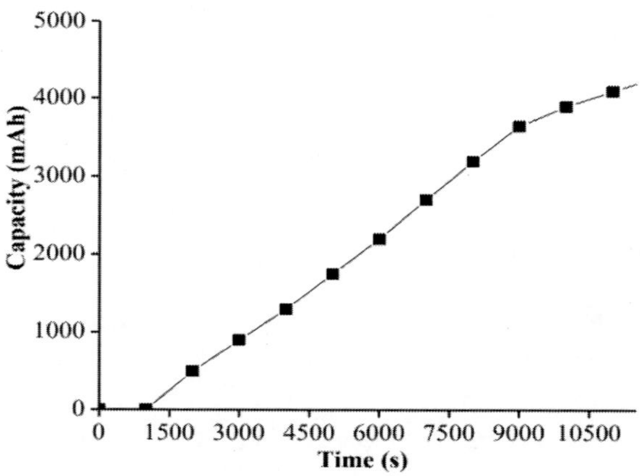

Figure 11.4. The capacity of battery packs (estimated).

During the charging phase of LiPo batteries, the SoC estimate approach and real-time transmission of battery characteristics were evaluated. The effective capacity calculated using the coulomb counting method is shown in Figure 11.4.

Rechargeable batteries are required for any portable equipment. Lead-acid and nickel-metal hydride batteries are still the most widely used battery chemistries today, however lithium-ion batteries are steadily displacing them. This is due to the fact that lithium batteries have a larger energy density, a greater capacity for current discharge, a longer cycle life, and a lower self-discharge rate. Because of their high energy density and low cost, lithium batteries are becoming increasingly popular. Clean energy technologies, such as electric vehicles and energy storage, have fueled technological advancements. Figure 11.5 depicts the performance and cost of lithium batteries from 1990 to 2015.

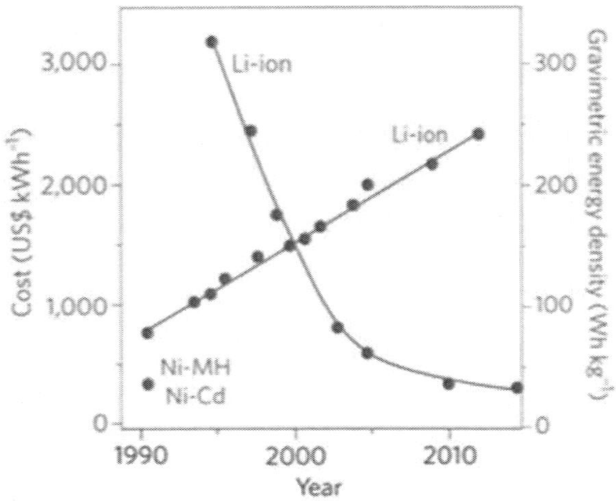

Figure 11.5. Lithium battery energy density and cost over time.

Lithium batteries have a significantly tighter operating tolerance than lead-acid batteries and are more prone to dangerous circumstances like overvoltages (OVs) and under voltages (UVs). As a result, lithium batteries are more dangerous and require constant monitoring by a battery protection circuit. These protective circuits will be referred to as a BMS in this chapter. Because battery voltage cannot be used as the only parameter for predicting capacity, lithium battery capacity is more difficult to estimate than lead-acid battery capacity. Lithium-ion batteries have a far flatter discharge profile than alkaline or lead-acid batteries, which means that the voltage decreases far less for a given amount of capacity lost. When simply voltage is employed as a measure for battery capacity, this leads to incorrect evaluations of battery capacity. Over 80 percent of their SOC, lithium-ion, lithium iron phosphate (LFP) batteries, for example, have almost no voltage change (within low-cost quantifiable limitations).

According to a recent WHO survey, India has 14 of the world's top 20 most polluted cities. The majority of drivers and car owners are personally affected by bad air quality, according to the report. 76 percent said they, their neighbours, acquaintances, or family, suffer from bad air quality on a daily basis or are beginning to exhibit signs of being harmed by pollution. The city of Delhi looked to be the most affected, with 91 percent of respondents claiming to have been affected by poor air quality themselves or someone they know. Similarly, Hyderabad had a score of 78 percent, while Chennai had a score of 75 percent. In addition, Mumbai received a score of 74 percent, which is only slightly lower than Chennai. Bangalore had a lower rate of 71 percent, which was lower than Chennai and Mumbai. Kolkata has a proportion of 71 percent, which is lower than other cities, which have high percentages of 70 percent. The same is explained in Figure 11.6.

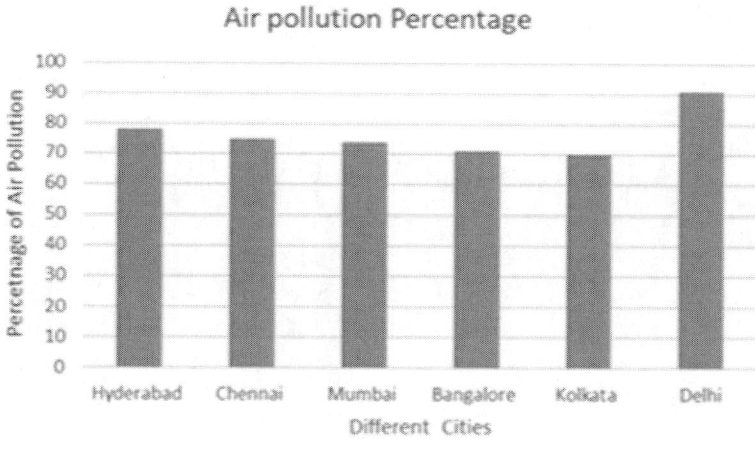

Figure 11.6. Air pollution affected in different cities of India.

Since they produce no exhaust, electric vehicles contribute to the improvement of urban air quality. With a target of 30% EV penetration by 2030, the Indian government has gained momentum through its Faster Use and Manufacturing of (Hybrid &) Electric Cars efforts. These initiatives encourage, and in some cases even mandate, the use of electric vehicles (EV). The programme fosters demand for electric vehicles while also promoting the establishment of charging infrastructure and stations in metropolitan areas. If these objectives are reached by 2030, they may be able to avoid up to 474 million tonnes of oil equivalent (Mtoe) and 846 million tonnes of net CO_2 emissions during the course of their lifespan.

Conclusion

Using electric cars and giving up your natural gas-fired boiler won't cut down on CO_2 emissions, and they might even make the pollution problem worse. According to the International Energy Agency's annual International Energy Outlook, which also illustrates many scenarios of future energy use, higher electrification may cause oil demand to peak by 2030. However, the greater use of power plants to charge them will negate any reduction in emissions from electric vehicles. According to the Paris-based organisation that provides governments with energy policy advice, electrification will need to be a part of a comprehensive package of measures to lower power sector carbon emissions and increase energy efficiency by 2040 in order to considerably cut hazardous pollution. As a result, this research suggests energy as a service business model to investigate how Electric Vehicles might help smart cities attain sustainable energy. This research also looks into the rising role of energy informatics, the importance of smart grids and sustainable energy systems, the benefits of integrating electric vehicles into the energy grid as flexibility, and the implications of electric vehicles for offering incentives to their users.

References

Asaad, M, Ahmad F, Alam MS, Rafat Y. *IoT enabled Electric Vehicle's Battery Monitoring System.* EAI SGIOT (2017).

Bellavista, P, G. Cardone, A. Corradi, and L. Foschini, "Convergence of MANET and WSN in IoT urban scenarios," *IEEE Sens. J.,* vol. 13, no. 10, (2013) pp. 3558–3567.

Bui, A, Zanella, N, Castellani A, Vangelista L, Zorzi M. Internet of Things for Smart Cities. *IEEE Internet Things J.* (2014), 1, 22–32.

Cassandras, CG. "Smart Cities as Cyber-Physical Social Systems. Engineering," 2, (2016), 156-158.

Farmanbar, Mina, Kiyan Parham, System Arild, and Chunming Rong. "A Widespread Review of Smart Grids Towards Smart Cities" *Energies* 12, no. 23: 4484, (2019).

Helmy, M, Wahab A, Imanina N, Anuar M, Ambar R, Baharum A, Shanta S, Sulaiman MS, Fauzi SSM, Hanafi HF. IoT-Based Battery Monitoring System for Electric Vehicle. *Int. J. Eng. Technol.* (2018), 7, 505–510.

Joshi, Sujata, Saksham Saxena, Tanvi Godbole, Shreya, Developing Smart Cities: An Integrated Framework, *Procedia Computer Science,* Volume 93,(2016), Pages 902-909, ISSN 1877-0509.

Malmö Stad, *"Climate-Smart Hyllie — Testing the Sustainable Solutions of the Future,"* Swedish Energy Agency, (2013).

Mills, D, Pudney S, Pevcin P, Dvorak J. *Evidence-Based Public Policy Decision-Making in Smart Cities.* (2022).

Peris Ortiz, Marta, Bennett, Dag R. Yábar, Diana Pérez-Bustamante, *Sustainable Smart Cities: Creating Spaces for Technological, Social and Business Development.* Springer. (2020).

Philips, "Connected Lighting System," press release, 2014. Retrieved From: www.newscenter.philips.com/main/standard/news/pre ss/2014/20140327-philips givesworkers-smartphonecontrolofoffice-lighting-with-ground breaking connected-lighting-system.wpd#.VL46kS5rNow.

Savari, GF, Krishnasamy V, Sathik J, Ali Z.M, Abdel Aleem SHE. Internet of Things based real-time electric vehicle load forecasting and charging station recommendation. *ISA Trans.* (2020), 97, 431–447.

Seuwou, Patrice, E. Banissi, George Ubakanma, *The Future of Mobility with Connected and Autonomous Vehicles in Smart Cities,* January 2020.

Sousa, RA, V. Monteiro, JC. Ferreira, AA. Nogueiras Melendez, JL. Afonso and JA. Afonso, "Development of an IoT System with Smart Charging Current Control for Electric Vehicles," *IECON 2018 - 44th Annual Conference of the IEEE Industrial Electronics Society,* (2018), pp. 4662-4667.

Teng, R, Li H, Zhang B and Miura R. "Differentiation Presentation for Sustaining Internet Access in a Disaster-Resilient Homogeneous Wireless Infrastructure," *IEEE Access,* 4,(2016), pp 514-528.

Top 6 Characteristics to Understand the Concept of Smart City, (2018). Retrieved from https://ecmapping.com/2017/03/31/top6characteristics-to-understand-the-concept-of-smartcity.

Vilajosana, I, J. Llosa, B. Martinez, M. Domingo-Prieto, A. Angles, and X. Vilajosana. "Bootstrapping smart cities through a self-sustainable model based on big data flows," *IEEE Communications Magazine,* vol. 51, no. 6, (2013), pp. 128-134.

Yao, L, Chen, YQ, Lim WH. Internet of Things for Electric Vehicle: An Improved Decentralized Charging Scheme. In *Proceedings of the 2015 IEEE International Conference on Data Science and Data Intensive Systems,* Sydney, Australia, 11–13 December (2015), pp. 651–658.

Zanella, N. Bui, A. Castellani, L. Vangelista, and M. Zorzi, "Internet of Things for smart cities," *IEEE Internet Things J.,* vol. 1, no. 1, (2014), pp. 22–32.

Chapter 12

Certichain: A Blockchain-Based Framework for Document Verification

A. Kamerish[1], R. Manimegalai[2,*] and R. Kingsy Grace[3]

[1]Springfield, Illinois, USA
[2]PSG Institute of Technology and Applied Research, Coimbatore, Tamilnadu, India
[3]Sri Ramakrishna Engineering College, Coimbatore, Tamilnadu, India

Abstract

Rapid advancements in image processing have led to a lot of constructive and optimal solutions for societal applications and they have led to some serious security issues as well. One such problem is the forging of documents and certificates by employing techniques such as image-warping, compositing and morphing. This problem leads to criminal activities such as tampering the original documents and generating bogus documents, especially the documents issued by central and state governments, educational universities, passports, Aadhaar-cards, passbooks, degree certificates, mark-sheets, etc. With the advent of image-processing tools, the adversary and evil users can easily forge the documents and submit them as legal documents. Blockchain technology solves this problem by empowering organizations, institutions and employers to verify documents instantly with minimal effort. With the advancements in blockchain technology, there are multiple solutions to the document verification problem. This work focuses on employing block-chain technology effectively to verify the authenticity of the

* Corresponding Author's Email: drrm@psgitech.ac.in.

In: Blockchain Enabled Secure Big Data Computing …
Editors: L. Ashok Kumar, D. Karthika Renuka, V. Chandrasekar et al.
ISBN: 979-8-89113-243-6

documents and certificates. The data stored on a blockchain network is immutable as the ledger has a copy in all the peers participating in the network. Peers are categorized as committing peers and endorsing peers. The committing peer only can update the ledger while the endorsing peers act as an endorser as they must verify and accept the transaction as mentioned in the endorsing policy and can also commit to the ledger. The transaction made in a blockchain is recorded by all the peers participating in the network and can be verified by all the peers. A smart contract is an important feature of the Hyperledger platform which provides the ability to implement business logic in the network. A smart contract which is also called chaincode in Hyperledger fabric is the core of this blockchain-based solution to reduce the time required in verifying the documents. The proposed blockchain-based framework, for document verification, is implemented using hashing and the details mentioned in the document or certificate by the original document-provider. An asset is created in the blockchain network so that the document or certificate is stored in the ledger. The QR code of the hash is attached to the certificate to verify the same without much time and effort. The given documents are verified easily by scanning the QR code in the corresponding client application. The proposed work focuses on providing an efficient solution for the document verification problem using block-chain technology with enhanced authentication, authorization and privacy.

Keywords: hyperledger fabric, blockchain, distributed ledger, document verification, enhanced authentication

Introduction

The process of validating the legitimacy of a document is known as document verification. Nowadays, documents play a crucial part in everyone's life and hold monetary value as well. Documents serve as evidence of a person's identity and properties. Digital versions of documents may be modelled and replicated with appropriate alterations thanks to advances in artificial intelligence and image processing. For example, a person can change his final grade point by forging the certificates. This is a major issue that affects everyone working in recruiting across the world and also a serious problem when medical certificates are forged. Manual approach for verifying the certificates by the certificate providers is time consuming and errors may occur most of the time. E-certificate providers have a portal to verify the certificate, but service cost is high. There are many problems faced by the certificate providers and third-party people who verify the certificates.

Searching through the manual-ledger is time consuming. Certificate issuers do not always have a way to check their certifications. For example, property documents are at high risk of being forged as there is no proper method of verification and tracing history is difficult. It is not economical to provide a cloud-based solution for the e-certificate providers, to make the certificate verifiable because the service for every certificate generated needs to be verifiable for at least 15 years. Providing all the details to the third-party verifiers is not necessary.

Blockchain technology addresses the problems of user privacy, cost, tampering, and time-required for certificate verification. Blockchain is essentially a digital ledger which is copied and distributed to all the participants in the network. The blockchain system of recording information makes it difficult to change, hack the system as it is not a single server instead it is a group of servers which needs to be hacked simultaneously to alter the ledger. There are three types of blockchain such as public, private i.e., permissioned and consortium. Blockchain works on the underlying principles of P2P protocol and cryptography. Public Blockchains are the blockchains where it is open to all where anyone can make a transaction, whereas private blockchain is restricted to a group of persons and there is an organization to monitor the blockchain to play the administrative role. Private blockchain is mainly used for privacy reasons. A consortium blockchain is semi-private as it is used and controlled by a group of organizations rather than a single organization in a private blockchain. The decentralized database is called distributed ledger technology. All the transactions taking place in the blockchain is recorded and cryptographically stored in the ledger. The main features of Distributed Ledger Technology (DLT) are: it is programmable, time-stamped, immutable, secured, and digitally verifiable. As there is no single server to regulate the data it is not possible to tamper with it. Most Blockchain systems are open source. It allows participants to modify the code and the technology in ways that best suits their problem and does not give them the ability to modify a running blockchain as it needs to be accepted publicly. Every blockchain has an underlying consensus governing all the transactions taking place. It validates every transaction before it reaches the ledger. There are many consensuses available and the most popular one is Practical Byzantine Fault Tolerance (PBFT). Each block in the blockchain contains five elements, namely, *i*) the main data; *ii*) the hash of the previous block; *iii*) the hash of the current block; *iv*) the timestamp and *v*) other relevant information. Blockchain can solve the problem of Document verification because it takes very less time to verify the certificate. The certificate issuers only need to create an asset and make a transaction in the blockchain so that it is recorded and can be verified later. In this case, there is no need for a cloud solution to verify the certificates. Blockchain is programmable so the users have complete control of the information and choose who to share the certificate with.

Related Work

(Raghav et al., 2019) have used Hyperledger fabric blockchain technology to solve the problem along with InterPlanetary File System (IPFS). The hash value of the certificate is computed and encrypted using asymmetric encryption which is stored in IPFS. Finally, a transaction that contains metadata of the certificate is done. The issue is that IPFS is still in its infancy and has to be further improved before being included as a widely used application. (Liu & Guo, 2019) have used Hyperedge fabric as the underlying blockchain technology. The entire methodology is divided into four layers namely application layer, business layer, smart contract and Hyperledger fabric layer. The more advanced feature Kafka clusters is used to improve the throughput of the entire system. Kafka is a distributed, partitioned, and replicable message system to improve the transaction processing speed. The proposed method did not have improvements in privacy, chain data load, database schema, and throughput. The Transactions Per Second (TPS) is 180-250, whereas the TPS of the latest version of Hyperledger fabric is up to 2500 (Liu & Guo, 2019).

(Badr et al., 2019) have proposed a solution with the Hyperledger Fabric blockchain technology. The Hyperledger composer is used to model the current business architecture and create a Business Network Archive (BNA) file for integrating the existing system into a blockchain. The proposed model converts the existing use case into a BNA file for integrating it with the blockchain. Arenas and Fernandez have proposed a methodology for the forging certificate problem using Credence ledger (Arenas & Fernandez, 2018). Credence Ledger is the name of the architecture developed using multi-chain which is an extended fork of bitcoin. The proposed work focuses on user privacy as it limits the amount of information the third-party certificate verifiers can view. Even though it can be made to work as the permissioned blockchain. It still needs to deal with cryptocurrency and mining and also it has very low transactions per second when compared to Hyperledger fabric. (Marella et al., 2020) have proposed a solution with the Hyperledger fabric blockchain platform. The proposed work is an early prototype showing an application of blockchain and it does not deal with user privacy, chain data load. (Hamitha et al., 2019) have proposed a methodology for the document verification problem based on the Ethereum blockchain platform. The review presents the usage of smart contracts in the Ethereum blockchain platform for verification of digital certificates.

(Neethu Gopal & Vani V Prakash, 2018) have used the Ethereum blockchain platform for verification of digital documents. The smart contracts represent the business logic of the system and sit on the Ethereum public blockchain. All the verification and enrollment take place in the cloud as every new verifier does not need to set up an Ethereum virtual machine which consumes a lot of data and storage. Privacy and scalability are real issues. From the literature, it is clear that blockchain based document verification either lacks privacy or fails to achieve reliable throughput which can be scaled in the future (Nakamoto, 2008). To overcome these issues, a solution is developed with the optimal database schema, which strictly maintains user privacy and is also scalable by using the Hyperledger fabric.

Certichain: Proposed Methodology for Document Verification

The literature survey indicates that there is scope for improvement in user privacy, connecting multiple channels, throughput and performance of the document verification. The proposed methodology, Certichain, mainly deals with the idea of filling these gaps for a more practical blockchain based document verification framework. Hyperledger Fabric is a blockchain framework in which Certichain is built upon. Hyperledger fabric has introduced a new architecture for transaction. It is called execute-order-validate. Also, it is different from the order-execute model in terms of resiliency, flexibility, scalability, performance and confidentiality. Hyperledger fabric executes a transaction first to check its correctness and if it is legal transaction, it endorses the same. It then orders the transactions using a consensus protocol i.e., Pluggable and then validates the transaction based on an application-specific endorsement policy written on the blockchain before sending it to the committing peers to be written on the ledger.

Hyperledger Fabric is an open-source distributed ledger technology designed for use in an enterprise context. It is an open-source initiative from the Linux foundation. It has some key differentiating capabilities over other blockchain platforms, such as (i) developed under open governance and by maintainers from multiple organizations (ii) fabric is highly configurable and modular architecture (iii) it is the first blockchain platform to support smart contract written in general purpose programming languages such as go, java and node.js (iv) the fabric platform is permissioned unlike the public permissionless network the participants are known to each other. The fabric has a pluggable consensus protocol that enables the platform to be more effectively customized to fit particular use cases and trust models. Fabric does not require cryptocurrency to fuel the smart contracts and thus does not require costly mining.

At the core of the blockchain is a distributed ledger that records all the transactions taking place in the network. This technology is often called decentralized because it is replicated across all the nodes which are part of the network. The append-only is the main feature of distributed ledger in blockchain. The feature of immutability makes it a standout feature in business use cases and makes it an enterprise solution. The distributed ledger is a sequenced record of all state transitions in the fabric. State transitions take place as a result of chain-code invocation as transactions are submitted by participating parties. There is only one ledger per channel and every peer in the channel has an exact copy of the ledger. Main features of Fabric Ledger are: the ledger can be queried using key based and ranged queries and composite key queries. Transactions contain signatures of endorsing peers and are submitted to the ordering service. The transaction is ordered into blocks and delivered from an ordering service to peers on a channel. Smart contracts contain the business logic which contains the lifecycle of a business object. A blockchain network uses smart contracts to restrict access to the ledger. It has functions for querying, transacting, appending data to the ledger.

Smart contracts allow participants to automate certain functions based on the state of the ledger. In Hyperledger fabric, the terms chaincode and smart contract are used interchangeably. A chaincode is used by administrators to group related smart contracts for deployment in a channel. Smart contracts need to be deployed on a network with permission from the peers according to the endorsement policy. Smart contracts are one of the core features of blockchain technology because they can represent an infinite array of business use cases relating to the immutability of data in multi-organizational decision making. A consensus is a fault-tolerance mechanism used in blockchain technologies to achieve the necessary agreement on a single data value or a single state of network among distributed nodes. It keeps the ledger transactions synchronized with the network, it ensures the ledgers update the transactions only when approved by the participants of the network and it also ensures they all update in the same order in all the nodes. Consensus plays its role in the entire transaction flow from proposal and endorsement to ordering, validation and commitment. A multitude of endorsement, validity and version check take place along with identity verification happening in all directions on the transaction flow. Hyperledger fabric is a permissioned blockchain network and it needs a method to provide identities to all the devices participating in the network to enable the transactions on the network. Certificate authorities work based on Public Key Infrastructure (PKI) and it provides all the entities with a pair of public and private keys. A peer uses its private key to endorse and sign a transaction since the private key cannot be shared anywhere. An MSP holds the peer's public key, and it is used to verify the transaction as only the matching public key can decrypt the message. Thus, the MSP is the mechanism that allows the identities to be trusted and recognized by the rest of the network without ever revealing the member's private key. MSPs occur at two domains in a blockchain network, locally on actor's node and in the channel configuration. The key difference between them is their scope; it lists roles and permissions at a particular level of administration.

Hyperledger fabric has a peer called orderer or ordering node. This orderer node does the ordering transaction, which along with other orderers, forms the ordering service. Orderer maintains a list of channels that are allowed to create channels called the consortium. They enforce basic access control for channels restricting read and write operations on them. From the first phase of the transaction, the client application receives an endorsed transaction proposal response from a set of peers. This transaction response is now sent to the ordering service. The ordering service creates blocks of transactions which will ultimately be distributed to all peers on the channel for final validation and commit. It received a transaction proposal from many different clients concurrently and it orders them on a first come first serve basis and creates a block. These blocks will become the blocks of the blockchain. The ordering service ensures that there are no ledger forks in the peers resulting due to inefficient ordering service. There is a different implementation of the ordering service in Hyperledger fabric as it is a pluggable consensus. They are Raft, Kafka, and Solo.

The basic transaction flow of an asset transfer is explained below with the following assumptions, the network is entirely setup and running and chaincode for asset transfer is instantiated in the network with the endorsing policy requiring both the parties to endorse every transaction. Now, Client A is sending a request to buy an asset from them to all the peers. In this example, the peers are client A and client B, and both must endorse the transaction as stated in the endorsement policy. Now using an appropriate fabric SDK, a transaction proposal is created with a request to invoke chaincode function with specific input parameters with the intent of updating the ledger. The SDK acts as a shim to package the transaction Proposal to the protocol buffer over gRPC format and uses the user's cryptographic material to produce a unique signature for this transaction proposal.

As the endorsement policy states both the peers must endorse the transaction proposal it is sent to both the peers. Now both the peers check the integrity of the transaction proposal and pass the inputs to the invoked chaincode's function. The chaincode is executed against the current state database and the transaction results along with the response, read/write set are signed by the peer's cryptographic material and passed back as proposal response by the SDK. No changes are made to the ledger at this point. Now the peer which initiated the transaction receives all the proposal responses and checks the integrity and determines whether the proposal responses are the same. If the responses match the endorsement policy of the channel, then, the application broadcasts the transaction message to the ordering service. The ordering service collects all such transaction messages and orders them in a block. Now, these blocks are sent to all the peers. These peers append the block to the channels ledger and an event is emitted by each peer to the client application stating the success message of the transaction. The transaction flow of Hyperledger Fabric is shown in Figure 1.

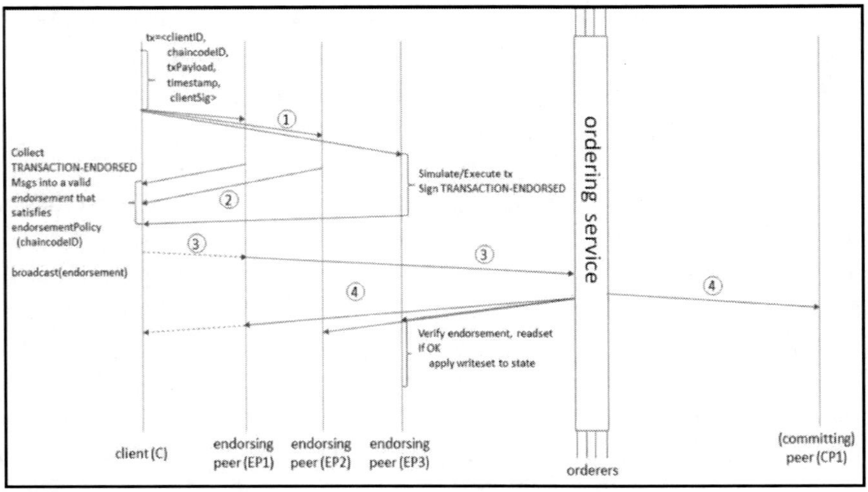

Figure 1. Transaction Flow of Hyperledger Fabric.

In the proposed work, the participants are the certificate issuer, third party verifiers and documents owners. Here in this model, the digital document is the asset in the blockchain network. The entire prototype contains a front-end web portal backed by a cloud, offering blockchain as a service for handling all the work of the three types of entities participating in the network. The backend in Certichain is the blockchain network itself. The blockchain network is designed in such a way that every organization issuing certificates will form a channel, for example in our case Aadhar documents will form a channel, PAN documents will form a channel etc. Client web portal for verification of documents is depicted in Figure 2 and network design is shown in Figure 3.

The web portal is used to add a new document, modify/update an existing document, and delete a document. Every action done on a document will notify the corresponding asset owner as it is available on a shared basis based on the endorsement policies. The Document owners in this case will be added to all the channels the documents are issued. The third-party verifiers will be part of every channel automatically but the ability to verify the documents will be restricted based on the permission network admin. All the participants of the network can access the service with any device having a connection to the internet as it is backed by the cloud, so, they need to pay for the service provided or else they can setup their peer node for all the transaction purpose. All the permission is written in the endorsement policy and chaincode completely so that the data is confidential and can be accessed by permissioned people.

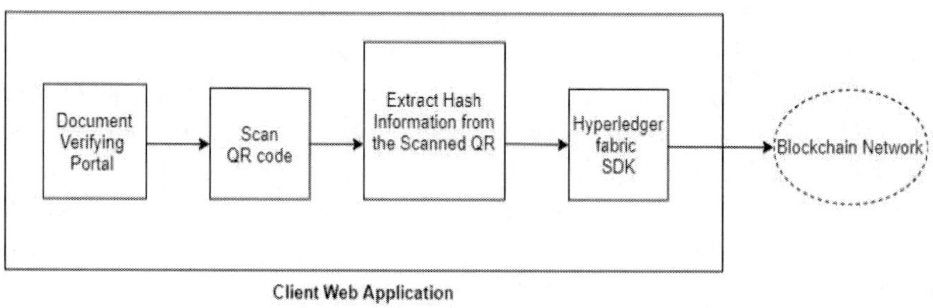

Figure 2. Client Web Portal for Verification of Documents.

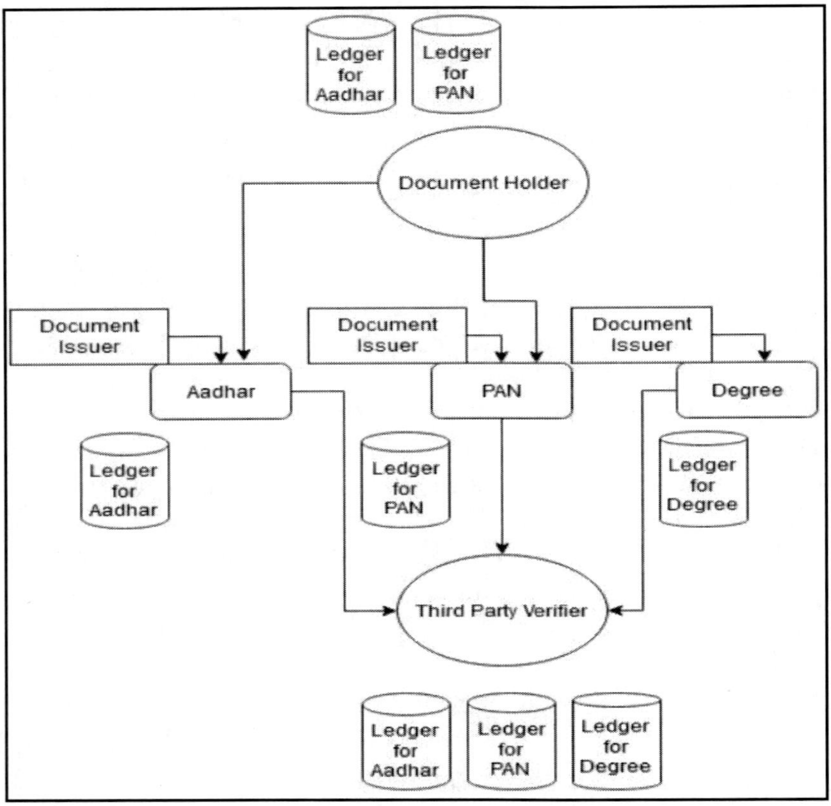

Figure 3. Network Design.

Every verification done triggers an event that will notify the document owner. The document owner can restrict the third-party verifiers from verifying the documents using the smart contracts. In Certichain, the document issuer gets first enrolled from the blockchain network administrator and starts adding all the documents in the blockchain network in the respective channels. These documents automatically get shared between the people to whom the document is issued to. Once this process is completed the document owner can enable third-party verification for their documents.

The certificates are designed in such a way that required details are clubbed in a specific order and then hashed to get the key. The certificate is stored in the blockchain with the hash key to map it in the database. Blockchain uses NoSQL as database in the backend as it needs to store more tables and club them together for scalability. The certificate indicates the values that are part of the hash so that it can be regenerated using the portal provided by the blockchain network. The web portal has the fabric SDK at its core, and it sends all the API calls to the cloud where the peers are setup and running.

The Chaincode is the group of smart contracts for the document verification system Certichain. It has different smart contracts for document owners, document issuers, and third-party verifiers based on their roles. Certichain is different from the previous solutions because it is an end-to-end prototype. It covers the scalability issue by using cloud peer deployment so that it can be easily managed from anywhere in the world. The data stored in the ledger is cryptic and the third-party verifiers can view it only with the permission from the document owners thus enabling privacy. Using a cloud solution is always an optional one as peers can also be set up locally which is an inexpensive solution.

Conclusion

The blockchain based document verification solution, *Certichain*, is powered by Hyperledger fabric and deals with user privacy upfront. It enables instantaneous access to document verification by providing blockchain as a service in the Cloud. The smart contracts are designed based on roles and packed into a chain-code and deployed in the network to ensure confidentiality. The data stored in the blockchain is also encrypted hence it is secure. Interplanetary File System (IPFS) can be incorporated if it becomes widely used and accepted technology in the security perspective which in turn reduces the data load on the chain for faster access.

Conflits of Interest

None.

References

Arenas R. and P. Fernandez, "CredenceLedger: A Permissioned Blockchain for Verifiable Academic Credentials," *IEEE International Conference on Engineering, Technology and Innovation Stuttgart*, 2018, pp. 1-6, doi: https://doi.org/10.1109/ICE.2018.8436324.

Badr A., L. Rafferty, Q. H. Mahmoud, K. Elgazzar and P. C. K. Hung, "A Permissioned Blockchain-based System for Verification of Academic Records," *10th IFIP International Conference on New Technologies, Mobility and Security*, Spain, 2019, pp. 1-5, doi: https://doi.org/10.1109/NTMS.2019.8763831.

Gopal N., Vani V Prakash "Survey on Blockchain Based Digital Certificate System," *International Research Journal of Engineering and Technology,* Volume: 05 Issue: 11, 2018.

Hamitha Nasrin M., S. Hemalakshmi, G. Ramsundar, "A Review on Implementation Techniques of Blockchain enabled Smart Contract for Document Verification," *International Research Journal of Engineering and Technology*, Volume: 6 Issue: 2, pp.412-416, 2019.

Liu D. and X. Guo, "Blockchain Based Storage and Verification Scheme of Credible Degree Certificate," *2nd International Conference on Safety Produce Informatization,* China, 2019, pp. 350-352, doi: https://doi.org/10.1109/IICSPI 48186.2019.9095961.

Marella V. and Vijayan, Anoop, Document Verification using Blockchain for Trusted CV Information, Americas Conference on Information Systems, *2020 Proceedings.* 12.

Nakamoto S., "Bitcoin: A Peer-to-Peer Electronic Cash System," *www.Bitcoin.org*, p. 9, 2008.

Raghav N., R. Andola, S. Verma, S. Venkatesan and S. Verma, "Tamper-Proof Certificate Management System," *IEEE Conference on Information and Communication Technology,* India, 2019, pp. 1-6, doi: https://doi.org/10.1109/ CICT48419.2019.9066236.

Chapter 13

Blockchain and TinyML for Low Powered IoT Devices in Smart City Applications: Current Trends and Future Prospects

Kavi Priya Santhosham*, **PhD**
Bala Ganesh Nallathambi[1], **PhD**
and Pon Karthika Kasirajan[2]
Department of Computer Science and Engineering, Mepco Schlenk Engineering College, Sivakasi, Tamil Nadu, India

Abstract

By 2024, Smart cities sectors will reach nearly 54 million low powered IoT (Internet of Things) connections, as stated in statistical reporting. The primary motive of encapsulating Tiny Machine Learning (TinyML) into IoT framework is to achieve effective bandwidth utilization, ensure data security and privacy, low latency and minimal cost. Moreover, the integration of blockchain and tinyML is expected to enhance the capabilities of smart city applications by enabling secure and efficient data processing and analysis on low-powered devices. This will help to optimize resource usage and reduce energy consumption, leading to more sustainable and environmentally friendly smart cities. This study introduces the definitions and background information of Low powered IoT devices, TinyML and Blockchain technology. It also exhibits how TinyML and Blockchain technology can be implemented in smart city applications containing low powered IoT components. Moreover, it

* Corresponding Author's Email: urskavi@mepcoeng.ac.in.
* Corresponding Author's Email: urskavi@mepcoeng.ac.in.
* Corresponding Author's Email: urskavi@mepcoeng.ac.in.

In: Blockchain Enabled Secure Big Data Computing …
Editors: L. Ashok Kumar, D. Karthika Renuka, V. Chandrasekar et al.
ISBN: 979-8-89113-243-6

discusses the current trends, challenges, and prospects of sustainable smart city development. This study will impart valuable knowledge to the research community for future research direction in the area of TinyML and Blockchain in various IoT enabled smart city applications.

Keywords: internet of things, blockchain, tinyml, smart city

Introduction

Cities around the world are experiencing rapid urbanization and are becoming hubs for businesses and top talent, leading to economic growth. However, this growth also leads to increased congestion and environmental challenges. As populations in cities continue to grow, there is a higher demand for resources, urban infrastructure, and public services, which puts pressure on the environment. The need for Smart Cities, enabled by connected technologies, is becoming increasingly important. The COVID-19 pandemic has demonstrated the need for more efficient and reliable urban management.

The concept of a smart city (Arroub et al., 2016) involves the integration of various systems and services, such as transportation, energy management, waste management, public safety, healthcare, and communication, to create an efficient and sustainable urban environment. The idea of Smart Cities embodies an ideal urban environment that combines information and communication technologies (Camero and Alba, 2019), facilitating the generation and exchange of real-time data among citizens, governments, and organizations. A smart city ecosystem refers to the complex network of interconnected components, stakeholders, and technologies that work together to create a smart city. The ecosystem involves a variety of actors including government agencies, private sector companies, citizens, and various service providers who collaborate to deliver services and optimize urban living.

Government agencies play a crucial role in the smart city ecosystem, as they are responsible for creating policies and regulations that enable smart city initiatives to thrive. They are also responsible for providing public services such as transportation, healthcare, and waste management. In a smart city, government agencies work closely with other stakeholders to identify opportunities for improvement and implement innovative solutions. Private sector companies are key players in the smart city ecosystem, as they provide technology solutions, infrastructure, and services that support smart city initiatives. Companies such as IBM, Siemens, and Cisco provide technologies and services that enable smart city solutions such as IoT, AI, and cloud computing. They also invest in research and development to drive innovation in the smart city space. Citizens are an important component of the smart city ecosystem, as they are the end-users of smart city solutions. Citizen engagement is critical for the success of smart city initiatives, as it helps to identify needs and preferences, and ensures that solutions are designed with the needs of the community in mind. Service providers play a key role in delivering services in a smart city ecosystem. They include transportation providers, energy providers, waste management providers, and other service providers that collaborate with government agencies and private sector companies to deliver efficient and sustainable services. In a smart city, data is a critical resource that enables real-time monitoring and analysis of urban systems. Data providers include sensors, cameras, and other IoT devices that collect and transmit data, as well as data analytics companies that process and analyze this data to provide insights and inform decision-making. The physical infrastructure of a smart city includes buildings, roads, transportation systems, energy grids, and other infrastructure that supports urban living. In a smart city ecosystem, infrastructure is designed to be interconnected and intelligent, enabling data sharing and optimization of urban services.

Overall, the smart city ecosystem involves a complex web of actors, technologies, and systems that work together to create a more sustainable, efficient, and livable urban environment. Collaboration and innovation are critical for the success of smart city initiatives, as they require coordination across multiple stakeholders and the integration of advanced technologies and systems. The vision of smart cities in India is to create modern urban areas that are sustainable and provide an enhanced quality of life for citizens. The mission is to deliver basic urban infrastructure services such as water, sanitation, and transportation while simultaneously leveraging advanced technologies such as IoT, cloud computing, and data analytics to optimize resource usage and promote economic growth. The objective is to make cities more livable, environmentally sustainable, and inclusive by developing citizen-centric solutions that enhance safety, security, and accessibility. This mission aims to establish a template for urban development in India that can be replicated across the country and globally. Smart cities are built on a foundation of advanced technologies that are used to optimize various aspects of urban management and improve the quality of life for citizens.

These technologies include:

- Internet of Things (IoT) - This technology is used to connect various devices, sensors, and systems within a city (Treiblmaier et al., 2020), enabling real-time data collection and analysis for better decision-making.
- Big Data Analytics - By analyzing large and complex data sets, smart cities can gain insights that can be used for better urban planning, traffic management, public safety, and other areas.
- Artificial Intelligence (AI) is used to enable predictive analytics, automate routine tasks, and optimize resource usage. For instance, waste management, parking, and public transportation can be optimized using AI.
- Blockchain - Blockchain is a technology that is utilized to create secure and transparent transactions and establish a decentralized infrastructure for smart city services (Hammi et al., 2017; Alasbali et al., 2022).
- TinyML – TinyML (Schizas et al., 2022) is an emerging technology that involves deploying machine learning (ML) algorithms on small, low-power devices such as sensors and microcontrollers. The goal of TinyML is to enable intelligent data processing and decision-making at the edge of the network, without the need for constant connectivity to the cloud. This is especially relevant in the context of smart cities, where numerous IoT devices are deployed throughout the urban environment to gather data on various aspects of city life such as traffic, air quality, and energy consumption. By leveraging TinyML, these devices can perform local analysis of the collected data and provide insights in real-time, without relying on a centralized data processing infrastructure. This can lead to increased efficiency, better resource allocation, and improved quality of life for citizens. TinyML can also help overcome the challenges of limited bandwidth and high latency in communication networks, while reducing the overall energy consumption and cost of smart city deployments.
- Renewable Energy is promoted in smart cities to facilitate sustainable urban development and reduce their carbon footprint by employing renewable energy sources such as solar, wind, and geothermal power.
- 5G Networks are utilized to provide high-speed and low-latency connectivity, facilitating new services such as remote healthcare and autonomous vehicles, while also enabling the adoption of other advanced technologies such as IoT.

Smart cities continue to adopt and explore new technologies that can enhance their efficiency and sustainability, while improving the quality of life for citizens. The integration of all these pioneering technologies brings up new research challenges and opportunities to build an optimized and efficient Smart City.

The remainder chapter content is composed as follows. Section 2 provides a brief literature review. Section 3 introduces a five-tier architecture of smart city applications combining IoT, blockchain and big data technology. The possible research challenges and future opportunities are discussed in Section 4. Section 5 presents different case studies of smart city applications in real time. Finally, section 6 concludes the chapter with a neat summarization.

Literature Review

Henkel et al., (2017) summarized the IoT energy consumption methodologies and approaches to design energy efficient IoT devices. Kadam et al., (2020) presented a survey on integration of blockchain with low powered IoT devices for IoT applications. Hakak et al., (2020) enumerated the necessary requirements of integrating blockchain with smart cities. The authors proposed a theoretical architecture interpreting the security of smart city using blockchain technology with a case study. Deepa N et al., (2021) provides a detailed survey on blockchain for big data in this work. They briefed overview of blockchain for big data along with the motivation for integration. The survey about various services in blockchain for big data such as blockchain for secure big data acquisition, data storage, data analytics and data privacy preservation were discussed. At last, the latest studies which use the blockchain in domains such as smart city, smart healthcare, smart transportation, and smart grid were reviewed. Pajooh et al., (2021) proposed the design and implementation of layer-based distributed data storage in blockchain-enabled large-scale IoT system. This system has been developed to mitigate the trust issues caused by the dependencies in centralized server solution by using Hyperledger Fabric (HLF) platform for distributed ledger. The transaction verification and audits records are performed by using HLF peers in the big data system with the help of blockchain technology. The porotype was implemented for deploying the proposed solution in IoT edge computing and big data ecosystem. The proposed work has been evaluated based on the metrics such as throughput, latency, communication, and computation costs. Bauer et al., (2021) delivered a review on overview of IoT enabled smart cities, detailed, and highlighted the impacts of embedding IoT with smart city services. Bellini et al., (2022) reviewed the research literature on IoT-enabled smart cities. This work focused on main open challenges and trends in adopting IoT technology for development and efficient smart cities. In this paper, the technology proposed for IoT framework implementation was reviewed and the main smart city methods and frameworks has been reviewed based on classification into eight domains such as governance, living and infrastructures, mobility and transportation, economy, industry and production, energy, environment, and healthcare. Schizas et al., (2022) provided a detailed overview of TinyML, covering various aspects of research related to this rapidly growing field. Hardware innovation is crucial for the development and stability of TinyML, as it requires ongoing advancement to keep up with its progress.

The significant contributions of the present book chapter are:

- Introduces a five-tier architecture of Smart City applications using Low-powered IoT and Blockchain technologies.
- Provides the basic introduction of the rudimentary technologies in developing Smart City frameworks.
- Presents the insights of current trends, practical and technical difficulties, and the future direction of research to address the issues in implementing smart city applications.
- Emphasizes the case studies of worldwide profitable smart city solutions.

Smart City Architecture and Solutions

Smart City application framework comprises of five layers: Physical Sensing layer, Network layer, Middleware Processing layer, Application Layer and Business layer. Figure 13.1 depicts the five-tier architecture introduced for the implementation and deployment of Smart city applications combining multiple rudimentary technologies. A detailed description of the functionalities and technologies involved in each layer of the architecture is presented in the following subsections.

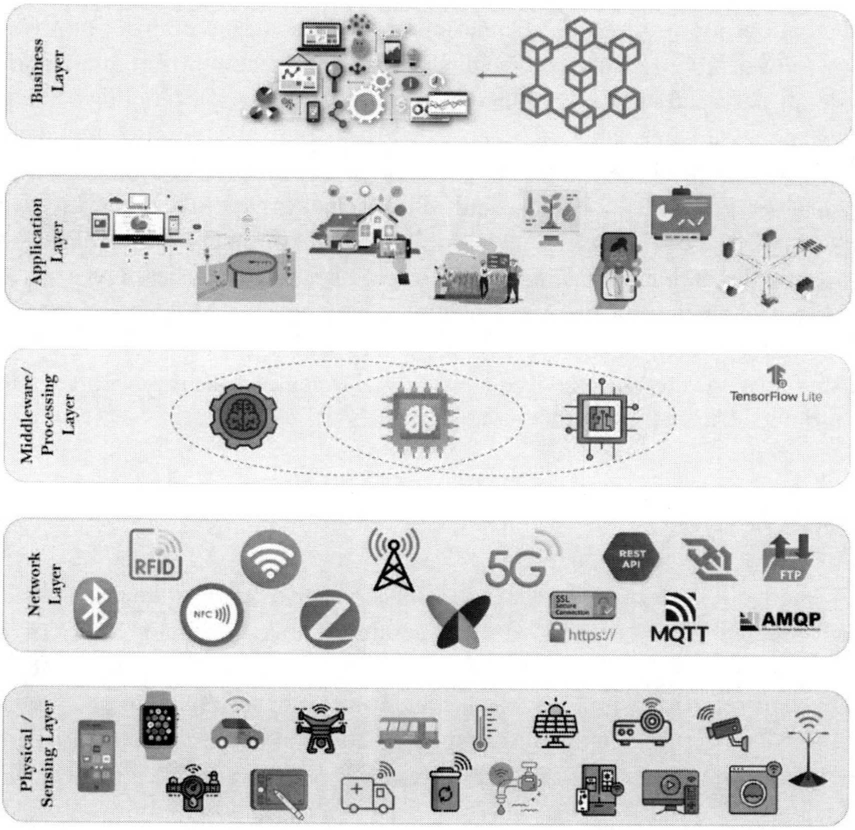

Figure 13.1. Five-tier Architecture of Smart City Application Framework.

Physical/Sensing Layer

The amalgamation of Internet of Things (IoT) and Internet of Everything (IoE) technologies forms the crux of Smart City solutions. By the year 2025, the worldwide count of connected IoT devices will raise about 75 billion and the commercial impact of IoT will reach about $11.1 trillion. The naïve beauty of IoT technology is that allows the devices to share data irrespective of the location of devices and can be governed and controlled in multiple ways. Nowadays, the IoT devices are expected to be very small, spontaneously responsive, and compatible considering various design features including security, connectivity, battery life, personalization, and cost. The critical design goal of IoT devices is to reduce the power consumption for enhancing the battery life to meet the current market demands. The ultimate motive of low-power designs is to optimize the lifespan of the battery by minimizing the overall static and dynamic power consumption of a device leveraging the latest design techniques and approaches. From the smart city application perspective, heterogeneous low-powered edge devices including sensors, actuators, gadgets, wearables, and various interactive modules which can exchange data over wireless networks, collectively forms the physical/sensing layer of the architecture. The two different aspects of physical layer are: (i) sensing and gathering environmental information, and (ii) taking actions with respect to the control signals. The challenges that need to be addressed while using low-powered IoT devices are safety, privacy and security, eco-friendly, on-device data storage, processing, and analysis.

Network Layer

The network layer facilitates data transmission and reception between physical and processing layer using wireless communication technologies. Network brokers/gateways serves as the bridge for communication between IoT nodes. The diverse network protocols and technologies are clustered into two groups: Limited coverage communication protocols and Broad coverage area networks. Bluetooth, RFID (Radio-Frequency Identification) tags and NFC (Near-field communication) have coverage over limited proximity. Wireless broadband technologies like Wireless fidelity (WiFi), Long Range Wide-Area Network (LoRaWAN), ZigBee, 5G and Sigfox have fairly wide coverage area.

Bluetooth is a short range, low energy and larger bandwidth wireless communication protocol specially designed for portable device applications. It can be configured easily with minimal setup and has the range of communication roughly about 10 meters (30 feet). BLE (Bluetooth Low Energy) is introduced to provide Bluetooth connection at reduced cost and power makes it suitable for low-powered IoT applications.

RFID is also a wireless communication protocol based on radio waves intended to detect, communicate, and track human and real time objects like vehicles, fruits, etc. The RFID technology comprises of readers and tags. The readers radiate radio waves to get identity information from the tags, which can be either passive or active.

NFC is two-way wireless transmission protocol with high proximity up to 4cm. NFC can be used in three distinct modes: P2P (Peer to Peer) mode where two or more NFC peers can share data and is applied in Android Beam,

Reader/Writer mode where the NFC device can able to read or write passive NFC labels and tags and Card emulation mode where the NFC device enacts as the NFC card to be read by other NFC terminal or reader.

WiFi is an array of IEEE 802.11 standard wireless network protocols, predominantly used for wireless local area networks (WLAN). It has a high frequency range from 2.4 GHz to 5 GHz establishing internet connectivity at high-speed. WiFi supports IoT applications with broad and limited bands with different range of signals and power consumption. The upcoming 5G standard model will give more priority to IoT domain with low latency (below 4 milliseconds) and high availability to handle Ultra-Reliable Low Latency Communications (URLLC) components.

LoRaWAN is a Low-Power, Low-cost and Wide Area wireless communication protocol, targeted to connect low battery powered devices to the local or global internet. It is implemented using a star topology for direct communication with the gateway node. It satisfies the IoT networking constraints such as mobility, security, localization and two-way communication.

Zigbee is also a low-power communication protocol having mesh topology supporting more flexibility, scalability, and reliability along with 'self-recovering' capability. Zigbee is more suitable for automation and control applications requiring high speed and reliability adopting battery operated components.

Sigfox is a wireless cellular network which supports IoT network with wide range, less power, minimum throughput, and data rate connectivity between M2M and other IoT devices. Sigfox technology is more suited to remote monitoring and control applications.

5G is the fifth-generation broadband cellular network technology with 10 times more speed that current LTE technologies. 5G enables an opportunity to realize the full potential of IoT. The significant improvements enforced in 5G network include ten times reduced latency, more connection density and improved throughput, three times more spectrum efficiency, hundred times more capacity to handle traffic and network efficiency. Almost all smart city applications will get profited from the 5G network enhancements.

The network layer is also in charge of formatting the data for presentation. There are two different schemes available for data transmission i.e., client-server architecture and distributed publish-subscribe messaging architecture. In client-server architecture, the subscription policy is followed by the IoT devices and agents and data sending/receiving is done using push and pull request respectively. HTTP/HTTPS, FTP, Web services and REST call are examples of pull protocols. Advanced Message Queuing Protocol (AMQP), Message Queue Telemetry Transport (MQTT) are examples of push protocols. Publish-subscribe messaging model is used to share data among multiple applications and devices. Some of the software platforms available for this distributed scheme are Apache Kafka, FIWARE-Orion Broker, etc.

Middleware/Processing Layer

The role of middleware/processing layer in Smart City is to provide numerous functionalities to integrate and process the data from heterogeneous sources via divergent protocols. The primary work of the middleware layer is to provide interoperability among multiple connected devices. Interoperability should be employed at three distinct levels: technical level, semantic level, and syntactic level.

IoT middleware acts as an interface between IoT components also ensuring the stability, scalability, reliability, and robustness of the system to manage and control the voluminous data generated by the increasing number of IoT connections and message transmissions. When the data is transmitted at the big data scale, the energy consumption of low powered IoT devices tends to increase and the battery lifetime will get reduced. Also, only the interested portion of the transmitted data will be useful for processing and remaining data will get flooded over the network. In order to reduce energy consumption and increase the lifetime of battery, the IoT devices needs to be intelligent to separate the significant and insignificant data and transmit only the important and interested data. To design smarter and intelligent ultra-low power devices 'TinyML (Tiny Machine Learning)' enters the frame and brings up ways to deploy ML models into the device hardware.

TinyML and Edge AI

TinyML brings the Edge Artificial Intelligence (EdgeAI) into a tiny microchip leveraging the full potential of edge computing to perform on-device computation in real time applications with the core benefits of:

- Minimal Latency
- Lower bandwidth cost
- Enhanced Security
- Exceptional Reliability

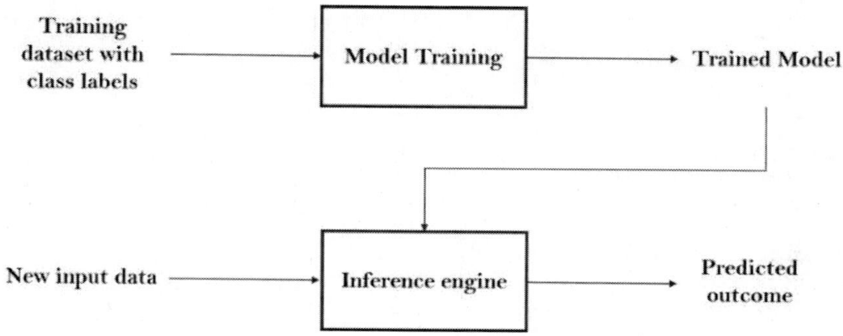

Figure 13.2. Machine Learning Model Training Vs Inference.

The typical machine learning model construction process consists of training phase and inference phase. In the training phase, the ML model tunes the hyper parameters based on the received data to attain better prediction accuracy. The backpropagation mechanism is applied to fine-tune the model using the error rate/loss value obtained from previous iteration/epoch. The training process is repeated for the specified number of epochs and is hard to execute even in regular computers or laptops. On the contrary, inference is the process of predicting the result using the pre-trained model. Figure 13.2 demonstrates the ML model training and inference process. To make the prediction feasible by the edge devices, only inference is executed for the new incoming data. The five-step flow diagram of TinyML model construction for low-powered edge devices is exhibited in Figure 13.3.

i. Model Training is the process of training ML model using the conventional ML learning techniques.
ii. Model Conversion is the process in which the neural network model is pruned and compressed to a tiny ML model adopting various TinyML model packages. Quantization technique is adopted to minimize the memory allocation for numerical values. Knowledge distillation is employed to detect and select only the relevant and significant features of the model.
iii. Model Optimization is carried out to optimize the parameters of the model to make it more fast, accurate, and efficient.
iv. Model Deployment is the process in which the optimized TinyML model is embedded into the edge device hardware.

v. Edge Inference Generation where the embedded edge computing hardware will execute the deployed TinyML model to make predictions on new real time data.

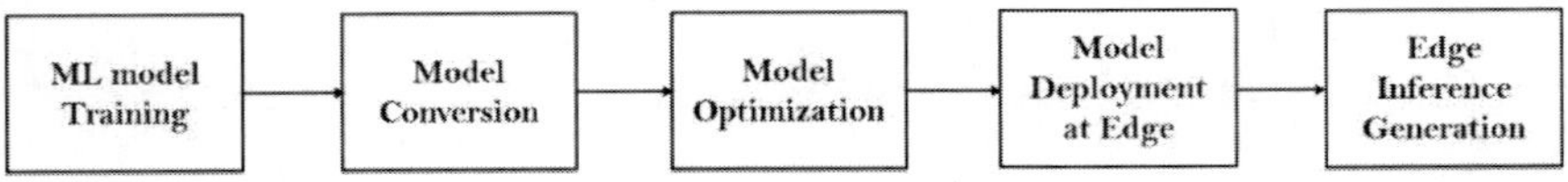

Figure 13.3. Process flow diagram of Tiny ML model.

The **Software Platforms and Frameworks** helps to develop and optimize TinyML models without compromising the efficiency of the model. The most popular and widely used TinyML software frameworks are:

- TensorFlow Lite Micro
- Edge Impulse
- OpenMV

Here are some of the **TinyML Hardware** modules available in the market:

- Arduino Nano 33 BLE Sense – It is the basic standard hardware from the Arduino Nano family equipped with 32-bit ARM Cortex-M4F microcontroller with 256KB RAM and 1MB flash, embedded with built-in sensors, Bluetooth and microphone.
- Arduino Nano RP2040 Connect – It is one of the latest embedded hardware with Raspberry Pi's RP2040 microcontroller.
- Wio Terminal – It is a complete AI framework built with ATSAMD51P19 and ARM Cortex-M4F at 120MHz, integrated with built-in WiFi, Bluetooth, IMU, buzzer, microphone, 2.4" LCD display, microSD slot, light and infrared sensor. It has authorized support of TensorFlow Lite and Edge Impulse.
- Seeeduino XIAO – It is the tiniest Arduino board of thump tip size and belongs to Seeeduino family, equipped with powerful SAMD21 microchip and a wide range of hardware interfaces. XIAO RP2040 and XIAO BLE SENSE are the two strong and potential siblings of Seeeduino XIAO. The latter is the perfect choice for TinyML.

Application Layer

Smart cities have multiplex smart and intelligent applications of diverse domains to address the challenges like overpopulation, insufficient resources, poverty, traffic congestion, etc. The synthesis of TinyML and IoT can revolutionize the Smart City application development with tiny ultra-low powered device that can be virtually deployed anywhere to collect and analyze the data. Smart city applications then make use of the collected data to elevate the public services and utilities, living and infrastructure and many more. This section composes the various smart city applications integrating low powered IoT, TinyML, big data and AI technologies.

Smart Water Management

The ultimate goal of Smart Water Management is to maintain and regulate the water consumption and water recycling to deal with the increasing population, environmental impacts and demands in agriculture and industry sectors. The key objectives of the system are: reducing the wastage of water, enhancing water quality, prevent and control water leakages, etc. Smart Water Management system is the integration of complex systems and parameters to observe, govern, control, and manage the quality and utilization of water resources along with the maintenance of hardware and software components connecting the people with the system. The hardware components include sensors, actuators, meters, pipes, mobile devices, etc. The software components include data storage, processing, and visualization frameworks. The adoption of leading-edge smart water technology improves the transparency and control over the complete water supply chain with more sustainability and reduced manual intervention at optimized cost. The end-to-end tracking process involves IoT devices and software toolkits to optimize the procurement, allotment, consumption, and smart water treatment systems. The sensors play a vital role in smart water management and are used to measure the quality and chemical composition of water at various stages, to alter the water quantity in storage tanks, to monitor the pressure in the distribution pipes, etc. Smart meters and monitoring nodes enable the system to measure and monitor the consumption, exorbitant usage, leakage, and wastage points and rectify the water usage routines and predict the future need and consumption. Dynamic water flow and supply can be achieved using Intelligent water distribution systems and ML models. IoT unlocks new possibilities to convert rainwater and storm water into one of the great resources of water by means of rain and storm water management system.

Smart Energy Management

IoT-powered smart energy management solutions integrate multiple sustainable and renewable energy sources with the IT sector to strengthen the effective energy distribution and consumption. Smart metering is the cornerstone of smart grid systems promotes efficient energy distribution at urban level. Smart meter automation system allows online tracking and monitoring of energy availability and consumption. It helps to lessen the impact of outages and highly responsive in detecting the reason and recover power without on-spot inspection. Smart energy grid solutions aim to encounter current demands through optimized energy production and distribution targeting larger audience. It incorporates renewable energy from heterogeneous sources and distribute energy in a flexible manner. Smart grids ensure reliability by supplying power with minimal outage impact and at reasonable cost.

Smart Healthcare

The new generation technology such as Internet of Things (IoT), big data, cloud computing and artificial intelligence provides the transformation of traditional medical system into more efficient, convenient, and personalized smart healthcare system (Islam et al., 2023). The common application of IoT device for smart healthcare system is remote patient monitoring. This device automatically collects the patient's health information such as heart rate, blood pressure, temperature, etc., These data are sent to software applications where healthcare professionals can view in order to recommend the treatment or generate alerts by analysing the data. The information about depression symptoms and patients' mood are difficult to collect continuously. This challenge can be overcome by depression and mood monitoring IoT devices which collects the information that infers the patient's mental state such as heart rate, blood pressure and even movements of patient's eye can be collected using advanced mood monitoring tool. Robotic surgeries are performed in the healthcare by using the small IoT devices which can reduce the size of incision required to perform surgery, leads to less invasive process and faster healing for the patients.

Smart Agriculture

Due to increase in demand and shortage of food and food products, the IoT smart agriculture provides the solution which enables the formers to reduce waste and increase productivity. Farmers can obtain the better control over the process of growing livestock, crops, making it more predictable and improving its efficiency by using various smart agricultural gadgets. The smart agricultural sensors can be used to monitor the crops which helps the farmers to decide the exact number of pesticides, fertilizers in order to reach the optimal efficiency. The IoT devices (sensors) used in monitoring climate conditions are located in various areas in crop filed. These sensors collect data from the field and sent to the cloud where the required measures are taken which helps to predict the climate conditions, choose particular crops for the climate, etc., In Green House Automation, the sensors collect the real-time greenhouse condition such as lighting, soil condition, temperature, humidity, etc., The crop management kind of IoT device that collect the data specific to the crop such as temperature, precipitation to leaf water potential and overall crop health. This helps to detect the diseases and infestation that harm the yield. One of the most advancements in the smart agriculture is agriculture drone otherwise known as Unmanned aerial vehicles (UAVs). Along with the surveillance task, drones can perform variety of other tasks such as planting crops, fighting pests and infections, agriculture spraying, crop monitoring, etc.

Smart Education

Internet of Things (IoT) plays an important role in the field of education. The adoption of IoT devices such smart phones, tablets in the educational field makes the traditional education system more efficient. Smart Education provides students the reliable access to the learning materials that enables better understanding of the subjects, and the teachers can measure the real-time students' learning process. The e-books provide zooming and saving feature. Instead of black boards, smart boards such as white boards are used to write with a marker and the topics related to the images and graphics can be displayed. This kind of devices are connected to central server which provides the control and monitor the syllabus wise or topic-wise categorization for the students. The other types of smart devices include voice command system for teacher, speech-to-text based note taking system for students, smart security cameras, school buses with GPS tracker, disaster alarms and tablets, and smartphones. Automated attendance recording enables through biometric attendance, barcode-based with the identity card of the student can be used when they entered the classroom. The Live classes, pre-recorded classes and online timer-based assessment questions are all very helpful for distance learners.

Smart Living and Infrastructure

In modern days, smart cities provide better quality of life. Smart Environment are designed for improving energy efficiency and environmental quality in cities. Smart mobility enables organization improvement and urban mobility optimization. The 24-hour software in the smart fire detectors makes the monitoring, detection, and fire prevention possible in urban areas in addition to that, it gives the methods to identify the different sources of fire. The intelligent video surveillance increases the public safety, and the predictive analysis can be established in order to optimize the traffic flow. In smart home, the home appliances can be controlled remotely by set them in such a way that can be automatize house maintenance. They can be combined as the single network. The control and monitoring in smart homes enables the control over the devices in the homes such as fridge can notify if the milk gets expired in three days or can add milk in the shopping list. The cost and energy savings helps to reduce the wastage of energy by shut down the devices which are not in use. Security camera, smart locks and other motion sensors, smoke sensors provide the security in the smart home system when the anomalous behaviour is sensed in the environment.

Smart Industry

Each industry in the market becomes smart industry by adopting the IoT technology with the smart solution in business process which helps the organisation to take their business to next level by using the opportunities in the IoT. The large availability of sensors, reliable connectivity, scalable platform, storage capacity for structured and unstructured data and enterprise IoT solutions makes the IoT has become very helpful for business success rate. Smart factory system provides certain features such as quality control and statistical analysis which are used in every station in production line. The detailed analysis can be performed by the supervisors with the smart factory solutions which provides real-time continuous and statistical task level data. Smart Asset Monitoring provides the automated management of industrial assets like manufacturing and industrial machinery, power and utilities, mining, oil & gas etc., Smart Connected Worker enables tracking and monitoring the safety of the worker as well as the activity of the worker.

Smart Environment

Smart Environment provides better understanding and control over the environment which are built with the various embedded sensors, displays and computing devices. Smart cities provide the control and monitoring of environment in terms of pollution, wastage management, water management, managing streetlights and traffic signals, etc. Pollution is controlled by ride sharing and bike sharing services in smart cities. Living conditions are monitored by using sensors and cameras which tracks the air quality, temperature, sound levels, traffic etc., IoT based trash collection procedure saves the money and manual work by utilizes the IoT smart devices and sensors and enabling the machine-to-machine communication. The ultrasonic and proximity sensors give alert when the trash bin full, assess segregation level and decide whether the driver should stop or continue. The full bin can be identified by the weight sensor. Smart dumpster provides efficient waste collection which collects information such as location, temperature, fill levels and then transmit it to garbage management firms. This information can be used to identify the best path for garbage collectors based on the prioritization of area which need the cleaning most and it reduce the missing garbage pickup. Smart waste management provides efficient way to recycle the both the valuable and hazardous components.

Business Layer

The business layer is responsible for administering the complete system architecture handling applications, business models, security, and privacy of the users. Blockchain and Big Data is a perfect tech combination to meet the business layer requirements of the Low-powered IoT enabled smart city applications. The two major purposes of combining blockchain and big data in smart city applications are:

- i. Efficient and secured data management
  - a. Big Data Acquisition – For secured big data collection and big data sharing and transmission
  - b. Big Data Storage – For securing file systems, databases and storage systems.
- ii. Effective and fast analysis of data and make accurate predictions.
  - a. Big Data Analytics – For secured data learning/training employing AI models.
  - b. Big Data Privacy – For preserving the privacy of big data processing.

Figure 13.4 illustrates the internal components of business layer and the connectivity between the application layer and business layer. Blockchain

chain technology adds additional layer of security to the big data generated in smart city applications by it's certain, distributed and decentralized nature. The subsequent section outlines the integration of blockchain and bigdata for Smart city applications.

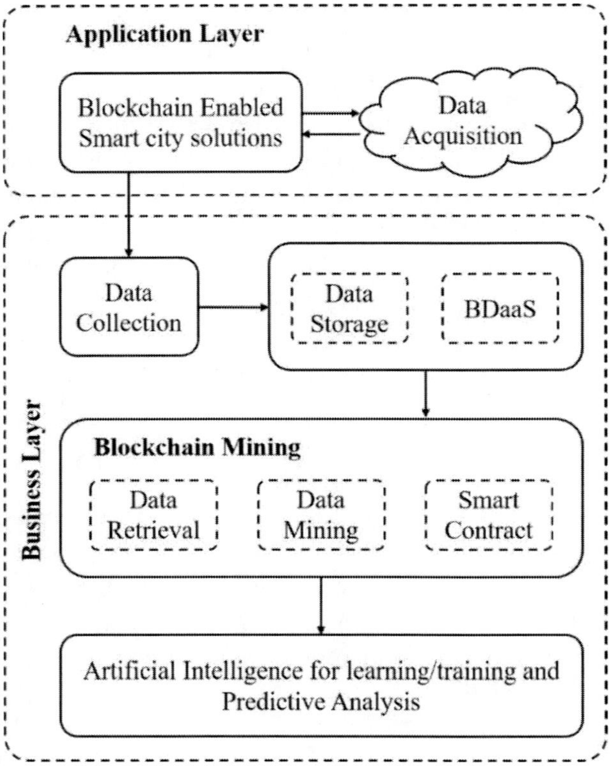

Figure 13.4. Overview of business layer integrating blockchain and big data.

Blockchain and Big Data enabled Smart City

Even though the integration of pervasive technologies such as IoT, Information and wireless communication technology (ICT) and big data makes the city life easier and smarter, there exists several issues such as security, maintenance, adaptability, affordability, and reliability. Blockchain technology overcomes these issues by allowing transparency, scalability, robustness, and sustainability of the edge devices.

Blockchain and Big Data enabled Smart Healthcare

Security and privacy of data transmission and data logging are the major concerns of the smart healthcare. Blockchain provides opportunities to enable security by encrypting the private, public, and other relevant data collected from sensors. Further, the direct access to data is allowed for authorized entities with the consent of the patient.

Blockchain and Big Data enabled Smart Transportation

Blockchain technology helps to develop transparent and secured intelligent transportation systems (ITS) to deliver remote services such as smart charging, smart insurance services, vehicle communication systems, etc. Security and transparency of transportation is assured by the secured key management system for vehicle data associated with the transport keys.

Blockchain and Big Data enabled Smart Grid

The conventional energy management system faces the consequences of radical energy shifts due to the progress in distributed sources of energy. The distributed consensus procedures can be implemented in energy generation stations. Blockchain provides security, scalability, and reliability to address the problems related to micro grids and distributed energy systems using distributed key management technique.

Research Challenges

Blockchain has tremendous advantages in solving large scale and complex real time problems. But there are issues that needs to be addressed when integrating blockchain with IoT and big data, and is discussed in the following subsections.

Challenges in integrating IoT and Blockchain

The major challenges addressed by blockchain when integrated with IoT devices are:

- Interoperability issues among various diverse IoT platforms, formats, and protocols.

- Resource constraints of IoT hardware.
- No software reusability compliance to specific IoT standards.
- Conventional programming software are still used for IoT applications.

Challenges in Integrating Blockchain and Big Data

The integration of blockchain and big data technologies uncovers new challenges in various smart city application sectors. Some of the key challenges are:

- Maintaining stability among accountability, privacy, and security.
- Blockchain implementation for mobile crowdsensing.
- Industrial commercialization.
- Energy and resource utilization while implementing large scale data.
- Possibility of double-spending attack on blockchain.
- Standardization of divergent transaction schemes followed by traditional systems.
- Handling diverse, complex, and dynamically streaming big data.
- Blockchain implementation is costlier when involving more computational complexity to deal with large scale applications.

Future Research Directions

The above listed research challenges create new opportunities for future directions of research to integrate blockchain in large scale smart city applications effective in terms of both computation and cost. Some of them are:

- Flexible and Customized blockchain for big data in cloud computing, SDN (Software defined networks) and mobile ad-hoc edge computing.
- Blockchain supporting the next generation network technologies.

Case Studies

Table 13.1 displays the successful compilation of worldwide real time (World Economic Forum - Global Future Council on Cities and Urbanization 2020) smart city solutions under various sectors.

Table 13.1. Case Studies – Smart City Applications

Sector	Project Name and Country	Description	Scale
Governance and policy	Smart Dubai Happiness Meter – Dubai, United Arab Emirates (UAE)	Live, powerful, and citywide tool that simply captures sentiment. It is developed to make Dubai the happiest city across the globe.	Deployed in 172 private and public sectors in Dubai.
	Urban Environment Quality Index – Russian Federation	Standard monitoring tool to improve and maintain the city infrastructure facilities.	Implemented national wide across more than 1000 cities.
	Open Data Roadmap – Maputo, Mozambique	Initiated to enhance the accountability and transparency for a well-designed city.	Mozambique
	Reinventing Paris – France	Scheme launched to renovate Paris maintaining the social and cultural features.	31 underused spaces across 12 cities across the world.
	New Urban Mechanics – Boston, USA	Public innovation group created to impart people-centric design ideas across the country.	Philadelphia, Utah
Society	Bájale al Acoso (Turn Down Harassment) – Quito, Ecuador	Implemented to fight against public transport sexual harassment via instant messaging system in SMS.	Quito, Buenos Aires in Argentina and other cities
	Estaciones Saludables (Health Stations) – Buenos Aires, Argentina	Scheme to support the healthy lifestyle by providing health consultancy services (HCS) to maintain diet via smart stations.	More than 5.5 million HCS are provided targeting than 1.2 million people
	Harnessing City Data – Singapore	Collecting and exploiting city data to improve services and generate economic value.	Singapore, Amsterdam, Barcelona, Copenhagen, Dubai, London, and New York

Table 13.1. (Continued)

Sector	Project Name and Country	Description	Scale
	Hush City Mobile Lab – Berlin, Germany	Free mobile to find nearby calm and peaceful areas.	Municipality of Berlin for the Berlin Noise Action Plan
	CareTech – Fukuoka, Japan	IoT network to track and monitor elder people suffering from dementia.	Start-ups, Companies, medical and nursing care takers, investors, and citizens.
Infrastructure and services	Copenhagen Cycling Infrastructure – Copenhagen, Denmark	Established an infrastructure for cycling with more than 450 km of lanes.	Approximately 36,000 cyclists per day.
	Santa Monica's Wellbeing Project – California, USA	Exhaustive collection and analysis of data to realize the demands for future planning and budgeting.	Bloomberg Philanthropies granted $1 million.
	Urban Solar – Singapore	Installed rooftop solar panels to meet the energy demands.	By 2030, it is planned to install 2 gigawatt-peak of solar capacity.
	1200 Buildings Program – Melbourne, Australia	Aims to provide financing for commercial buildings at low cost.	5.4 megawatts of solar has been installed across 60 businesses.
	Smart Seoul – South Korea	Enhance the people lifestyle by providing human-centric services.	Expanding continuously.
Environment	Hydrogen From Sewage – Fukuoka, Japan	Generated hydrogen from sewage of households to charge the vehicles.	Throughout the city.
	Ahmedabad Heat Action Plan – India	Innovative early heat warning system to warn the public during emergency heat events.	More than 30 Indian cities.
	Tenjin Big Bang Project – Fukuoka, Japan	Develop urbans resilient to disasters.	Private sectors and partnership with sustainable urban development.
	Water-Conscious Urban Development – Fukuoka, Japan	Implemented to conserve water and track the water supply chain.	Completely deployed across the city.
	EDGE Olympic – Amsterdam, the Netherlands	Promotes healthy environment fostering neutral energy and sustainable resource management.	Berlin (EDGE Grand Central), Rotterdam and Hamburg.

Sector	Project Name and Country	Description	Scale
Business and economy	The Platform Economy – Amsterdam, the Netherlands	Launched a platform economy to gain the trust of public.	More than 100 platforms in Amsterdam
	Dubai Blockchain Strategy – UAE	Implementing blockchain technology for multisector, multiparty use cases.	BaaS (Blockchain as a Service) for about 20 use cases.
	Repair Cafés – Amsterdam, the Netherlands	Free repairing centres for mechanical objects.	20 centres in Amsterdam, 1,500 places globally.
	Circular Glasgow – United Kingdom	Online platform for waste management.	More than 550 businesses have been committed.
	The Mayor's Civic Innovation Challenge – London, United Kingdom	Initiative to provide technical solutions to city level problems.	Can be deployed across London and other cities.

Conclusion

The growing urban sprawl resulted in smart cities development, which requires intelligent and effectual solutions for governance, environment, infrastructure, agriculture, industry, healthcare, energy, and water optimization. The fusion of influential technologies like IoT, big data, block chain, artificial intelligence and power efficient web technologies provides robust and reliable solutions to meet the demands of smart city sectors. The present chapter introduces the mission and vision of smart city followed by the basic concepts of the incorporated technologies and a short literature survey of related works. Then, the five-tier architecture of Low-powered IoT and blockchain enabled Smart City application is introduced with the detailed description of the layers. Further, various research challenges, future research directions are discussed. Subsequently, twenty-five case studies of smart city solutions are neatly presented. Therefore, it is evident that this chapter captures the important aspects of smart city implementation along with the challenges and opportunities.

Disclaimer

None.

References

Alasbali N, Azzuhri SRB, Salleh RB, Kiah MLM, Shariffuddin AAASA, Kamel NMINM, Ismail L. Rules of Smart IoT Networks within Smart Cities towards Blockchain Standardization. *Mobile Information Systems* (2022) 2022:11. https://doi.org/10.1155/2022/9109300.

Arroub A, Zahi B, Sabir E, Sadik M. A literature review on Smart Cities: Paradigms, opportunities, and open problems. *2016 International Conference on Wireless Networks and Mobile Communications (WINCOM)* (2016) 180-186. https://doi.org/10.1109/WINCOM.2016.7777211.

Bauer M, Sanchez L, Song J. IoT-Enabled Smart Cities: Evolution and Outlook. *Sensors* (2021) 21(13): 4511. https://doi.org/10.3390/s21134511.

Bellini P, Nesi P, Pantaleo G. IoT-Enabled Smart Cities: A Review of Concepts, Frameworks and Key Technologies. *MDPI in Applied Sciences* (2022) 12. https://doi.org/10.3390/app12031607.

Camero A, Alba E. Smart City and information technology: A review. *Cities* (2019) 93:84-94. https://doi.org/10.1016/j.cities.2019.04.014.

Deepa N, Pham Q, Nguyen DC, Bhattacharya S, Prabadevi B, Gadekallu TR, Maddikunta PKR, Fang F, Pathirana PN. A Survey on Blockchain for Big Data: Approaches, Opportunities, and Future Directions. *Future Generation Computer System* (2022), 131:209-226. https://doi.org/10.1016/j.future.2022.01.017.

Hakak S, Khan WZ, Gilkar GA, Imran M, Guizani N. Securing Smart Cities through Blockchain Technology: Architecture, Requirements, and Challenges. *IEEE Network* (2020) 34(1):8-14. doi: 10.1109/MNET.001.1900178.

Hammi B, Khatoun R, Zeadally S, Fayad A, Khoukhi L. Internet of Things (IoT) Technologies for Smart Cities. *IET Networks* (2017) 7. https://doi.org/10.1049/iet-net.2017.0163.

Henkel J, Pagani S, Amrouch H, Bauer L, Samie F. Ultra-Low Power and Dependability for IoT Devices (Invited paper for IoT technologies). *Design, Automation & Test in Europe Conference & Exhibition (DATE)* (2017) 954-959. https://doi.org/10.23919/DATE.2017.7927129.

Islam M, Nooruddin S, Karray F, Muhammad G. Internet of Things: Device Capabilities, Architectures, Protocols, and Smart Applications in Healthcare Domain. *IEEE Internet of Things Journal* (2023) 10(4):3611-3641. https://doi.org/10.1109/JIOT.2022.3228795.

Kadam SB, John SK. Blockchain Integration with Low-Power Internet of Things Devices. *Handbook of Research on Blockchain Technology* (2020), 183-211. https://doi.org/10.1016/B978-0-12-819816-2.00008-3.

Pajooh HH, Rashid MA, Alam F, Demidenko S. IoT Big Data provenance scheme using blockchain on Hadoop ecosystem. *Journal of Big Data* (2021) 8(114). https://doi.org/10.1186/s40537-021-00505-y.

Schizas N, Karras A, Karras C, Sioutas S. TinyML for Ultra-Low Power AI and Large Scale IoT Deployments: A Systematic Review. *Future Internet* (2022) 14(12):363. https://doi.org/10.3390/fi14120363.

Treiblmaier H, Rejeb A, Strebinger A. Blockchain as a Driver for Smart City Development: Application Fields and a Comprehensive Research Agenda. *Smart Cities* (2020) 3(3):853-872. https://doi.org/10.3390/smartcities3030044.

World Economic Forum - Global Future Council on Cities and Urbanization. Smart at Scale: Cities to Watch - 25 Case Studies. *World Economic Forum* (2020).

Chapter 14

The Conceptualization of Blockchain for Secure Storage in Smart Cities

R. Anusuya[1], M.Tech
D. Karthika Renuka[1], PhD
and L. Ashok Kumar[2], PhD
[1]Department of Information Technology, PSG College of Technology, Coimbatore, Tamil Nadu, India
[2]Department of Electrical and Electronic Engineering, PSG College of Technology, Coimbatore, Tamil Nadu, India

Abstract

Applications in sectors including healthcare, smart grid and supply-chain management are exploiting the advantages given by blockchain owing to its decentralized nature. The utilization of edge computing combined with blockchain in such applications might assist in improving latency which is very much essential for real-time data processing. Usage of edge computing also helps the scalability of blockchain implementation. In recent years, smart cities have emerged as a new paradigm for providing citizens with high-quality amenities by optimizing city resources in a dynamic manner. In terms of healthcare, transportation, energy consumption, and education, smart cities can provide the best possible services to improve the daily lives of their residents. Nevertheless, the concept of smart city is still in its early stages, and despite its visionary potential, there are a growing number of security issues. Due to its auditability, transparency, immutability, and decentralisation, blockchain technology has the potential to promote the development of smart cities. In our study, we examined blockchain technologies and feasible architecture for merging IoT devices, edge devices and blockchain. We

In: Blockchain Enabled Secure Big Data Computing …
Editors: L. Ashok Kumar, D. Karthika Renuka, V. Chandrasekar et al.
ISBN: 979-8-89113-243-6

have identified a few strategies for strengthening the security of blockchain. We have conceived an architecture using security for the integrated technologies stated above.

Keywords: edge computing, blockchain, IoT, security in blockchain

Introduction

A block chain is a technique of organizing data that separates it into units, or blocks, with each one including a collection of data (Nofer et al., 2017). When a block's storage limit is reached, it is linked to the preceding block, forming a data chain known as a "blockchain." The data after that is synthesized to a fresh block and then added to the chain. A block chain may be used to store a few data formats, although its most frequent use to date has been as a transaction ledger. Block chain is utilized in a decentralized manner in Bitcoin, which implies that no one person or organization has power, instead all users have collective control. A decentralized block chain's data is irreversible, which means it can't be altered. This means that all Bitcoin transactions are permanently recorded and available to everyone. Although the concept of block chain seems to be complicated, it is really very simple.

How a Blockchain Work ?

A blockchain database is a one-of-a-kind database. Block chains vary from traditional databases in the manner they store data. Data is stored in blocks that are then linked together. On the platform, block chain will store any type of data exchange (Zheng et al., 2018). Consequently, it functions similarly to a ledger system, with each data transaction being logged. The data exchanges in the system are referred to as transactions. Whenever a transaction is verified, it is stored in the ledger system as a block. No one can remove or change anything after it's entered the ledger. Block chain is built on a peer-to-peer distributed network, which guarantees the technology's decentralized nature. Every device connected to the network is referred to as a node. Blockchain Technology is built on the foundation of using “keys”. Furthermore, the keys provide network security. A network user will create key pairs known as private and public keys to do this. Once you’ve started utilizing the keys, you'll have a one-of-a-kind credential that no one else has.

New information is added to a new block when it becomes available. The block is chained onto the preceding block once it has been filled with data, resulting in a data chain in chronological sequence.

Storage in Block Chain

Typically, to store data in block chain technology, the data is modified into the form of transactions. The data on the block chain is transformed into a time-stamped network of secure records of data (Kurt Peker et al., 2020). Consequently, the data saved in a blockchain is typically immutable and very secure. In certain instances, it may not be possible to modify the position of data recorded in the block chain in a transaction form and subsequently save it. When block chain technology does not allow for this, data is stored in tiny bits on the chain using addresses. The receiving system's address is automatically encoded with the storage location. This guarantees that the data is safely transferred. The way data is organized in a conventional database versus a block chain varies considerably. A block chain is a data storage technique that separates data into blocks, each one including a set of data. When a block hits its storage limit, it is connected to the preceding block, forming the so-called "block chain." A database stores data in tables, while a block chain stores data in linked chunks (blocks). As a result, all block chains are database structures; yet not all databases are block chains. As a result, all block chains are also databases; however, not all databases are also block chains. When a block is filled, it becomes permanent and becomes part of the chronology. It is necessary to assign a specific timestamp to each block when it is added to the chain.

Security in Blockchain

Security and trust issues are addressed in a number of ways by block chain technology. To start, all new blocks are always maintained in a sequential and linear order. Each block on the Bitcoin block chain is identified by a unique position on the chain, referred to as a height. The block height will have reached 656,197 blocks by November 2020. It is impossible to edit the content of a block after it has been put to the block chain unless a majority agrees. Each block has its own hash in addition to the preceding block's hash and a time stamp. Digital data is converted into a series of numbers and characters using a mathematical formula. When data is changed in any manner, the hash code is also changed. When the rest of the copies are compared, the hacker's version is different and discarded as fake. To achieve this compromise, the attacker must simultaneously hold and change fifty-one percent of the block chain copies, because of which their changed copy becomes the majority copy and therefore the approved chain. An attack of such type would be very costly in terms of resources since it would require rewriting every block with new time stamps and hashes. Considering the scale and pace of Bitcoin's network, any such task will be prohibitively expensive. Not only would such an endeavor be extremely costly, but it would also very certainly fail. An operation like this would not go undetected, since such large modifications to the block chain would be seen by network users. Members of the network would then branch off to a chain that was unaffected. This would result in the targeted Bitcoin's value falling, thus making the attack pointless, since the malicious person would now possess a worthless commodity. The same thing would occur if a malevolent individual tried to attack Bitcoin's recently forked version. It is designed in such a way that participating in the network is much more beneficial economically than attacking it.

Ethereum

Ethereum is often referred to be the second biggest cryptocurrency in terms of market capitalization, after Bitcoin. Ethereum is a distributed ledger technology built on top of a block chain network. A block chain is a distributed public ledger that is decentralized and validates and records all transactions. All the participants on the Ethereum network have an exact copy, which allows them to view all previous transactions. It is decentralized in that it is not regulated by a single entity, but by all the distributed ledger holders. Participants get compensated in the form of cryptocurrency tokens. In the Ethereum system, these currencies are referred to as Ether (ETH). Ether, like Bitcoin, may be used to purchase and trade goods and services. Furthermore, its value has risen dramatically in recent years, making it a constitutionally risky investment. Using Ethereum, users can create applications to execute in blockchain. Personal data might be stored and transferred and complicated financial transactions can be handled using these apps.

Ganache Ethereum

Ethereum smart contracts are programs that run in the context of Ethereum blockchain transactions. Truffle Suite includes Ethereum Ganache, a developer tool that allows users to test smart contracts and reproduce block chain settings locally. The execution of smart contracts on the Ethereum block chain differs significantly from that of other kinds of software. To begin with, the on-chain environment makes communication with the outside world difficult. Transactions are the only way to connect with the block chain, thus simple input/output operations, such as writing to a console, are not possible. Furthermore, since the block chain is transactional, all state-changing transactions with a smart contract are asynchronous. This means that transactions have no effect unless they are verified by inclusion in a block. Lastly, the block chain architecture puts some restrictions on the kind of code that may be executed, the majority of which are related to the transaction cost. Smart contracts are irreversible once deployed, and each deployment incurs a fee. To debug and test smart contracts prior to deployment, it is critical to enable developers to replicate block chain environments locally, free of provisioning fees and transaction latencies. The Truffle Suite is specifically developed for this purpose.

Raiden Network

The Raiden Network's underlying principle is to circumvent the block chain consensus bottleneck. This is accomplished via the use of a network of payment channels that enables safe off-chain value transfer, i.e., without engaging the block chain in each transaction. The Raiden Network project token (RDN) is a decentralised, general-purpose coin that is intended to be interoperable with and utilised on the Raiden Network. It may be used to establish micropayment channels on Raiden, which is presently deployed on the Ethereum mainnet at 0x1440317CB15499083dEE3dDf49C2bD51 D0d92e33, and to conduct transactions across these channels.

Related Works

Xiong et al. (2018) demonstrate blockchain in resource constrained edge devices. They demonstrated mobile devices' ability to access and use CPU power on the edge network, which enables them to integrate blockchain technology into their applications. Their work demonstrated that when additional miners join the network, integration benefits both miners and service providers. According to Samaniego et al. (2016), blockchain technology has the potential to enhance IoT devices by allowing for storage without tampering and authorized transactions. The main barrier to successfully integrating blockchain in IoT devices, according to them, is infrastructure. They then conduct assessments of the fog and cloud as potential deployment platforms, revealing that the fog outperforms the cloud. Liu et al. (2019) established a resource allocation model for streaming video systems using edge computing to enable blockchain adoption. In an edge computing environment, they created a reward system and tested two methods for offloading transcoding tasks. The findings confirmed that the approach produces favorable outcomes when it comes to optimizing average earnings in offloading and adaptive resource allocation schemes. Conoscenti et al. (2017) performed a comprehensive A research of the literature on the blockchain's usage in IoT applications was conducted to determine the possibility of merging block chain with peer-to-peer (P2P) to build a decentralized model for IoT applications. The article revealed three impediments to successful integration: anonymity, adaptability, and integrity. They discovered that pseudonyms are the only method to secure anonymity, that integrity is contingent upon the difficulty of PoW and the presence of many trustworthy miners, and that compliance is constrained by the complexity of PoW. Different research studies have examined mitigation strategies for such problems (Sasson et al., 2014).

Anusuya et al. (2021) discussed the role of blockchain in adding security to edge computing storage requirement.

Design and Implementation

Despite ongoing efforts to provide a useful framework for blockchain implementation in Internet of things, critical issues such as anonymity, adaptability, and data integrity remain unresolved, making safe data storage impossible. The suggested idea is to use a Proof-of-Space solution in combination with a smart contract-based data integrity service to authenticate stored transactions and ensure data integrity in the absence of a central authority. Zerocash, zerocoin and Linkable ring signatures were determined to be sufficient methods for hiding user identity and obscuring information included in transactions to guarantee anonymity.

Setting up a personal Ethereum blockchain using Ganache Ethereum, which you can use to conduct tests, issue commands, and examine state while maintaining control over the chain's operation. MetaMask is the most straightforward and safe method to connect to blockchain apps. When engaging on the new decentralized web, you are always in command. The Raiden Network's main goal is to bypass the blockchain consensus bottleneck. This is accomplished by using a network of payment channels that allow for safe off-chain value transfers, i.e., without engaging the blockchain in every transaction.

The Raiden Network's project token (RDN) is a decentralized coin that is intended to be interoperable with and utilized on the Raiden Network. It may be used to make payments via Raiden micropayment channels, which are presently implemented on the Ethereum mainnet at 0x1440317CB154990 83dEE3dDf49C2bD5d92e33.

Zerocoin and zerocash are two different types of cryptocurrencies. This uses non-interactive zero knowledge proof to build ideal tools for obfuscating transactional data. An attacker will be unable to destroy anonymity by altering the data bits in a transaction to an incoherent random bit sequence that has no relation to the original data. The protocols are more appropriate for ensuring anonymity in our model because of their robustness in preventing ill-intentioned individuals from recovering information regarding the transaction.

Design of Software Architecture

The following is the experiment's software architecture design shown in Figure 14.1.

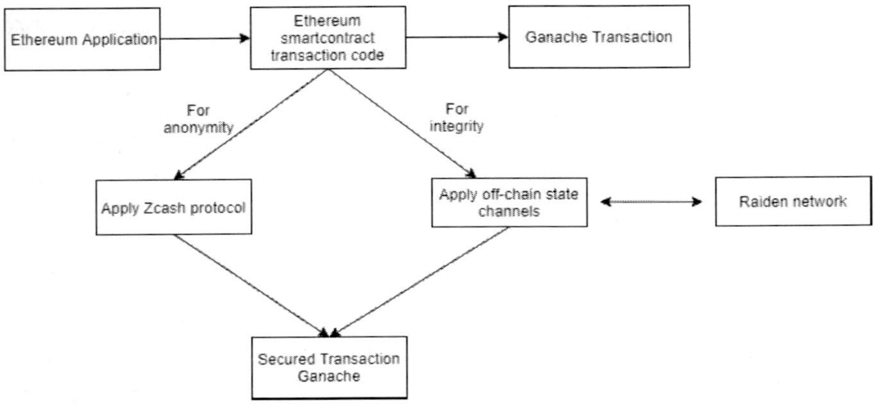

Figure.14.1 Software Architecture Design.

Module Description

Blockchain Data Integrity Service

The Data Integrity Service (DIS) is offered as a significant data integrity approach for blockchain. In DIS, there are data owner applications (DOA) for owners and data consumer applications (DCA) for consumers. In practice, the cloud storage service may be utilized either as a stand-alone cloud service or as a blockchain node. Owners and consumers were individually identified in the blockchain system by their respective public keys. When the DOA and DCA joined the blockchain network, a key pair was established for them consisting of a private key and a matching public key. The public key is used to identify the account of each node, while the private key is needed to obtain access to the account. All transactions in the system were restricted to nodes with adequate money in their accounts. Although both DOAs and DCAs had the ability to join the network as miners at any moment, DOAs found it difficult and often unnecessary to earn their deposit as miners owing to their poor processing capacity. DCAs, on the other hand, may serve as miners, depending on their hardware and financial resources.

Using a mix of Ethereum and smart contracts, we were able to provide storage which maintains integrity for our integrated framework. This method enforces end-to-end encryption to enhance security.

Proof-of-Space (PoSpace) is a mechanism that peers in a P2P network use to verify their claims of depositing and committing space. In this context, PoSpace refers to a method by which a prover might demonstrate genuine interest in a service by devoting a substantial portion of memory or disc space to fulfilling a challenge set by a verifier. It is critical to note that to overcome the PoSpace problem, a large quantity of data must be shared between the prover and verifier, rendering this solution unfeasible. However, security always comes at a cost.

Before peers may build and connect a transaction in the blockchain, they must first register and confirm their transactions via proof of space verification activities. IoT users utilized a smart contract to store transactions in this framework. After then, the data was encrypted locally to prevent it from being stolen.

Unauthorized access resulted in the creation of a transaction, which was then publicized to the P2P network by owner clients, who filed claims for needs and asked about charges to be paid. Miners (peers on a peer-to-peer network) responded by analyzing customers' needs and existing services in transactions to provide clients with the required storage for rent. Therefore, IoT devices may outsource data storage to a decentralized peer-to-peer storage system that enforces appropriate incentives and penalties through smart contracts. To secure the integrity of outsourced data, Internet of Things users issue a new challenge transaction that needs data miners to compute and maintain a proof of work (that users may validate). While the calculated evidence does not pass verification, data hosts are paid for the deposit made while enrolling as IoT users. Miners may remove the committed space at any time by issuing a cancelling transaction and withdrawing the payment made upon registration.

Anonymity Service for Block Chain Data

While the computed evidence fails verification, the data hosts are penalized by being paid for the deposit they made when signing up as IoT users. This service is used to provide network control mechanisms for data transit between devices through intermediate nodes. This also applies to the smart contracts' internal messages, which contain rules governing various transactional parts. Contracts may be transferred between nodes that are presumably on different sides of the network, and this transmission must be secure. To ensure dependable and efficient network coordination, it is critical that both data and contract communications be secured. These messages provide information about privileges, user addresses, cryptographic keys, and the duration of a transaction. Security risks at different levels must be handled in the design since communications are exchanged between devices in the network. Edge computing links the lower and higher levels of computing and interacts with several systems using cellular and wireless networks.

In this diverse setting, network administration unavoidably becomes a problem. The usage of software-defined networks (SDN) and its extension to SDN components (SDNC) is a successful method for this deployment because it separates the control plane from the data plane, allowing for better network visibility. Finally, in the context of SDNs, utilizing dynamic virtualization of network resources simplifies network management and provides network privacy and security through blockchain.

Methods for Preserving Blockchain Anonymity

We found three methods for maintaining blockchain anonymity: zero-knowledge proof, ring signatures and mixing services. By concealing connections between senders and recipients in a transaction, mixing services may defend against anonymity attacks. The obfuscation conceals the identity of the entity involved in the connection, as well as the data being sent. A ring signature happens when a group of individuals forms a ring without the intervention of a central entity, and one of the members anonymously signs the message on behalf of everyone. The ring signature produces a valid but anonymous digital signature from a collection of individuals without disclosing the originator's identity. ZKP (zero-knowledge proof) is a cryptography-based solution for validating a transaction without revealing any additional information. Due to the spontaneity with which blockchain transactions are combined, mixing services suffer significant latency when participants find one another prior to their transactions being combined. Without going into detail, the ring signature utilizes a public key infrastructure (PKI) to produce valid signatures for all ring members that the verifier may verify without knowing the signer's real identity. When an entity in the ring may sign the same message again with the same tag, but the resultant signatures can be connected through PKI procedures, maintaining the signer's anonymity, the linkable ring signature method is created.

Offchain State Networks

By integrating a layered architecture into the blockchain system, we were able to overcome the scalability and interoperability limitations imposed by the heterogeneity of IoT devices and the many protocols utilized in the edge computing environment. This is accomplished by stacking the blockchain deployment and decoupling it from the application layer, enabling less competent IoT devices to retain just the piece of the blockchain necessary for their own transactions. Thus, it will be possible to avoid storing whole blocks, which is prohibitively costly for low-capacity devices. The suggested framework section described the different entities' operations and interactions at various levels. As previously stated, the off-chain state channels architecture enables some blockchain activities to be performed off-chain. This allows for the separation of blockchain-intensive tasks into a distinct layer. Our solution enables the off-chain storage and processing of enormous amounts of data from IoT devices to off-chain resources at the edge layer and other peers in the system. With enormous quantities of storage and computing capacity collected in a peer-to-peer network, the framework may provide sufficient storage and processing capacity to allow blockchain's adaptation to IoT applications. Using off-chain storage and computation in our solution, large data storage and heavy transactions from IoT devices may be offloaded to off-chain resources at the edge layer and other peers in the system.

Proofs of Zero Knowledge

On the other hand, ZKP provides an appropriate protocol for anonymously verifying blockchain transactions through its non-interactive zero-knowledge proof (NIZK) version. The link between the prover and verifier is severed in NIZK proofs to ensure transactional anonymity. Without going into depth about how they work, both zerocoin and zerocash rely on cryptographic processes known as NIZK proofs to avoid transaction analysis and increase anonymity. Zerocoin was created to combat visual analysis of transactions and to enable completely anonymous money transfers. It uses NIZK proof techniques to validate generated coins before redeeming them with new coins of identical value that have no previous information, separating transactions from their payment sources. Additionally, zerocash enhances blockchain privacy by allowing participants to make direct payments to one other without revealing the source of the transaction, destination, or value. In a transaction, shielded addresses and created ZKP are used to transfer and receive Zcash, letting other participants to validate encrypted data without disclosing the address.

Design of Prototypes

We provided a comprehensive experimental design and the tools necessary to build a prototype of our suggested framework in this part.

For edge computing server, a powerful workstation was utilized, such as one equipped with an intel Xeon Processor (E5450 12M Cache, 3.00 GHz, 1333 MHz FSB). Although a peer-to-peer network of edge servers was intended for real operation, this configuration should be sufficient for our first testing. On the other hand, IoT end nodes were deployed using android-based devices. We also developed, installed, and tested our framework on Microsoft Azure.

The goal was for IoT end nodes to act as miners, processing data on an edge computing server whose compute and storage capabilities could be enhanced using Azure cloud services. Using the Ethereum JavaScript API web3 (https://github.com/ethereum/web3.js/), we developed a data storage decentralized application (DApp) whose client module could be loaded and performed on IoT nodes and controlled by the Ethereum platform running on the edge workstation. Three essential technologies were needed to create this application: TestRPC, Truffle, and MetaMask. TestRPC, as one of the Ethereum clients, may be used to run and test smart contracts on a local Ethereum network.

TestRPC may be used as an Ethereum client to locally imitate an Ethereum network for the purpose of executing and testing smart contracts prior to their deployment on the main network. It is accessible through the npm package manager, which is Node.js's default package management. Using Node.js, the Ethereum network may then be established and deployed locally, replete with default accounts, private keys, and a listening port. Truffle is an Ethereum development toolkit for solidity-based smart contracts. Solidity is a high-level object-oriented programming language with a syntax like JavaScript that is widely used in the construction of smart contracts. Truffle comes with a compiler for smart contracts, as well as testing and deployment tools. It also has a JavaScript abstraction layer that makes communication between smart contracts and user interfaces easier. It's also available on npm and works with TestRPC and the web3 API without any further installation or configuration. After that, the truffle service may be started in the newly established project tree to host smart contracts, handle deployment files, and store test files for contract and application testing throughout development.

Finally, MetaMask offers a browser based Ethereum lightnode client that exposes the web3 API to the DApp and enables users to communicate with the Ethereum network through the browser. MetaMask is a Chrome addon that allows users to connect to many Ethereum networks, including the main network and various online test networks. The user may connect MetaMask to the Ethereum network operated locally by TestRPC by switching networks from the given port. We built our decentralized application using the environmental setup, combining it with the zerocash project (https://github.com/zcash/zcash) and the linkable ring signature method (https://github.com/sorrge/LSAG) for increased anonymity. The Raiden network, which is an extension of the Ethereum network, was used to facilitate scalable and rapid transactions. To provide the data integrity service, the DApp's smart contracts were constructed on the Ethereum network. Mobile devices may connect to the edge server through a wireless access point (AP) and conduct mining operations via the DApp. In this situation, miners were able to request storage and processing resources from the edge server using Ethereum services. The Ethereum network may then access and distribute the newly minted transaction blocks. We evaluated the framework's performance using a mix of the TestRPC running on Node.js to imitate the Ethereum network and our own DApp for smart contracts and other integrated features. We started by using Node.js to create an initial batch of blocks, say 5000, and then utilized mobile nodes to begin block mining on top of the main blockchain. We examined three distinct scenarios by varying the number of mining devices in each case. We set the number of transactions in each mined block (say, 20) and modified the "difficulty" objectives for PoW puzzles using smart contracts. Following that, we looked at CPU use to see whether there were any trends in a miner's computation requirements in all three cases. We concentrated only on successful assessments at this early stage of our model's development. By altering the number of miners and their associated demands, we were able to assess computer service requirements using the CPU utilization of edge servers. The probability of success in mining the blocks was reflected in the demand for services. Similarly, we intended to conduct our tests to assess energy and memory usage, as well as to simulate a full-scale deployment of incentivized outsourced storage with optimal offloading models.

Results

The experiment to determine if block chain technology might be used to fulfil our framework's design objectives of anonymity, integrity, and flexibility revealed possible solutions. These solutions must be integrated into a single decentralized application built on the blockchain network to ensure its seamless functioning. Our study's objective was to develop a secure data storage system and a high-performance IoT architecture by combining edge-computing with blockchain. We suggested a combined Internet of Things architecture based on peer-to-peer networks to allow edge computing and block chain to complement one another in terms of storage capacity and security. To ensure that our model meets the security and performance criteria, we examined current technologies that solve the critical problems of integrity, flexibility, and anonymity associated with the combination of block chain and edge computing. To validate recorded transactions and guarantee data integrity without the participation of a central authority, we utilized a proof-of-space solution for a data integrity service based on smart contracts. We utilized a layered architecture in our design to separate the block chain from the application layer, which is implemented via the Raiden network, which offers fast and scalable transactions on top of Ethereum, to improve scalability. This split allows resource-constrained Internet of Things devices to store just the parts of the block chain required for their own transactions. We found that linkable ring signatures, zerocoin, and zerocash are all adequate techniques for concealing user identity and obscuring transaction data in terms of anonymity. In the prototype of our system, we utilized zerocash and linkable ring signatures to increase anonymity.

Performance Evaluation

The performance of ZKP module was analysed using Hyperledger Calliper and the results are depicted in Figure 14.2. We can observe that for sending 560 successful transactions, in Ethereum the TPS (Transactions Per Second). We conduct tests in which we raise the length of secrets from one to ten, hundred, thousand, and ten thousand characters and then measure the outcomes. The findings indicate that regardless of the secret length, both proof production and verification durations stay consistent, at about 30 ms and 240 ms, respectively. Because a secret message m is hashed to a fixed length of 256 bits prior to proof production using the SHA256 technique, our ZKP module can offer constant proof creation and verification timings. Therefore, the time required to generate and verify proof is independent of the length of secret messages.

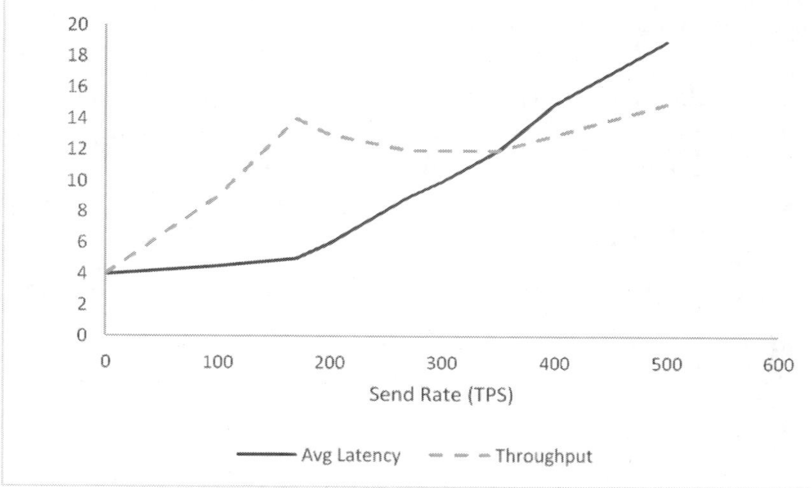

Figure 14.2. Graphical Representation of Performance Analysis.

Conclusion

Our research resulted in the development of a functional system prototype that integrates edge computing and blockchain technology for secured decentralized data storage. Theoretical assessments of the integrated system imply that they are capable of effectively integrating in such a way that huge quantities of data transferred across various IoT devices may be securely stored. The variety of IoT devices and the numerous protocols utilized in the edge computing infrastructure presented scalability and interoperability challenges, which were overcome by integrating a layered architecture into the blockchain system. It involves segmenting the blockchain deployment and isolating it from the application layer, enabling less capable IoT devices to store just the blockchain components necessary for their transactions. The cost of storing all the blocks is too costly for low-capacity devices, and so will be avoided. The proposed framework section detailed the activities and interactions of the various entities at the various levels.

We have presented just an abstract design for a system prototype in this article and a succinct assessment based mostly on the merits of the specific solutions chosen.

Future Work

Our future efforts will center on evaluating the effective functioning of our suggested system and optimizing the many parameters involved utilizing the decentralized application created. We envision a full-scale deployment across all levels by starting with fundamental CPU, memory, and power utilization patterns on the edge server.

Disclaimer

None

References

Anusuya, R., Renuka, D. K., & Kumar, L. A. (2021, January). Review on Challenges of Secure Data Analytics in Edge Computing. In *2021 International Conference on Computer Communication and Informatics (ICCCI)* (pp. 1-5). IEEE.

Conoscenti, M.; Vetro, A.; De Martin, J.C. Blockchain for the Internet of Things: A systematic literature review. In *Proceedings of the 2016 IEEE/ACS 13th International Conference of Computer Systems and Applications (AICCSA),* Agadir, Morocco, 29 November–2 December 2016

Kurt Peker, Y., Rodriguez, X., Ericsson, J., Lee, S. J., & Perez, A. J. (2020). A cost analysis of internet of things sensor data storage on blockchain via smart contracts. *Electronics,* 9(2), 244.

Liu, M.; Yu, F.R.; Teng, Y.; Leung, V.C.M.; Song, M. Distributed Resource Allocation in Blockchain-Based Video Streaming Systems With Mobile Edge Computing. *IEEE Trans.* Wirel. Commun. 2019, 18, 695–708.

Nofer, M., Gomber, P., Hinz, O., & Schiereck, D. (2017). Blockchain. *Business & Information Systems Engineering,* 59(3), 183-187.

Samaniego, M.; Deters, R. Blockchain as a Service for IoT. In *Proceedings of the 2016 IEEE International Conference on Internet of Things (iThings) and IEEE Green Computing and Communications (GreenCom) and IEEE Cyber, Physical and Social Computing (CPSCom) and IEEE Smart Data (SmartData),* Chengdu, China, 15–18 December 2016.

Sasson, E.B.; Chiesa, A.; Garman, C.; Green, M.; Miers, I.; Tromer, E.; Virza, M. Zerocash: Decentralized anonymous payments from bitcoin. In *Proceedings of the 2014 IEEE Symposium on Security and Privacy,* San Jose, CA, USA, 18–21 May 2014.

Xiong, Z.; Zhang, Y.; Niyato, D.; Wang, P.; Han, Z. When Mobile Blockchain Meets Edge Computing. *IEEE Commun. Mag.* 2018, 56, 33–39.

Zheng, Z., Xie, S., Dai, H. N., Chen, X., & Wang, H. (2018). Blockchain challenges and opportunities: A survey. *International Journal of Web and Grid Services,* 14(4), 352-375.

Chapter 15

Smart Transportation for Bus Travellers and Identifying the Optimal Insurance Premium Using Internet of Things

A. K. Kowsalyadevi[1]
M. Jayasanthi[1,*]
C. V. Pavithra[2]
R. Divya[2]
and D. Archana[1]

[1]Department of ECE, PSG Institute of Technology and Applied Research, Coimbatore, Tamilnadu, India
[2]Department of EEE, PSG Institute of Technology and Applied Research, Coimbatore, Tamilnadu, India

Abstract

Using infrastructure development and connectivity technologies, Smart Cities are designed to increase the value of your life through monetary expansion and improved quality of life. A Smart City would consist of a wide range of groundwork components, including ample water supply, reliable electricity, education, public health, solid waste management, resourceful urban mobility, and civic transport, affordable housing for the poor, good governance, including e-Government and citizen participation, sustainable environment, robust IT connectivity and digitalization, and safety and security for citizens, especially for children and women. A safer, greener and more convenient next generation of transportation goes beyond simply reaching A to B. The future of

* Corresponding Author's Email: jayasanthiranjith@gmail.com.

In: Blockchain Enabled Secure Big Data Computing …
Editors: L. Ashok Kumar, D. Karthika Renuka, V. Chandrasekar et al.
ISBN: 979-8-89113-243-6

intelligent transportation is close at hand when intelligent processors are connected and distributed inside vehicles. The system alarms the usual travelers by intimating the heavy traffic zones at the time to plan the travel. Insurance premium details are calculated by checking not only the current condition of the vehicle but also by analyzing whether the traffic rules and operating rules of the vehicle are obeyed/followed or not. Then the premium amount to be paid is sent to the vehicle owner. Aiding people with these practices can save public time, promote traffic knowledge, utilize transportation efficiently, and brand the city even smarter. It is desirable to transform transportation in profound and unimaginable ways. Moreover, seamless connectivity will be available across all modes of transportation. Public bus information is provided to frequent travelers by the Smart Transportation System. As the bus moves around a city, the system keeps track of real-time information including timings, seat availability, the bus's location, the time it takes to travel to a specific destination, and the density of people inside the vehicle. The method is not only limited to providing significant information. On top of it, it serves to assess the performance of the driver that tends to decide the vehicle insurance based on run-time data. The system can be implemented in two phases namely - the Data Acquisition phase which tracks the inline details of the buses using sensors including the location and seat availability of the bus. The second phase, Data Analytics, acquires these details and provides the information to the appropriate users through a mobile application.

Keywords: smart transport – data acquisition – data analytics – optimal vehicle insurance

Introduction

Shortly, buses could connect wirelessly to the internet, intelligent traffic systems, and even other vehicles through the next generation of transportation. In effect, the bus becomes a digital lifestyle extension. Existing and emerging wireless standards will enable vehicles and devices across the transportation infrastructure to interact with one another, facilitating automated intelligent real-time travel decisions. For example, installed safety cameras will automatically detect congestion or accidents and alert emergency personnel. Moreover, traffic can be redirected by notifying approaching cars and updating nearby electronic roadside signs.

Intelligent traffic signals will become more common and intelligent than they are now. Detecting real-time traffic flow from all directions and changing signals automatically to improve it. Distributed intelligence is secure as well as reliable. A significant service is the transformation of the complete traffic infrastructure from passive to practical. A connected intelligent response system improves travel in all aspects. Furthermore, to make this transportation vision a reality, the infrastructure has to be built on an established universal standard-based architecture.

Smart Transportation for Bus Travellers using IoT (Internet of Things) refers to the use of connected devices, sensors, and data analytics to enhance the safety, efficiency, and convenience of public transportation systems (Jana, M., & Biswas, S., 2021). IoT technology can be used to track the location and condition of buses in real-time, allowing for better scheduling and route planning. It can also provide information to passengers about bus arrival times, route changes, and other relevant updates via mobile apps, digital displays, or voice assistants.

IoT can also be used to monitor the health of buses, detecting maintenance issues before they become major problems and improving overall safety. IoT sensors can also provide insights into passenger behaviour, such as in crowded areas, and enable operators to adjust accordingly (M Nesheli, A Ceder, et al., 2017). Overall, Smart Transportation for Bus Travellers using IoT can improve the quality of public transportation, increase passenger satisfaction, and reduce the environmental impact of transportation systems.

Flexible premium methods for monitoring bus conditions throughout the year using IoT technology have several benefits (M Syafrudin, G Alfian et al., 2018), (A P Murdan, V Bucktowar et al., 2020), including:

- Early detection of maintenance issues: By monitoring bus conditions in real time, operators can identify maintenance issues early, reducing the need for costly repairs and downtime. This can save money on maintenance costs over time.
- Improved safety: Regular monitoring of bus conditions can help identify potential safety hazards, such as worn brakes or faulty electrical systems before they cause accidents. This can improve the safety of passengers and other road users.
- Better reliability: By monitoring bus conditions, operators can proactively address issues that could lead to breakdowns or delays. This can improve the reliability of bus services and increase customer satisfaction.

- Enhanced environmental performance: By optimizing bus performance through real-time monitoring, operators can reduce fuel consumption and emissions, leading to improved environmental performance.
- Improved customer experience: Real-time monitoring of bus conditions can enable operators to provide passengers with accurate and up-to-date information about bus arrival times and service disruptions. This can enhance the customer experience and increase passenger satisfaction.

Overall, flexible premium methods by monitoring bus conditions throughout the year using IoT technology can help operators optimize their operations, reduce costs, and provide better service to passengers while improving safety and environmental performance.

The paper discusses the literature and relates its significance to smart bus travellers and their flexible payment plans. In the next section, a proposed technique is presented for transforming Public Transportation with IoT and identifying the optimum insurance premium. Following that, existing performance metrics were used to evaluate the performance of IoT-based smart transportation. The work concludes with a description of the benefits to society and the scope of its future applications.

Related Works

The challenges in the transportation system due to the increasing urban population in major cities of developing countries are investigated in (Vidya Bhilawade, and L. K. Ragha, 2018). Several advanced traveller information systems were reported earlier in many research works. The advanced traveller information system presented in (S. Pal, and V. Singh, 2011) has integrated Global Information System (GIS) and World Wide Web (WWW) for spatial analysis of data. The authors in (Bhupendra Singh and Ankit Gupta, 2015) have highlighted the conclusions extracted from the studies of different systems and have given the future scope in the field of Intelligent Transportation Systems (ITS) to make it more user-friendly and accessible. The significance of driving behaviour characterization which the insurance companies have benchmarked to quantify accident risk and provide personalized rates is discussed in (National Association of Insurance

Commissioners, 2017). The driving behaviour analysis done through CAN Bus in an uncontrolled environment is discussed in detail in (Umberto Fugiglando et al., 2019).

In (Miraal Kamal, Manal Atif, Hafsa et al., 2019), IoT-based smart bus stops with a system prototype were introduced to test and verify the design and operation of the proposed solution. It has been proposed that bus stops be outfitted with a standalone microcontroller with WiFi access, which would be linked to sensors and actuators. This controller transfers readings from sensors to a cloud-based real-time data acquisition hub. In the event of unusual measurements or breakdowns, mobile application software is integrated into the system and transmits the information to the operators or technicians. This application also contains a map interface that enables operators to remotely monitor a bus stop's temperature, humidity, estimated occupancy, and air pollution levels.

A Smart Public Transportation system which is the subsystem of Intelligent Transportation Systems (ITS) that controls the public transportation systems in a greatly clever manner is reviewed in (J. Jalaney, Dr R. S. Ganesh et al., 2019). It keeps up the execution and gives information on excursions and working conditions of the system to travellers and leaders. The review showed that IoT utilization tends to give priority to safety to avoid road accidents. But it has not yet been discussed how intelligent systems can be developed by integrating bus scheduling, bus availability information, and advance payment by the passengers to book seats so that congestion can be limited and the waiting time of the passengers can be reduced. The paper features the investigations of numerous works and provides a possible future extension to utilize the IoT concept in a full-fledged manner.

The paper (A J Kadam, Virendra Patil, Kapish Kaith et al., 2018) proposes an android-based system to track buses using Global Positioning System. It also provides information about the number of passengers on the bus and also the estimated time of arrival of the bus to the users' location. The location of the Bus can be tracked by the commuters using mobiles with Android Applications. This also contains all the details of the bus like Bus number, Bus routes, Bus Stops, Bus timings and frequency of the bus on a particular route.

The development of an urban IoT framework that is utilized to construct an Intelligent Transportation System (ITS) is proposed in (Parag Gawade, Prof. Meenakshi A et al., 2017). IoT-based intelligent transportation systems are suggested with a vision of a Smart City, which can use sensors and other communication systems to manage the smart city and the inhabitants. The development of smart products and services is made easier by the advent of smart sensors, single system-on-chip computers, the Internet of Things (IoT), and cloud computing. These consist of intelligent public lighting, intelligent petrol stations, intelligent parking lots and intelligent bus stops. With multiple devices connected to the cloud, an extensive amount of information is created widely. Therefore, this vast amount of information needs to be managed and transformed into effective frameworks. In today's public transportation, reliability in the bus arrival is crucial, and the suggested technique makes it possible.

Significance of Smart Transportation for Bus Travelers

With the increasing population, the management of city transportation is an important aspect of the establishment of a smart city that will replace the age-old transportation framework (Y. Kuo, J M Leung, et al., 2023). Nowadays, people prefer to use public transportation as the key mode of commuting to avoid traffic congestion and parking issues.

Even with the development of transportation infrastructure over the last few years, it is found that there is a lag in safe and comfortable transportation. Capacity inadequacies in public vehicles and poor traffic management are the key issues that demand smart transportation (X Qin, J Ke et al., 2022). A smart transportation system integrates various controls such as vehicle sensing, providing an optimum route to travelers, avoiding traffic congestion and so on.

In this proposed work, smart transportation is planned for bus travelers to enable them to identify the bus location, information on bus timings at each stop, the time required for the bus to reach the particular destination, the route of travel of the bus from starting point to the destination and also the information on the density of passengers inside the bus is planned to be assessed. Furthermore, based on the run-time data of the bus and the performance of the bus driver in terms of reaching the location on time and public safety, vehicle insurance is determined.

Though full implementation of smart transportation necessitates the deployment of Internet of Things (IoT) devices for the above-mentioned process which may take years to establish, it is one of the optimum solutions for an increasing population. Thus smart transportation will enhance the quality of life of people, enable quick response to emergencies and disasters, and attract more tourists. Overall, the significance of smart transportation lies in resolving the uncertainties in travel by implementing better travel plans and enhancing the mobility of people in an improved manner.

Significance of Identifying the Optimum Insurance Premium

Uncertainties are part and parcel of the journey of life, which is not under our control but the risks of uncertainties are within our limit. On travel, bus drivers are bounded to ensure the safety of passengers. Hence, they should adhere to the transportation policies and procedures of the government.

Diving behavior explains how the driver has operated the vehicle in the context of external surrounding influences in preventing accidents. The hours of driving, period of rest and the tiredness experienced by the drivers, characteristics of the driving environment are important aspects of the safe operation of the vehicle. However, insurance companies provide a personalized rate to quantify the emergency risks based on the driver's behavior. Hence, identifying an optimum insurance premium is important for the drivers to get the maximum benefit in case where some untoward incidents take place beyond their control.

Smart transportation thus allows assessing the driver behavior by incorporating the vehicles with needed sensors and control units to track real-time data like vehicle speed, time the vehicle has been trapped in traffic congestion, energy consumption, smoke emission, brake pedal pressure, steering wheel angle and drivers' potentiality. These signals are either directly or indirectly communicated between the vehicle and the driver using IoT-based devices. Thus with the information obtained from the IoT devices, the driver's vehicle insurance is decided in the smart transportation system.

Proposed Methodology

The public transport systems are not as effective as they could be worldwide. Even though more people are using public transportation, relatively few of them are satisfied with the standard of the services they receive. These individuals frequently encounter bus arrival time delays because of maintenance and other unexpected situations. Transports become busy in some locations and vacant in others as a result of inefficient vehicle deployment across different routes. The adoption of a highly effective and practical transportation system made possible by the Internet of Things has the potential to reverse the trend of growing private vehicle ownership. The proposed public transport systems can provide a convenient, comfortable travelling experience for people travelling from one location to another.

Transforming Public Transportation with IoT

The management of public transportation must have a significant impact on both commuters' and non-commuters' lives. For instance, effective bus route planning can alleviate the issues related to traffic jams and bottlenecks on roadways, making life easier for those who travel in private vehicles. When insufficient buses are allocated to high-traffic routes, it also results in fewer people clustered at bus stops. The government can employ a smart, unified system to make sure that the operations of the transit system are well-coordinated and efficient by applying smart strategies to prevent these challenges and other public transportation-related concerns.

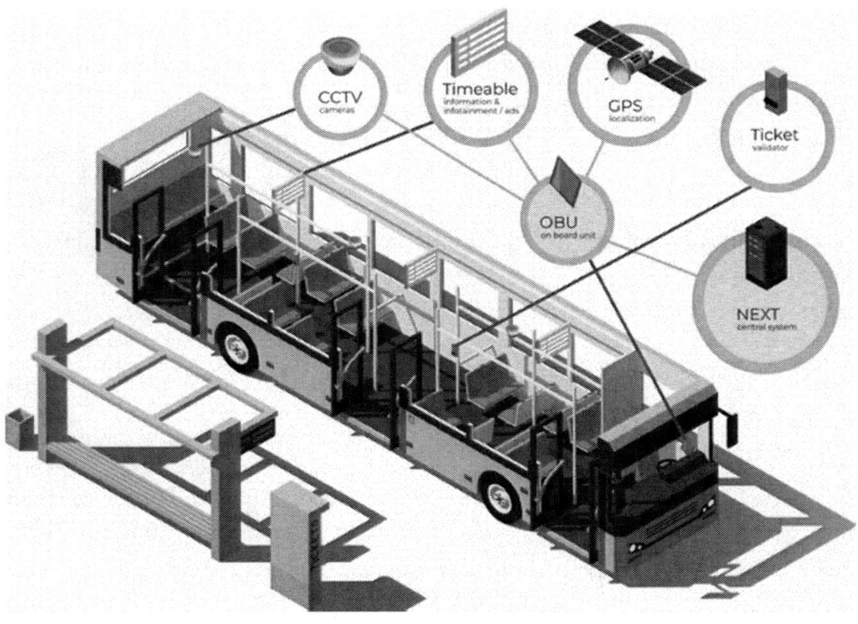

Figure 1. A model of a smart transport system.

Such a system can offer hassle-free journeys to commuters while assisting operators in cost reduction and maximizing the efficiency of transportation systems. Such a system can be created with the aid of IoT in the transit sector. A model of the automated transportation system is shown in Figure 1. The implementation of IoT in the transportation sector can assist in building a system with a variety of interface units, such as CCTV cameras, GPS for tracking, onboard units, ticket information, and delivery systems, to allow smart functioning.

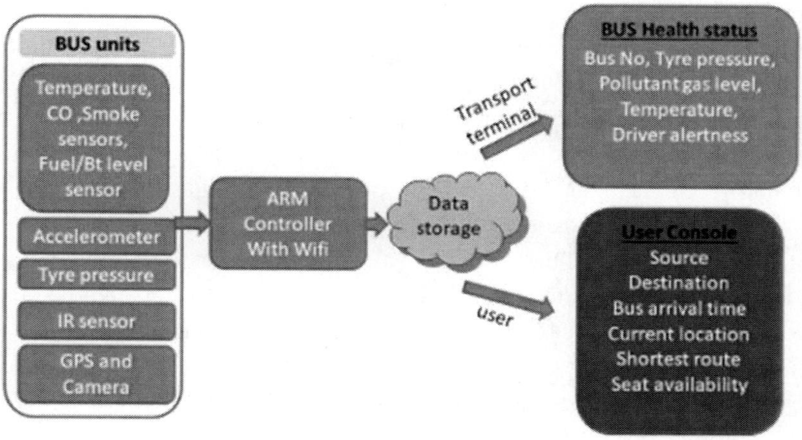

Figure 2. Block diagram - Implementing IoT in transportation.

Figure 2 displays the block diagram of the recommended transportation networks. Because of their high reliability and low propensity for delays, passengers can schedule their trips without accounting for delays. This implies that commuters do not need to leave their homes or places of employment early and wait at the bus stop or train terminal. Furthermore, commuters can see whether buses are full or vacant. They can also learn how many people are waiting for the transport. They will be able to determine the optimal time to start their commute from home or the workplace. As a result, commuters can depart from their homes to catch the next empty bus by knowing the precise Estimated Time of Arrival. Having immediate detail on the current state of buses in addition to their regular schedules enables commuters to plan for their travel flexibly. Using IoT-enabled GPS monitoring of buses provides commuters with up-to-the-minute status updates. Passengers can receive this information via specialized smartphone apps.

The use of IoT in transportation systems to continuously track the various health parameters of vehicles and transportation infrastructure aids in the early detection of potential issues. Engine temperatures, tyre pressure, pollutant petrol levels, fuel and battery levels, and other related health variables may be among the parameters. IoT sensors can be used to continuously check these parameters for anomalies and alert the appropriate officials to start maintenance procedures. Greater passenger safety and highly effective services with few interruptions will result from this. To address the dissatisfaction with most public transportation systems caused by frequent infrastructure or vehicle breakdowns, modern maintenance techniques like preventive and predictive maintenance must be adopted. This ends the brief service interruption that left the travellers stranded. Thus, it is possible to prevent the inconvenience for the passengers and distraction of the complete transportation system by postponing more people's commutes. This has a cumulative impact on how well a city or area that depends on the transportation system operates overall.

Transportation authorities can evaluate the processes taking place at predefined locations using infrared sensors at bus stops. They can determine how many passengers can be picked up at each location. They can use this information to permit the right number of vehicles to travel on various routes. They can also determine the quickest routes by using the traffic data collected by the sensors placed along the roads, at signals, and on streetlights. Bus drivers can receive continuous updates on the fastest path between two locations, which they can use to get their passengers to their destinations as soon as possible.

Transport regulators could be able to maintain continuous sight into the public transportation system because of sensors and cameras on highways, streetlights, bus stops, and other elements of travel routes. Transportation authorities may currently offer improved services to their citizens without compromising operational effectiveness due to this and the data collected from passengers' devices.

To automatically warn emergency personnel and divert traffic by notifying approaching vehicles and updating surrounding digital roadside signs, cameras can be used to monitor traffic congestion or accidents.

In the long run, new transit routes and bus services can be planned to utilise the data obtained by this system. Making decisions involving infrastructure is made easier with the use of this information, which is useful for city planning and management. People will prefer using public transportation more as it gets more dependable, not simply for financial reasons but also for reasons of convenience in general. This will decrease the number of private vehicles on the road and, as a result, increase the speed and reliability of public transportation. A city that relies primarily on public transportation also significantly reduces its carbon footprint, which has positive impacts on the environment. Hence, the application of IoT in transportation can lead to the creation of really smart and sustainable cities.

Identifying the Optimum Insurance Premium

Estimation of the optimal insurance premiums to be paid for the vehicle can be performed from the data obtained from the sensors placed on the bus. The following parameters and procedures can be applied for identifying the Optimum Insurance Premium using IoT.

The first step in identifying the optimal insurance premium using IoT is to deploy devices that can collect data on different risk factors. These devices could include the sensors mentioned above in addition to sensors to monitor traffic, weather, and other conditions that could affect the likelihood of accidents.

Once the IoT devices are deployed, data is collected on the different risk factors. Using machine learning algorithms, data are analyzed, and patterns are identified that can help predict the likelihood of accidents or other incidents.

Based on the analysis of the data, the level of coverage needed to adequately protect against different risks needs to be determined. This may involve identifying specific policies that can be tailored to individual vehicles based on their unique risk profiles. Once the coverage needs have been assessed, the appropriate premiums to charge can be determined using machine learning regression models. This involves using predictive models that take into account the different risk factors and the potential cost of claims. We may also need to consider the need to remain competitive in the marketplace while offering appropriate coverage at a fair price.

This mode of calculation of premium amount thus helps the vehicle owners and government to pay the optimal amount without the need to pay a fixed amount every year.

IoT is not the only solution to the problems with today's transportation systems, even though it has the potential to be revolutionary in terms of its strengths and applications. To enhance the public transport system, other established and emerging areas, such as machine learning, must also be considered. IoT devices can be deployed in the city's infrastructure as well as on transportation modes like buses and trains.

Performance Metrics

Smart transportation for bus travelers can be evaluated using various performance metrics, including (Tao Liu, Oded Cats, Konstantinos Gkiotsalitis (2021), Laura Eboli, Gabriella Mazzulla, (2012), and B Aniley, R Negi, (2015)):

- On-time performance: This metric measures the percentage of buses that arrive at their destination within the scheduled time. A high on-time performance indicates that the transportation system is reliable and efficient.
- Average journey time: This metric measures the average time taken for a bus to complete a journey. A shorter average journey time indicates that the transportation system is faster and more efficient.
- *Passenger satisfaction:* This metric measures the level of satisfaction among bus travellers. It may be measured through surveys or feedback mechanisms. A high level of passenger satisfaction indicates that the transportation system is meeting the needs of its users.
- *Capacity utilisation:* This metric measures the percentage of available seats that are occupied during peak hours. A high capacity utilisation indicates that the transportation system is efficiently using its resources to serve as many passengers as possible.
- *Safety and security:* This metric measures the number of accidents or incidents that occur within the transportation system. A low number of accidents or incidents indicates that the transportation system is safe and secure for passengers.
- *Environmental impact:* This metric measures the number of emissions generated by the transportation system. A lower environmental impact indicates that the transportation system is environmentally friendly.
- *Operational costs:* This metric measures the total cost of running the transportation system, including fuel, maintenance, and labour costs.

A lower operational cost indicates that the transportation system is efficient and cost-effective.

- *Service availability:* The span of transit service refers to the hours in a day when it is available, which can differ based on the day of the week, route, or stop. To assess this, an indicator can be computed as the average of the daily hours during different periods of the year or for various routes and stops. The duration of service within a day affects the ease of utilizing the transit system and may limit the types of journeys passengers can take by transit.
- *Cleanliness:* Transit agencies typically clean the interiors of their buses daily, even if it's just to remove larger items like bottles and newspapers. In addition, they could perform more comprehensive cleanings of the entire bus periodically, which would include the interior. The frequency of these detailed cleanings can vary depending on the agency, occurring monthly, quarterly, or annually, and their extent is determined by the amount of daily cleaning done as part of regular service. Lastly, transit agencies could take measures to prevent graffiti and vandalism, such as defining procedures for graffiti cleaning and removal. One way to assess the cleanliness of buses is to calculate indicators such as the frequency of interior cleaning and exterior washing.

Table 1. Comparison of smart transportation with existing systems

Performance Metrics	Existing System/ Methods	Smart Transportation
On-time performance	70%	85%
Average journey time	45 mins	35 mins
Passenger satisfaction	75%	85%
Capacity utilization	80%	90%
Safety and Security	5 accidents per month	2 accidents per month

By evaluating these performance metrics, transportation operators and stakeholders can identify areas for improvement and implement strategies to enhance the performance of smart transportation systems and it is shown in Table 1.

To evaluate the performance metrics for optimal insurance premiums using the Internet of Things, some of the key metrics that can be considered are (Vaida Vabuolytė Rasa Ušpalytė-Vitkūnienė (2018) and Eva Celcer (2009)).

Efficiency: This metric measures the extent to which the system is optimally utilizing its resources. In smart transportation, it could be measured by the number of passengers transported per hour or the distance travelled by the bus fleet in a day. For IoT-based insurance, it could be measured by the percentage of claims processed and paid out within a certain time frame.

Reliability: This metric measures the consistency and predictability of the system's performance. In smart transportation, it could be measured by the frequency of bus breakdowns or delays in arrival times. In IoT-based insurance, it could be measured by the speed of claims processing or the accuracy of the data collected by IoT devices.

Customer satisfaction: This metric measures the level of satisfaction of customers with the system or service provided. In smart transportation, it could be measured by feedback received from passengers, while in IoT-based insurance, it could be measured by the number of customer complaints or the percentage of policyholders who renew their policy.

Safety: This metric measures the degree of protection provided by the system. In smart transportation, it could be measured by the number of accidents, while in IoT-based insurance, it could be measured by the frequency and severity of claims related to accidents.

Cost-effectiveness: This metric measures the extent to which the system or service provides value for money. In smart transportation, it could be measured by the cost per passenger or kilometers, while in IoT-based insurance, it could be measured by the premiums charged compared to the coverage offered.

Table 2. Comparison of performance metrics for the optimal insurance premium with existing systems

Performance Metrics	Existing System/Methods	Smart Transportation and IoT-based Insurance
Efficiency	Manual ticketing and payment systems	Automated ticketing and payment systems using IoT
	Limited use of real-time data for bus routing and scheduling	Real-time monitoring of bus routes and passenger demand to optimise bus scheduling and routing

Performance Metrics	Existing System/Methods	Smart Transportation and IoT-based Insurance
	Limited use of data analytics to optimise operational efficiency	Data analytics used to optimise bus fleet maintenance and management
Reliability	Inconsistent arrival times and limited communication with passengers	Real-time updates on bus arrival times and service disruptions are provided to passengers through mobile apps and other channels
	Limited maintenance and monitoring of bus fleets	Continuous monitoring and predictive maintenance of bus fleets through IoT devices
Customer Satisfaction	Limited communication and interaction with passengers	Improved communication with passengers through real-time updates and personalised recommendations
	Limited access to information on bus routes, schedules, and amenities	Enhanced access to information on bus routes, schedules, and amenities through mobile apps and other channels
Safety	Limited safety features on buses	Advanced safety features such as collision avoidance and driver monitoring systems
	Limited use of data analytics to predict and prevent accidents	Predictive analysis and prevention of accidents using real-time data from IoT devices
Cost-effectiveness	High operational costs due to inefficient resource utilisation	Lower operational costs due to optimising resource utilisation and targeted marketing
	Limited use of data analytics for cost optimization	Use of data analytics to identify areas for cost optimization and revenue generation
	Conventional insurance policies with fixed premiums	Usage-based insurance policies with premiums based on actual risk and usage patterns

The above comparison Table 2 shows the performance metrics of the existing system and the proposed system for insurance offerings. This comparison helps to evaluate the effectiveness of smart transportation and IoT-based insurance solutions and identify areas where further improvements can be made. By evaluating these performance metrics, stakeholders can identify areas for improvement and implement strategies to enhance the performance of the smart transportation system for bus travellers and IoT-based insurance offerings.

Social Benefits

Smart cities embrace smart transportation enabled by infrastructures. With India marching towards a developed nation, intelligent transportation forms an integral part of reaching its boundary. Smart transportation enriches users with prior information about traffic and reduces travel inconveniences thus ensuring safety and quick mobility.

A quality public transportation system ensures closed-loop monitoring of public transport by the administrators, smart traffic light control for enabling quick movement of emergency vehicles, recognizing the vehicle location, identifying the density of passengers in the vehicle, and smart insurance. Thus, the proposed work has plans to achieve smart transportation by incorporating IoT devices for data acquisition making surface transportation more reliable and efficient.

Smart Transportation for Bus Travellers using IoT (Internet of Things) has numerous social benefits (S Porru, F Misso et al., 2020) and (L Raju, S Rani et al., 2017), including

- Improved accessibility: By providing real-time information about bus schedules and routes, Smart Transportation using IoT technology can improve accessibility to public transportation, particularly for people with disabilities or those living in remote areas.
- Reduced traffic congestion: Smart Transportation can contribute to minimizing traffic congestion and enhancing traffic movement by optimizing bus schedules and routes. This can result in a reduction of air pollution and an improvement in the environment.
- Enhanced safety: By monitoring bus conditions in real-time, Smart Transportation can help reduce accidents and improve overall safety for both passengers and other road users.
- Increased social inclusion: By providing accessible and reliable transportation, Smart Transportation can help increase social

inclusion, allowing people to access employment, education, and social activities that they may otherwise not be able to reach.

- Improved economic development: By improving transportation access, Smart Transportation can help stimulate economic development by connecting people to job opportunities, markets, and services.
- Reduced carbon footprint: By optimizing routes and reducing idling time, Smart Transportation can reduce fuel consumption and emissions, leading to a smaller carbon footprint and contributing to a healthier environment

Overall, the social benefits of Smart Transportation for Bus Travellers using IoT can contribute to a more sustainable and equitable society, improving quality of life and supporting economic growth. To conclude the socio-benefits of smart transportation, it keeps passengers connected, information enriched, and safe and ensures hassle-free journeys.

Future Scope

Looking into the future, vehicles could potentially come equipped with intelligent processors capable of handling multiple tasks concurrently, such as enhancing navigation, communication, and entertainment for passengers. This would allow for safer and more efficient travel while also making it possible to indulge in activities like downloading music, movies, and TV shows from the cloud.

Passengers can mark destination locations after receiving full routing information via rich visual maps and satellite imagery. When the vehicle's fuel level and battery power begin to deplete, it automatically searches for fuel or recharging stations, providing directions to the lowest rates.

Sensors inside as well as outside the car will improve safety. Since an interior camera recognises the driver and passengers, it is possible to customize seating navigation and entertainment preferences for ease. A lane exit can be detected by outside sensors. Detect and respond to possible hazards, such as failing to slow down for an approaching stop sign and failing to identify objects in front.

Smart Transportation aids in preventing collisions by enabling features such as automatic braking, stopping at a safe distance behind other vehicles, and adjusting cruise control to maintain a safe following distance.

Connected intelligence has the potential to stretch beyond vehicles. Intelligent travel can be multimodal, enhancing all modes of transportation, including car, train, and air travel.

Conclusion

This paper has discussed the need for smart and intelligent transportation for the next generation which would considerably save the time of the public. The public transportation discussed in the paper is passenger bus transportation that connects people at different locations. The paper includes two different sections. The first section deals with the Data acquisition phase, where the location of the bus and the seat availability is studied and the next section deals with the Data analytics phase, which provides appropriate information to the user through a mobile application. This enables the public to plan their travel. Also, uncertainties in travel are inevitable. The proposed work also concentrates on identifying an optimum insurance premium using machine learning regression models for the drivers in case some untoward incidents take place beyond their control. An IoT-enabled system with sensors and monitoring devices are deployed to track real-time data and thus the driver's vehicle insurance is decided in the smart transportation system itself. Finally, the existing transportation is compared with smart transportation in terms of performance metrics like journey time, passenger satisfaction, safety and security. It is found that the continuous monitoring and predictive maintenance of bus fleets incorporated through IoT devices, and usage-based insurance policies with premiums based on actual risk and usage patterns, have enhanced public transportation.

References

Kadam A J, Virendra Patil, Kapish Kaith, Dhanashree Patil, Shambhavi Bendre, Smart Bus for Smart City using IOT Technology, *International Journal of Advanced Research in Computer and Communication Engineering*, Vol. 7, Issue 1, January 2018.

Murdan A. P., V Bucktowar, V Oree, M. P. Enoch, Low-cost bus seating information technology system. *IET Intelligent Transport Systems*, 14(10), 1303-1310, 2020.

Aniley B, R Negi, "Service quality attributes affecting passengers' satisfaction with HIGER city bus" African *Journals Online (AJOL),* 2015.

Bhupendra Singh, Ankit Gupta, Recent trends in intelligent transportation systems: a review, Journal of Transport Literature, April 2015.

Celcer Eva, "Public transport fixed route quality indicators" *Digital Repository UL FGG*, 2009.

Jalaney J., Dr R. S. Ganesh, Review on IoT Based Architecture for Smart Public Transport System. *International Journal of Applied Engineering Research*, Volume 14, Number 2, pp. 466-471, 2019.

Raju L, S Rani, G Ram, B A Dannison, Smart Information System for Public Transportation Using IoT. *IJRTER,* 3, 222–230, 2017.

Eboli Laura, Gabriella Mazzulla, *Performance indicators for an objective measure of public transport service quality*, 2012.

Jana M, S Biswas, Intelligent and smart enabling technologies in advanced applications: Recent trends, *Recent Trends in Computational Intelligence Enabled Research*, 355-365, 2021.

Nesheli M, A Ceder, F Ghavamirad, Thacker, S, Environmental impacts of public transport systems using real-time control method. Transportation Research Part D: Transport and Environment, 51, 216-226, 2017.

Syafrudin M, G Alfian, N. L. Fitriyani, J Rhee, Performance Analysis of IoT-Based Sensor, *Big Data Processing, and Machine Learning Model for Real-Time Monitoring System in Automotive Manufacturing*, 18, 2946. Sensors 2018.

Miraal Kamal, Manal Atif, Hafsa Mujahid, Tamer Shanableh, A. R. Al-Ali, Ahmad Al Nabulsi, IoT Based Smart City Bus Stops, *Future Internet* 2019, 11, 227; doi:10.3390/fi11110227.

Parag Gawade, Prof. Meenakshi A, IOT Based Smart Public Transport System, *International Journal of Scientific and Research Publications*, Volume 7, 390, Issue 7, July 2017.

Porru S, F. E. Misso, F. E., Pani, Repetto, C. Smart mobility and public transport: Opportunities and challenges in rural and urban areas. *Journal of Traffic and Transportation Engineering (English Edition)*, 7(1), 88-97, 2020.

Pal S., and V. Singh, GIS Based Transit Information System for Metropolitan Cities in India, *Proceedings of Geospatial World Forum*, January 2011.

Liu Tao, Oded Cats, Konstantinos Gkiotsalitis, "A review of public transport transfer coordination at the tactical planning phase" *Transport Research Part C* 133 pg. No. 103-450, 2021.

Umberto Fugiglando, Emanuele Massaro, Paolo Santi, Sebastiano Milardo, Kacem Abida, Rainer Stahlmann, Florian Netter, Carlo Ratti., Driving Behavior Analysis through CAN Bus Data in an Uncontrolled Environment, IEEE Transactions on intelligent transportation systems, Vol. 20, February 2019. Usage-Based Insurance and Telematics. [Online]. Available: http://www.naic.org/cipr_topics/topic_usage_based_insurance.html, January 2017.

Vaida Vabuolytė Rasa Ušpalytė-Vitkūnienė, "Indicators of service quality of public transport" VGTU Journals (Vilnius Gediminas Technical University - Vilnius Tech), 2018

Bhilawade Vidya, L. K. Ragha, Intelligent Traffic Control System, *International Journal of Scientific and Research Publication*, vol. 8, no. 2, 571-574, Feb. 2018.

Qin X, J Ke, X. Wang, Y Tang, H Yang, Demand management for smart transportation: A review. *Multimodal Transportation*, 1(4), 100038, 2022.

Kuo Y, J.M. Leung, Y Yan, Public transport for smart cities: Recent innovations and future challenges. *European Journal of Operational Research*, 306(3), 1001-1026, 2023.

Chapter 16

Merging Blockchain and IoT Techniques to Provide an Effective Supply Chain Network for Logistic Business

S. Suhasini*
and D. Sterlin Rani[1]
Department of Computer Science and Engineering, R.M.D. Engineering College, India

Abstract

Delivery delays are one of the main problems in the supply chain and logistics business since a supply chain network comprises so many different parties. To address this problem, Blockchain and IoT came into this world. Blockchain is a term used to describe a distributed, decentralised, encrypted computer filing system intended to make it easier to create real-time, tamper-proof documents. IoT is a term used to describe how computing devices that are integrated into ordinary objects are connected to one another over the Internet so they may transmit and receive data. Combining these two technologies results in a prototype for a dependable, secure, and long - lasting method of storing data processed by intelligent, or more precisely "smart" devices. Blockchain technology has a promising future ahead of it, according to experts, as its application seems to be expanding. While IoT- enabled devices will let businesses monitor the progress of shipments at every stage, Blockchain will make the entire process transparent. IoT sensors (such as motion sensors, GPS, temperature sensors, etc.) can provide information on the status of shipments. For more transparency, this data is then recorded on the

* Corresponding Author's Email: ssi.cse@rmd.ac.in.
* Corresponding Author's Email: ssi.cse@rmd.ac.in.

In: Blockchain Enabled Secure Big Data Computing …
Editors: L. Ashok Kumar, D. Karthika Renuka, V. Chandrasekar et al.
ISBN: 979-8-89113-243-6

Blockchain network. After this is done, all supply chain participants named in the Smart Contracts have real-time access to the data. Blockchain and IoT working together can increase the traceability and dependability of the supply chain network.

Keywords: supply chain network, iot, blockchain, logistic business

Introduction

Supply chains rely largely on a range of external counterparties, become more dynamic, and incorporate a number of stakeholders. In order to untangle all data, documents, and communication transactions that exist within the supply chain network, blockchain has emerged as a strong contender. Massive amounts of data are transferred between different supply chain stages in any supply chain. Supply chain may use IoT and blockchain to handle massive amounts of data and secure their security. Nychas, G.-J.E. et al., (2016). The supply chain's security, output, efficiency, and profitability will all be improved as a result. The integration of interconnected devices is known as the Internet of Things (IoT), which enables data collection and informed decision-making. Shih, D.-H. et al., (2019) Because there are no built-in security or privacy safeguards, IoT is susceptible to security and privacy issues. IoT security requirements can be greatly reduced thanks to blockchain thanks to its Safety by Design. IoT's architectural shortcomings can be partially addressed by the immutability, openness, traceability, data encryption, and organisational flexibility of the blockchain. Dabbene, F. et al., (2011) With the help of IoT, it would be possible to monitor both the technical state of the machinery and the workers' safety and health. Hackius, N. et al., (2019) Additionally, it can increase in-transit visibility and offer more communication instructions to all parties in the supply chain. Wang, Y. et al., (2018) Businesses can benefit greatly from the integration of this cutting-edge technology, also known as the Blockchain of Things (BIoT), which includes both blockchain and the IoT solutions. Marsden, T. et al., (1999) The report condenses information on the many security risks that the supply chain faces as a result of the tremendous complexity and vast amounts of data. Zhao, J. et al., (2016) Information is frequently dispersed, challenging to retrieve, structured erratically, or challenging to evaluate. He, X. et al., (2018) The thought of promoting innovation in any industry may be intimidating. Park, K. et al., (2022) The good news is that even the most fundamental uses of blockchain technology have the potential to significantly raise profitability, efficiency, and production. This research illustrates Verhoeven, P. et al., (2018) how different security challenges can be managed by integrating blockchain technology with IoT into the established supply chain network.

Related Work

Bosona, T et al., (2013) investigated the idea of food traceability in terms of gathering, storing, sending, and maintaining information about food products throughout the various levels of the FSC using blockchain technology. This gives FSCs even more control over food quantity, quality, and safety. Lin, Q. et al., (2019) created a method for tracking food security and safety based on blockchain technology and the GS1 worldwide interoperability standard (EPCIS). It is specifically employed in the online gathering, management, and exchange of product information. Srivastava, H.S. et al., (2015) Customers can use this system to track information about their food through the consumer traceability client application. Alonso et al., (2020) launched a cutting-edge platform that combines artificial intelligence with the Internet of Things, employing edge computing and blockchain technologies to manage agriculture. Mastos, T. et al., (2021) With utilising blockchain technology, fragmentation and preserving food traceability from beginning to finish are no longer issues. Mastos, T. et al., (2022) It develops a standard platform for data collection. With all the blocks relating to a product, users can quickly trace food both forward and backward using blockchain technology. Tian, F. et al., (2016) Recently, FSCs are centralised and depend on centralised authorities to manage the flow of data and information pertaining to food. Numerous studies have discussed the FSC's inability to predict the food's freshness and its lack of ongoing observation. Gay, P. et al., (2008) Similar issues plague the traditional food monitoring system, including data discrepancies, interoperability with limited resources, fragmentation, and a lack of transparency. FSC practitioners have embraced several blockchain-related technologies in the food business to help allay the aforementioned worries. Bazoni, C.H. et al., (2017) A quick method for assessing physicochemical changes in foods that have been preserved, such soybeans, is near-infrared spectroscopy. David, et al., (2022) demonstrates the usefulness of the potential of blockchain technology in FSCM. Organizations can use this to maintain proof of work, requirements, traceability, innovation to reduce the need for middlemen, and other things. This chapter goes into great detail about the present state of blockchain technology and its applications in supply chain commerce. Ioannou, I. et al., (2022) The areas of the blockchain are highlighted that could decide the value of trading and supply chain finance. With the use of sensor devices and IoT approaches, the predicted acceleration in the world's population and the corresponding need for food from farms to markets have now also been detected. Jeppsson, A. et al., (2017) Low-power wide-area technologies are growing in popularity as a result of the IoT's quick development and the devices' power requirements. The emerging technologies known as material conscious information networks (MCIN) establish the architecture of smart agriculture and address issues like management and commerce that have an impact on the supply chain system. The difficulty of combining different technologies also calls for clever communication between these gadgets.

Challenges over Supply Chain Management

The complexity of global supply chains is being exacerbated by some of these macrotrends, like globalisation and connectivity. Aiming to lower operating costs, others are ingrained in the never-ending pursuit of efficiency. Figure 16.1 shows the complete flow of SCM.

Margin Deterioration and Abrupt Shifts in Demand

Businesses must respond swiftly to sudden demand changes in surroundings that are changing quickly and more frequently. Globalization brought more intense competition, which impacted profits.

Impact of the Extended Value Chain Ripples

Large supply networks put businesses at risk from significant fluctuation. The supply chain risk is considerably increased by a number of layers of stakeholders, including suppliers, distributors, and customers.

Poor Supply Chain Risk Management

Effective supply chain risk management programmes are essential for effectively tracking and forecasting risk so that appropriate responses can be made.

Lack of Total Transparency

Having a clear understanding of their supply networks, both internally and outside, is a challenge for businesses. As a result, they are exposed to a variety of hazards, including fraud and breaking the code of conduct.

The Ageing of Technologies

Modern supply chain management technologies demand a large and ongoing investment. This also has built-in implementation risks, which provide a significant hurdle for most businesses.

Overview of Supply Chain in Logistics

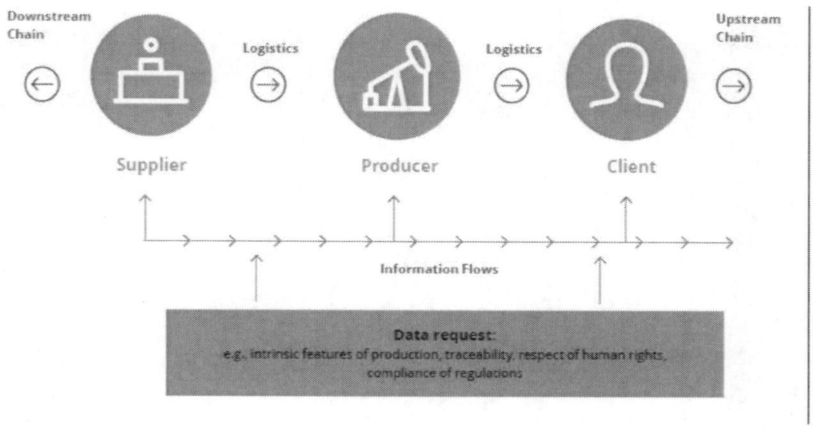

Figure 1. Stretch of Supply Chain between Upstream and Downstream Chain Layer.

Information flows require the effective data gathering and availability from the numerous stakeholders' information systems. As a result, the entire system depends on:

Faith: Every stakeholder must have faith in one another's willingness to share accurate information. The effectiveness of the lowest common denominator: This depends on systems' ability to accurately and effectively collect data, store it securely, and produce correct data for usage by other systems. Information flow along a supply chain is influenced by the weakest or "most wicked" connection.

Reconciliation: Because supply chains are not connected from beginning to finish and because there are few standards, there is no consistency in the information delivery. Each stakeholder must reconcile the data in order to verify accuracy and effectiveness, which lowers efficiency.

Restricted information flow manifestation: The ability of the data to accurately and effectively represent real-life items or events is constrained by the technologies deployed in supply chains to monitor and track the supply.

Sequential Structural Representation of Supply Chain works in Logistic Business:

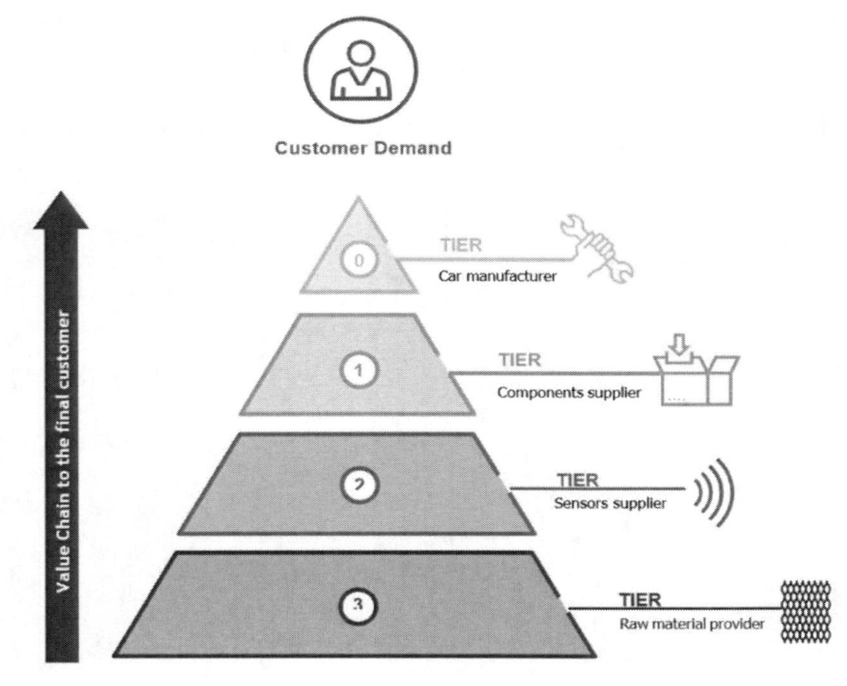

Figure 2. Structural Representation of Logistic Business in Supply Chain.

Figure 2 displays the Structural Representation of Logistic Business in Supply Chain which Oversee the whole supply chain Visibility is necessary for tracking supply chain activities Folkerts, H. et al., (1997) and processes, determining trouble spots, and even proactively recognising difficulties. This idea is essential for improving supply chain efficiency and ensuring traceability. Bazarsukh, R. et al., (2018) To remain competitive in the market, you must have the flexibility to quickly respond to problems and adjust your strategy without having a material influence on your operational costs. Instead of having to rely on one another, the stakeholders should have faith in the system's security to gather and deliver truthful and correct data Nizamuddin, N. et al., (2018). Policies, monitoring tools, and control mechanisms—all of which are of utmost importance—are needed for controlling systems.

Merging of Blockchain and IoT in Supply Chain Management

Supply chains suffer significant problems with information exchange and trust. Blockchain and IoT have promising prospects to address these issues thanks to their inherent qualities. To ensure traceability and lower inherent risks, it will be crucial to provide information continuity through the irrevocable and irreversible qualities of blockchain and effective information interchange among the many entities taking part in the global supply chain. Since they will eventually give people the access to information, they need to properly capitalise on the enormous volume of data generated along the supply chain, blockchains should make information quickly and openly available. The relationship between physical and information flows illustrates how data will be connected to the goods and materials at supply chain phases when matter is physically transformed. To reduce reputational risk, fraud detection and code of conduct violations will make sure that human rights and codes of conduct are upheld throughout the chain. To reduce business risk, effective fraud detection methods supported by suitable technologies will become increasingly crucial. The transparency and auditability of the blockchain will make this possible. This is an example of the life cycle of a vehicle along with some of the problems that may be encountered in relation to the four hurdles to further illustrate these risks and the places where they have an impact on the supply chain. Because each member has their own information system, it is difficult for the chain to share information, which prevents it from tracking information. It is difficult to properly track and link information to material without bias because of the intricate linkages among the stakeholders. Due to insecure sharing and the involvement of numerous stakeholders, all parties involved experience difficulty at some point in the process trying to obtain information properly. Recent fraud incidents, as was already said, have raised awareness of business risks throughout the whole global supply chain. One of the most popular concerns right now that needs to be considered is finding code of conduct violations. Figures 3 and 4 show the IoT Representation terms and structure.

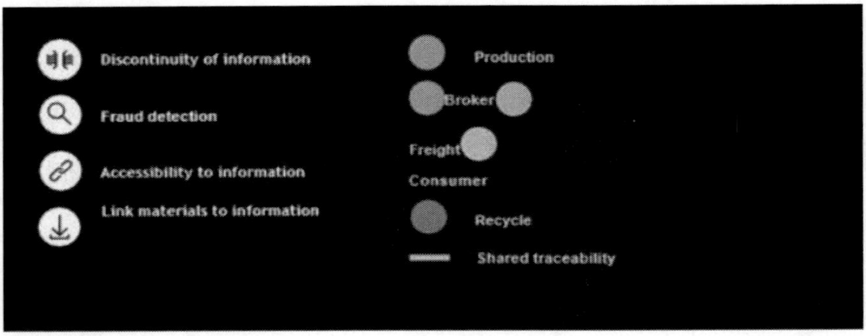

Figure 3. Representation of IoT Terms.

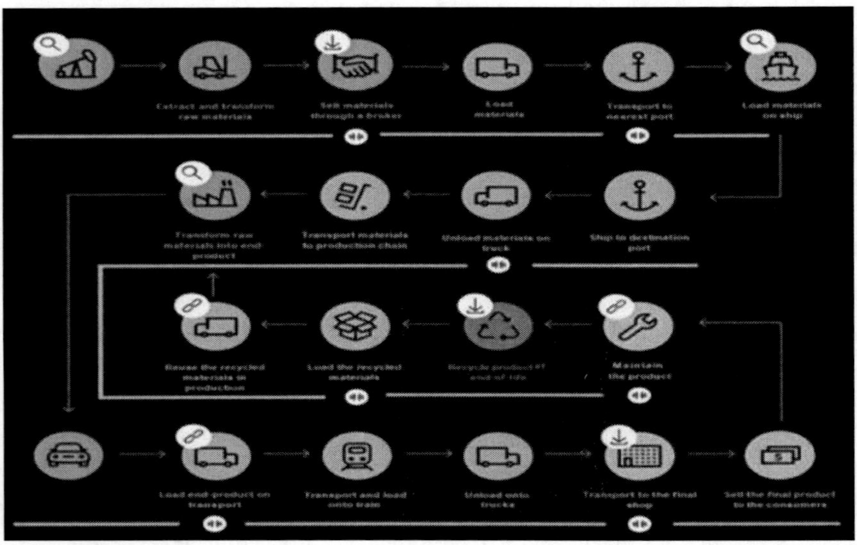

Figure 4. IoT Architectural Representation.

Stratergies of IoT and Blockchain Supply Chain Management

1. Recognize the risks and weaknesses in your supply chain.
2. Examine your supply chains as far upstream as you can to find any flaws or potential dangers.
3. Determine and categorise those flaws and dangers according to their type as well as their likelihood and effects. You will receive a heat map of your supply networks' weak points as a result.
4. Determine quickly whether blockchain and IoT can be of assistance for each of the problems.
5. Begin modestly. Choose a challenge that can be solved without a costly or complicated effort while yet demonstrating the added value. Apply this initially to a small area.
6. Attempt and fail until you are certain it is an effective and long-lasting fix.
7. Team up. Don't work alone; involve your consumer base and supplier network. Collaborate with young companies, technologists, and advisors that have a fresh perspective. Include your rivals who work with the same suppliers as you in this as well. You will split the costs and have your suppliers stick even more closely to your project.
8. Scale up gradually. As soon as people put the learning into practise, they begin growing the solution and increasing its application.

Conclusion

We have observed that the supply chain is currently dealing with significant difficulties and will continue to do so. The numerous food and auto tragedies that have shaken their respective industries are directly responsible for the recent rise in demand for transparency and traceability. Sources In that regard, we also made a point that future IoT and blockchain integration will be able to assist the various stakeholders in maintaining traceability along the chain and lowering risks. A single blockchain will be shared by all supply chain participants, guaranteeing that exchanged and altered goods are validated, lowering the risk of fraud, and making information access easier. By providing the blockchain with reliable data on the materials and goods, IoT, on the other hand, will serve as a bridge between the physical and digital worlds. IoT and blockchain technologies must develop in order to support these kinds of difficulties in the modern world. Nonetheless, players are actively helping to integrate these technologies into the realm of supply chains and offering fresh ideas for how to enhance and consistently secure information sharing. The numerous parties must, however, come to an agreement to invest in these technologies in order for this transformation to take place. To encourage the adoption of new technologies, the regulatory environment will also need to change. Ultimately, these technologies will transform how information is captured, shared, and accessed by the various participants in a supply chain on a safe, open, and transparent platform.

Review Comments

To remain competitive in the market, you must have the flexibility to quickly respond to problems and adjust your strategy without having a material influence on your operational costs. Nizamuddin, N. et al., (2018) Instead of having to rely on one another, the stakeholders should have faith in the system's security to gather and deliver truthful and correct data. (what does "Nizamuddin, N. et al., (2018)" belong to?)

Justification

For the above question, Nizamuddin, N. et al., (2018) is the reference of that point mentioned.

References

Alonso, R. S.; Sittón-Candanedo, I.; García, Ó.; Prieto, J.; Rodríguez-González, S. An intelligent Edge-IoT platform for monitoring livestock and crops in a dairy farming scenario. *Ad. Hoc. Netw*. 2020, 98, 102047.

Bazarsukh, R. Global Risks 2018: Fractures, Fears and Failures. Available online: https://reports.weforum.org/global-risks-2018 /global-risks-2018-fractures-fears-and-failures (accessed on 10 May 2022).Bazoni, C.H.; Ida, E.I.; Barbin, D.F.; Kurozawa, L.E. Near-infrared spectroscopy as a rapid method for evaluation physicochemical changes of stored soybeans. *J. Stored Prod. Res*. 2017, 73, 1–6. Bosona, T.; Gebresenbet, G. Food traceability as an integral part of logistics management in food and agricultural supply chain. Food Control 2013, 33, 32–48.

Dabbene, F.; Gay, P. Food traceability systems: Performance evaluation and optimization. *Comput. Electron. Agric.* 2011, 75, 139–146.

David, A.; Kumar, C. G.; Paul, P. V. Blockchain Technology in the Food Supply Chain: Empirical Analysis. *Int. J. Inf. Syst. Supply Chain. Manag.* (IJISSCM) 2022, 5, 1–12.

Folkerts, H.; Koehorst, H. Challenges in international food supply chains: Vertical co-ordination in the European agribusiness and food industries. *Supply Chain. Manag. Int. J.* 1997, 2, 11–14.

Gay, P.; Piccarolo, P.; Aimonino, D. R.; Tortia, C. Livestock identification and farm management by RFID systems. In *Proceedings of the International Conference on Agricultural Engineering, Hersonissos,* Greece, 23–25 June 2008; p. P-025.

Hackius, N.; Petersen, M. Blockchain in logistics and supply chain: Trick or Treat? In *Digitalization in Supply Chain Management and Logistics.* 2017. Available online:https://tore.tuhh.de/bitstream/11420/1447/1/petersen_hackius_blockchain_in_scm_and_logistics_hicl_2017 (accessed on 24 March 2019).

He, X.; Chen, X.; Li, K. A decentralized and non-reversible traceability system for storing commodity data. KSII Trans. *Internet Inf. Syst.* (TIIS) 2018, 13, 619–634.

Ioannou, I.; Demirel, G. Blockchain and supply chain finance: A critical literature review at the intersection of operations, finance and law. *J. Bank. Financ. Technol.* 2022, 6, 83–107.

Jeppsson, A.; Olsson, O. Blockchains as a Solution for Traceability and Transparency. *Master Thesis*, Lund University, Lund, Sweden, 2017.

Lee, Y.; Rathore, S.; Park, J. H.; Park, J. H. A blockchain-based smart home gateway architecture for preventing data forgery. *Hum.-Cent. Comput. Inf. Sci.* 2020, 10, 9.

Lin, Q.; Wang, H.; Pei, X.; Wang, J. Food Safety Traceability System Based on Blockchain and EPCIS. *IEEE Access* 2019, 7, 20698–20707.

Marsden, T.; Murdoch, J.; Morgan, K. Sustainable agriculture, food supply chains and regional development: Editorial introduction. *Int. Plan. Stud.* 1999, 4, 295–301.

Mastos, T.; Gotzamani, K. Sustainable Supply Chain Management in the Food Industry: A Conceptual Model from a Literature Review and a Case Study. *Foods* 2022, 11, 2295.

Mastos, T.D.; Nizamis, A.; Terzi, S.; Gkortzis, D.; Papadopoulos, A.; Tsagkalidis, N.; Ioannidis, D.; Votis, K.; Tzovaras, D. Introducing an application of an industry 4.0 solution for circular supply chain management. *J. Clean. Prod.* 2021, 300, 126886.

Nizamuddin, N.; Hasan, H.; Salah, K. IPFS- Blockchain-based Authenticity of Online Publications. In *International Conference on Blockchain* (ICBC 2018); Springer LNCS: Seattle,WA, USA, 2018.

Nychas, G.-J.E.; Panagou, E.Z.; Mohareb, F. Novel approaches for food safety management and communication. Curr. Opin. Food Sci. 2016, 12, 13–20.

Park, K. Cryptocurrencies, *Blockchain Is about to Revolutionize the Shipping Industry*, Bloomberg. 18 April 2018. Available online: https://www.bloomberg.com/news/articles/2018-04-18/drowning-in-a-sea-of-paper-world (accessed on 22 September 2022).

Queiroz, M. M.; Telles, R.; Bonilla, S.H. Blockchain and supply chain management integration a systematic review of the literature. Supply Chain Manag. *Int. J.* 2019, 25, 241–254.

Shih, D.-H.; Lu, K.-C.; Shih, Y.-T.; Shih, P.-Y. A simulated organic vegetable production and marketing environment by using Ethereum. *Electronics* 2019, 8, 1341.

Srivastava, H. S.; Wood, L. C. Cloud Computing to Improve Agri-Supply Chains in Developing Countries. In *Encyclopedia of Information Science and Technology*, 3rd ed.; IGI Global: Hershey, PA, USA, 2015; pp. 1059–1069; ISBN 978-1-4666-5888-2.

Tian, F. An Agri-food supply chain traceability system for China based on FRID & Blockchain technology. In *Proceedings of the 13th International Conference on Service Systems and Service Management* (ICSSSM 2016), Kunming, China, 24–26 June 2016.

Verhoeven, P.; Sinn, F.; Herden, T. T. Examples from Blockchain Implementations in Logistics and supply chain management exploring the mindful use of a new technology. *Logistics* 2018, 2, 20.

Wang, Y.; Han, J. H.; Davies, P. Understanding blockchain technology for future supply chains: A systematic literature review and research agenda. *Supply Chain Manag. Int. J.* 2018, 24, 62–84.

Zhao, J.; Fan, S.; Yan, J. Overview of business innovations and research opportunities in blockchain and introduction to the special issue. *Financ. Innov.* 2016, 2, 28.

Chapter 17

Location Tracking and Health Monitoring Systems During Disasters

K. Ramalakshmi[1]
L. Krishnakumari[2]
R. Rajalakshmi[2]
and D. Mahadevi[3]

[1]Department of ECE, P.S.R Engineering College, Sivakasi, Tamilnadu, India
[2]Department of ECE, Ramco Institute of Technology, Rajapalayam, Tamilnadu, India
[3]Department of Mechanical Engineering, P.S.R Engineering College, Sivakasi, Tamilnadu, India

Abstract

Natural disasters frequently occur in many different places in our daily lives. Because of the constant changes in geography, temperature, and nations that are populated close to the water, warnings and evacuation advice may save the lives of many people. Finding the safest site in that area will be a major challenge for travellers if disasters strike. Wearable technology, wearable fashion technology, tech togs, or fashion natural philosophy are terms used to describe intelligent electronic devices that can be integrated into clothing otherwise worn on the body as instill or garnishes. The Internet of Things in healthcare is demonstrated via wearable devices like Activity Hunter. Due to the fact that "things" like natural philosophy, software, sensors, and property are effectors that modify items to communicate information (including information quality) through the internet with a manufacturer, operator, and/or other connected devices without requiring human interaction. A sort of electronic gadget known as wearable technology can be attached to clothing, implanted inside of a user's body, or inked on the skin. The devices are hands-free gadgets with practical applications, powered by

In: Blockchain Enabled Secure Big Data Computing …
Editors: L. Ashok Kumar, D. Karthika Renuka, V. Chandrasekar et al.
ISBN: 979-8-89113-243-6

microprocessors and enhanced with the ability to send and receive data over the internet.

Keywords: wearable devices, activity hunter, accessories, microprocessors

Introduction

One of the global problems that people face is health. Taking care of a patient's health is a dangerous business. Telemedicine care is very important. A good healthcare system can help reduce the workload of doctors. The main goal is to create a reliable patient care system so that doctors can care for patients in the hospital or in daily life. Patients, on the other hand, frequently visit healthcare providers for necessary diagnosis and advice. Many natural disasters occur in various locations throughout our human life. If a disaster occurs in a location or in health, people face a major challenge in locating the safest place on that location.

This device service provides reliable health monitoring, position tracking at regular intervals to track the user's body conditions like temperature, pressure, heartbeat, and special objects, detects the user's relationship to his or her surroundings and takes protective measures as necessary, and notifies the base station when the signal is very low. These features are used to locate victims and also rescue people (El-Nour A. et al., 2019).

There are two major issues with this common procedure. Patients must at first spend some time in a clinic setting. Also, the patients should have access to the medical care specialists that are on site to assist them. A framework for remote patient monitoring has been provided to address these problems. The remote patient care center should store, send and display the patient's physical information anytime and anywhere. Typically, these frameworks have a worrisome framework that can alert patients and healthcare professionals to any unusual physiological data that is out of the regular reach. Due to the enhanced patient experience and cost effectiveness, mobile device based medical services (Abdullah, A. et al., 2015) framework have gained widespread recognition. The system combines an app, mobile phone and other apparatus to track the patient's physiological information and send it in the direction of doctors. Here, various modules are combined into a single Android application that is capable of taking a patient's temperature, heart rate, and blood pressure readings. The primary goal of the healthcare system is to help with the remote patient population's health monitoring. The method can thereby minimise the distance that patients must travel from the hospital to their homes or places of employment. IoT platform can be utilized in this framework (Kumar Yelamarthi et al., (2017). The IoT is a gifted technology for achieving the previously mentioned healthcare examination, furthermore it has the potential to improve medical service systems even further.

The traditional healthcare system is unable to provide patients with freedom and comfort due to a lack of monitoring setup. The advancement of wireless communication aids in the implementation of dynamic monitoring setups in today's and future healthcare systems, as well as the reduction of patient discomfort associated with extended stays in hospitals, at home, and so on. It also helps to achieve pervasive and mobile healthcare by lowering maintenance costs. Healthcare practitioners can employ ubiquitous and mobile computers, wearable biosensors, wireless communication technologies, and continuous monitoring and analysis of human vital signs to perform early diagnosis and treatment as well as create and validate new medicines. Wearable and non-wearable sensor devices can be used to track and monitor humans. Wearable technology is transforming telemedicine and remote healthcare.

Related Works

Tae-Wook Jo et al., (2008) displayed a bundle bridge concerning Bluetooth gadgets and in this way the UPnP network by implies of a virtual agent, with the objective of making a wearable environment. The bridge empowers a non-IP Bluetooth gadget to operate as a UPnP gadget based on informatics. When a Bluetooth gadget is identified, the bridge lists a virtual UPnP gadget and employs the extricated data to empower UPnP revelation. Test output comes about appears that Bluetooth gadgets will connect the UPnP network and give benefit with negligible overhead.

Cornelia Stez et al., (2010) created a private health system for stress detection. We intend to examine the discriminative control of electro dermal activity (EDA) in distinctive characteristic pressure from psychological feature weight in a workplace setting. During a laboratory intervention, a group of thirty-three subjects were exposed to a delicate psychological feature load as well as two stress elements that are significant in the workplace: mental stress from having to solve maths problems quickly and psycho-social stress from social- evaluative warning. Throughout the tests, a wearable gadget was employed to track the EDA as a real-time measure of each subject’s pressure response. The information analysis demonstrated that information about someone's level of strain is contained in the allocation of the EDA max out tallness and, consequently, the fast max out rate. The capacity of six classifiers to differentiate between psychological feature load and stress was examined. The highest accuracy for distinguishing stress from psychological feature load was 82.8%. Using a wearable EDA device, this could allow keeping track of nerve-racking phases throughout an operating day. Samuel Benjamin Scho et al., (2015) developed a wearable tip perception device that can make and break contact with the finger pad while also rendering shear and traditional skin deformations. The delta mechanism with new tuning spring and clamping mechanism repositioning technology facilitates the exploit of advanced motors and encoders for particular device manage: 10 Hz data is measured during testing and 0.255 metric linear units RMS detection error. In the first of the two experiments, participants pointed to a solid field in a fairly rigid virtual environment with a typical error of 7.6 degrees, alike to that set up in past tense learning by means of the past tense. In a subsequent experimentation, we assessed contributor's ability to interpret changes in friction. The maximum difference (JND) defined using the surface friction of our skin deforming device is 0 or 0.20, which corresponds to a constant friction of 0.5. Beyond the found abuse of proprioception feedback, this shows that users will comprehend variations in surface friction without proprioception forces that are grounded in the real world. These tests demonstrate that skin deformation in three dimensions (DoF) enables firmness and roughness unfairness even in the nonexistence of proprioceptive strength feedback. A wearable device that reads positive facial expressions by utilising physiological signals was developed by Anna Gruebler et al., (2014). We first analyzed 3D facial morphology and facial EMG of different regions of the face, which showed that we would find high EMG in the low facial mass region of the face, with effects related to the signals obtained from the electrodes. Ancient surface EMG of the image captures the location of the facial muscles of the face. To capture this signal, we use a multidimensional decision-making process to find the appropriate conductor position on either side of the face. Based on this analysis, we create and use reliable electronic devices. As the signals are recorded remotely, the device is ready to use free analysis and neural networks to examine them and obtain facial recognition results. Facial expressions identified by the wearable device will be recorded throughout the treatment, and facial expressions will be recorded over a long period of time to confirm and measure the user's mood. To encourage doctors Sudip Vhaduri et al., (2018) reported an automated product validation tool that uses a combination of three as well as coarse-grained minute-based biometrics: behavioral, physiological (heart rate) hybrid (calorie-burning and metabolic counterpart) activities. We prefer subjects that can show true accuracy around .93 (still) and .90 (still) with equal margin of error .05 Victim Binary SVM classifier is based on our analysis of a 17-month longitudinal health study of over 400 Fitbit users. Our results also show that integrated biological sciences are superior to other biological sciences, while behavioral biological sciences are not effective if not timely. Yoon Sang Kim et al., (2005) introduced SCURRY, a new data entry tool introduced by Samsung Advanced Institute of Technology. The device is supported by mechanophenotype sensors that allow the operator to select roles, conditions or activity level based on the need for spatial input with full force of the hand and finger click. It is similar to glove that preserve be wear on the human being hand. It has a core section, a controller, and two angle sensors (gyroscope), four ring modules (rings), and two sensor sensors on the back of the hand. Four finger Axis acceleration sensor (accelerometer). It is a combination of sub and ring modules, sensors, a transceiver or recipient for communiqué and a microcontroller, making the device squashed and lightweight. Two gyroscopes implanted in the base unit are responsible for determining the track of hand movement, hence the accelerometer is responsible for determining finger movement with finger touch. To have the appeal of finger clicks in SCURRY, the real finger tick identification process consists of three prepared elements such as feature extraction, quality click analysis and interference detection. Investigational outcomes and discussion are given. SCURRY will be utilized as a wearable mouse that allows three-finger operation spatially as left, middle, and right mouse buttons, and as a wearable keyboard that permits a person to set and pick features, events, or operate with gestures and finger taps work.

A sophisticated concern and attentive moveable tele medical supervision (AMON), a wearable medical observation and alert system for patients with compromised cardiac and respiratory systems, is described by Ya-Li Zheng et al., (2014). The system includes intelligent multi-parameter medical emergency detection, continuous collection and analysis of numerous important signs, and cellular affiliation to a centre. Continuous semi-permanent observation will be carried out without interfering with the patients' daily activities and without lowering the quality of those activities by grouping the entire system in an unassuming, wrist-worn attachment and using antagonistic low-power style methods. The delineated wrist-worn device, a communication link, and a comprehensive centre software system package have all been created, implemented, and tested by the Amen association. A medical study involving a group of 33 subjects has validated the system's effectiveness. The paper outlines the main concepts of the Amen system and provides information on each individual subsystem and solution, as well as the outcomes of the medical validation.

Shantanu A et al., (2019) describes a model of guidance called "finger vision." To help the visually impaired see and reach the objects around them. It has a small camera at the top and 4 equal parts of the tractor drawn in a circle attached to the index finger. Vibratory feedback, controlled by computer image analysis of the camera image, guides the user to move their hand towards a nearby target. The validity of the model was tested with two experiments. Preliminary studies have shown that participants were able to distinguish five vibrational zones (four different tactile receptors, all activated simultaneously) with an average accurateness of 88.9 percent behind receiving the initial instruction. Participants in the second experimentation were blind-folded and inculcated to shift the hand holding the appliance to one of four positions within arm's accomplish while the hand moved halfway through. The actuator is directed by two diverse approaches: (1) it repeats the axis with the greatest fault, and (2) it signals each axis alternately. Competitors begin to follow a direct path to the goal in each direction, but the physical parameters (speed, time) indicate fixed prizes along each axis. Minh Pham et al., (2018) was motivated by the desire to have a variety of home automation applications by the locals. For example, with people's location data, lighting and air conditioning can be better controlled, or home appliance users can improve serve residents. There is currently no good way for people to live in a closed environment. The theoretical method presented in this paper is to monitor the activity of the human body by combining a uncomplicated environmental sensor (submissive infrared sensor) and a light source. This approach reduces the burden on people even as keeping the charge of the process short. As a proof of concept, the hypothetical outline was tested and evaluated in a simulated building environment. We hope to do more accurate testing and in-depth analysis in preparation for the actual building transition. It is also essential to diminish the size and mass of the equipment used for measuring and to change the installation method of environmental sensors.

An emergency first aid delivery system that monitors patient health was proposed by Emre Oner Tartan et al., (2018). The services made available by mobile technology and location-based services. For advice, the mobile application sends an alarm message via notification, SMS, and email. The ECG is analysed on a mobile device, and if an abnormality is found, the data is sent to a server. The information that is transferred to the database in this system is used to assess the patient's condition. The heterogeneous system that M. Marzencki et al., (2011) proposed is capable of localization, environment monitoring, and the data is distributed to create context awareness. It is based on the ZigBee protocol and includes sensors and wearable nodes that can measure heart rate, body position, and location. The data is stored in the database. The data is sent to the destination via the wireless sensor network. Damian Kelly and others (2008). The importance of location-based systems is gradually increasing, particularly in the field of elder care. This paper focuses on the event of a single access point indoor location pursuit system. Room-level location accuracy is often faithfully achieved with a single access purpose localization system, it is believed. Damian Kelly introduces such a localization technique and provides the foundation for its operation.

An android-based energy-aware real-time location tracking system (EWAREL) using a GPS sensor was designed and implemented by Yunus Ozen et al., (2015). In order to save energy and be effective, its hybrid positioning technique makes use of additional inertial sensors and a data sending mechanism. When in danger, Junaid Ali et al., (2009) gave the option to the driver of the vehicle to activate a hidden switch at the monitoring station. To ensure proper system operation, the system also keeps an eye on the operating voltage and tracker temperature. A smart embedded system device that continuously monitors patients' health was proposed by Pratiksha WD. The patient's heart rate, body temperature, and saline liquid level are all tracked by this system. Pyeong-Gook Jung et al., (2015) developed a method based on air-pressure sensors and airbladders, for recognizing muscular activities. The change in air pressure in an airbladder contacting the interested muscle is used to detect muscular activity (s). Simone Farney et al., (2017) describe the W FYD architecture and detailed characterization of its narrow operating range, frequency response and dynamic response. She also wants to facilitate an understanding and the difference between materials used for active and inactive search, attack and non-hacking. The effect of weight was also taken into account. Mark Bachlin et al., (2010) developed a wearable assistive device for frozen walking (FOG) and Parkinson's disease (PD) patients. This wearable body uses a body acceleration sensor to measure the patient's muscle strength. FOG is automatically detected by identifying the frequency components found in these motions. When VOG is detected, the assistant sends a loud sound signal to the patient to encourage them to walk. Claudio Pacchirotti et al., (2017) presented a classification and analysis of effective fingertip and hand devices, focusing on systems that directly address the wear problem. The report also discusses key challenges and trends in tactile interfaces as well as future prospects for the industry.

Wearable Health Decices (WHDs)

Modern technology, called wearable medical devices (WHD), allows continuous physical monitoring of vital signs throughout a person's life (at work, at home, during physical activity, etc.). BHDs are important in intensive care settings as they reduce discomfort and interfere with human activities. WHD is division of the individual healthiness system, a perception related to doctors established in the 1990s with the aim of putting the person at the axis of health and managing their own health, the concept is called "Patient Empowerment." The aim is to increase satisfaction with their health, improve standards of care and encourage the make use of up-to-the-minute resources. Medical information, small and nanotechnology, resources engineering, electrical engineering and information and communication technology are all about this technology. The utilize of the WHD permits imperative signs collection and wellbeing observing over an expanded period (days/weeks) and exterior the clinic. This ability can make a valuable contribution to knowledge in many daily activities to become better, better, better, stronger, stronger, feel better, and/or function better and faster through treatment or recovery from physical injury. Sports/fitness to assess the performance of athletes or first responders or military personnel, to identify and monitor their body's response to various dangerous situations, and to better manage their energy and health. These products are used for daily physical care and all medicinal along with functional/
appropriateness/health intention.

Given this, the most straightforward nomenclature is "health," leading to WHDs. Regarding the areas to which they are applied, WHDs are more specialized. It has some distinctive qualities like low power usage, irresponsibility, security, comfort, and biotechnology.

Real-Time Observation of WHDs

Due to the time required, the patient needs to be followed up in the hospital for a long time. Use WHD. The clinical examination can be done outside the clinic, the patient may be reminded of any physical problem or asked to pay attention to himself and to know his vital signs while going about his daily business. In acute care settings, WHD allows patients to be seen anywhere in a designated area (usually a hospital) while transmitting their vital data wirelessly to a remote monitoring center. In a medical setting, some equipment can contaminate the patient's area. These options are not limited to one bed but allow patients to move around without the aid of equipment. This real-time system can generate reports for each patient to help identify necessary abnormalities at an early stage. Information about these cares is in the form of "field control" and "wide field" and increases as physicians inform patients when needed. Even medical records can be used to record vital signs for later analysis by doctors. The main advantage of WHD during the examination is that patients do not have time to look outside, all communication and telephone use, which is usually taken outside the reception, is at hand. This allows patients to lead a normal life during their care. His vital signs can be prolonged or sometimes forwarded to a foreign laboratory, where they can provide medical assistance and, if necessary, inform the patient about his condition. In order to understand the environment in which people live, they also need to have knowledge about the environment. Appropriate steps can be taken to prevent dangerous situations from occurring, such as exposure of the elderly to cold or air that can cause respiratory illness, dehydration or other conditions. Finally, essential signs be capable of also be conveyed via Bluetooth to transferable campaign or PCs to monitor and evaluate the patient. The current analysis will be used in sports competitions to examine the vital signs of athletes during everyday exercise or simple jog. These systems are referred to in the literature as "local" and "personal care" because there is no doctor involved. Another example is the use of this equipment when healthcare workers and firefighters are constantly monitoring mobile devices such as GPRS.

Remote Healthcare System

Figure 1 illustrates the system structural diagram of a common radio arrangement. The sensing devices are implanted to the patient's body. The sensors count the patient's vital signs such as body temperature, blood pressure and heart rate. These sensors are connected to the E-health shield along with the Bluetooth chip. This is an Arduino connected to the health protection shield. The Bluetooth chip transfers the input records to relevance on the patient's mobile phone. The system analyzes the sensor data and detects abnormalities in the information received from the patient's body. If the system finds a flaw in the data, it sends a message to the patient. The message also includes the address of the nearest medical center so that the patient can see a doctor for necessary tests. This app can footpath the patient's locality and launch messages to nearby doctors. This communication contains information concerning the idiosyncrasy in long-suffering information so that the adjacent doctor can follow up on the current patient and send the emergency team to the patient source (A. El-Nour et al., 2019).

This device is used to monitor the health condition, track the location, and find the safest location if a disaster occurs in the current location. It also sends status updates to the base station via SMS.

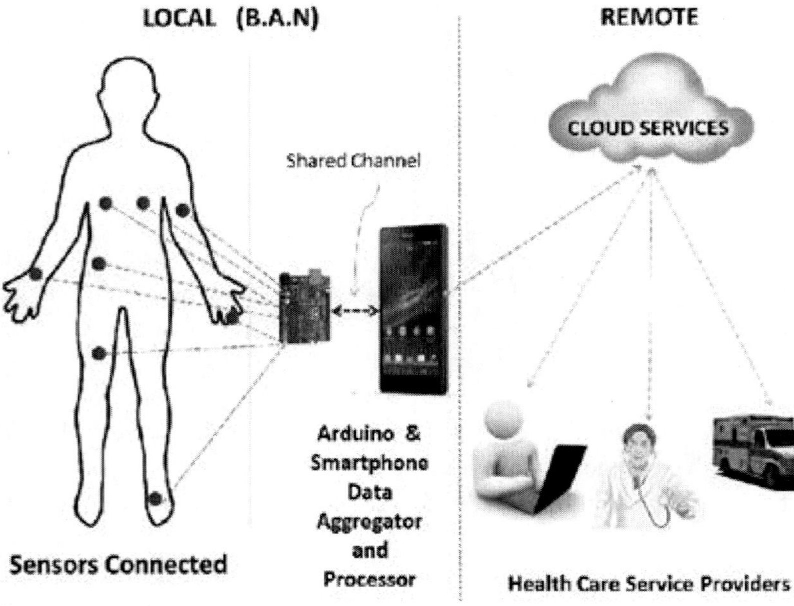

Figure 1. Structural diagram of remote healthcare system A. El-Nour et al., (2019).

The remote healthcare system includes the following components: Arduino Atmega328, GSM, GPS Module, Heart Rate Sensor, Temperature Sensor, MEMS, LCD Display, Power Supply, and Wearable sensing devices. Wearable sensing devices are becoming increasingly popular in the field of health monitoring. Wearable fitness trackers track the user's physical activity and heart rate. Intelligent health monitoring, heart and sleep monitoring, stress monitoring, monitoring the user's heart rate, blood oxygen level, recording ECG anytime and anywhere, and alerting the user when there is a problem. Wearable ECG monitors compute electrocardiograms, or ECGs, which help users follow their heart rhythm and rate, as well as other very important things such as blood pressure. Wearable blood pressure monitors track blood pressure as well as daily activity such as steps taken, and calories burned.

Wearable Biosensors

These are portable sensors that can be worn as gloves, clothing, bandages, or implants. They help to advise both users and doctors as well as regular and non-infectious and healthy maintenance of body movements and biofluids. Heart rate, skin temperature, respiratory rate, fall detection, body temperature, R-R interval (RR-I), single-lead ECG and tread reckoning are some measures.

Role of Wireless Communication in Healthcare System

Infectious disease-caused natural disasters have a devastating impact on people and societies. However, providing an appropriate response can prevent misfortunes from occurring as a result of infection spread. From the first flu epidemic in the year 1918 to the present year 2019 coronavirus illness, healthcare has played an important role in saving lives. Today, the development of smart healthcare is an admired leaning with wireless communication as the strength of character of the system. All hospitals should be equipped with technological equipment to provide diagnosis, treatment and various health services to patients in the hospital and remotely. An emergency response unit is also required to manage patient access and inquiries in the event of a disaster. Natural disasters such as hurricanes, earthquakes, tsunamis, nuclear accidents and epidemics have increased since the turn of the century. Online Web supported supervising Application: Doctors and nurses have an easy web-based submission to monitor and monitor their patients' health. The web application can be accessed via a standard web browser, smartphone or tablet device. The REST (RESTful) software architectural style is used to enable the resulting web applications to be adapted and modified. Also, JSON data representation is used for communication between web application modules. The working model includes the following components: LilyPad temperature sensor, heart rate sensor, GPS/GPRS/GSM MODULE V3.0 and

Arduino work together to make everything healthier. When the sensor is disconnected from the human body, the ECG displays the location (longitude and latitude) collected by the GPS/GPRS/GSM module. Lily Pad's thermometer is equipped with a digital thermometer and measures temperature in degrees Celsius and degrees Fahrenheit. Finally, the serial monitor shows all results. View the details of the sent short messages, together with the patient's name, body temperature, heart rate, and the latitude and longitude of the location. The patient's location can be determined using Google Maps and viewed via SMS. Wearable Personal Health and disaster Alert and pathway System: Meet with the development of personal health and emergency alert and tracking system, which can continuously monitor the user's body in real time and provide emergency medical aid reports.

The structural diagram of a wearable individual healthcare and urgent situation alert and tracking scheme is shown in Figure 2. This system is based on microcontroller board designs created by various vendors and employs various microcontrollers. These systems provide digital and analogue I/O pin sets that can be connected to various expansion boards ("shields") and other circuits. For loading programmes from personal computers, the boards have serial communications crossing point, including USB on some models. The Arduino mission provides an integrated development environment oriented on the dealing out project for programming microcontrollers, which consists of hold up for the C and C++ programming languages. The Universal Serial Bus (USB), which is enforced by utilizing USB-to-serial adaptor chips like the FTDI FT232, is used to programme Arduino boards. Other boards, like later-model Uno boards, substitute a separate AVR chip with USB-to-serial microcode that can be updated via its own ICSP header for the FTDI chip. Other variations, such as the Arduino small and consequently the unofficial Bardolino, eschew the Arduino IDE in favour of Bluetooth, a removable USB-to-serial adaptor board, or other approaches that were previously employed with outdated microcontroller tools. Compatible Arduino boards also use standard AVR ISP programming. The majority boards consist of a five-volt linear regulator and a sixteen-megahertz quartz oscillator or ceramic resonator in some alternates, though some styles, such as the Lily Pad, run at eight megahertz and do not include the aboard transformer due to form-factor constraints. When compared to other devices that require an external coder, Moreover, a boot loader that is pre-programmed into an Arduino's microcontroller makes it easier to upload programmes to the on-chip non-volatile storage. This makes utilizing an Arduino even easier by allowing the use of a standard PC as the coder. Opti boot bootloader is currently the default bootloader installed on Arduino UNO.

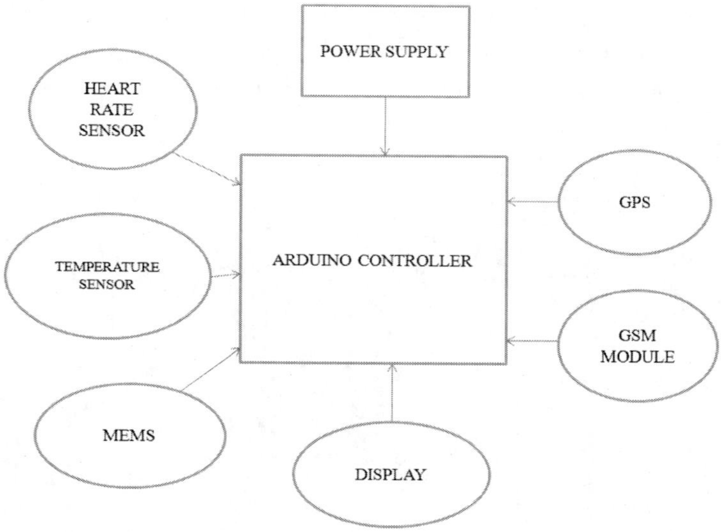

Figure 2. Wearable Personal Healthcare and Emergency Alert and Tracking System.

The experimental setup of the Wearable Personal Healthcare, Emergency Alert and Tracking System is shown in Figure 3. It describes the major components of the remote healthcare system and how the system is used to accurately analyse and notify health status and location tracking. The project's accuracy is achieved by utilizing modules such as an easy pulse sensor to analyse accurate heart beats and a GPS module to track the location. According to the programme, this system displays the status on the LCD and sends data to the base station. This is programmed and loaded into the Arduino microcontroller.

The research setup for wearable personal care and emergency alerts and tracking is shown in Figure 3. The projected scheme is designed to wrap end-to-end healthcare applications that can be created by two offices. Though, the most important utility of the first building block is to collect all the useful information about the person being cared for, while the second building block's role is to process, store and present the results of this phase to doctors and nurses. Personnel who follow the caregiver and take care of the patient. While the patient's heart rate transforms significantly, the Arduino, collecting the readings from the Pulse and LilyPad thermometers, instructs the GSM safeguard to launch a short message including the patient ID, readings and patient's place of residence. Sent to the doctor's phone, the doctor sends an ambulance to the patient's location. The Smart Embedded Board (SEB) section describes the hardware components used to create a connected smart board. The smart board uses a series of high-speed devices to understand people's health, and then sends the raw data via GSM SMS to the back-end server application.

Figure 3. Experimental setup of wearable personal healthcare and emergency alert and tracking system.

Classification of Major Application Areas of Healthcare System

The best healthcare system, from diagnosis to treatment, is a basic and prime need for society. Everyone, everywhere, should have access to high-quality healthcare. With the tremendous development of smart healthcare system applications, it is now possible. The healthcare system's foremost application areas are hospital surroundings, inaccessible healthcare, and disaster retort unit. Figure 4 illustrates the classification of major healthcare system application areas such as inaccessible healthcare, hospital surroundings and disaster retort unit.

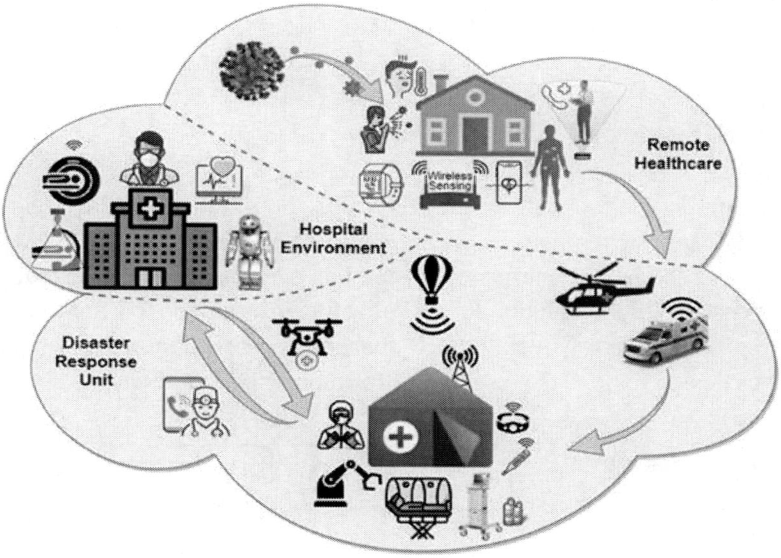

Figure 4. Classification of major application areas of healthcare system (Janjua et al., 2020).

Hospital Environment

A well-organized hospital environment with well-trained doctors, staff, and equipment for providing effective treatment is a must. In a smart healthcare system, effective information transformation among doctors, hospital administration, medical assistants and patients is critical. The Internet of Things is important in remote health monitoring and telemonitoring for tracking, monitoring, and maintaining records. The goal of the healthcare environment is not only disease management, but also patients' on the whole well-being. Sensing element-enabled IoT-utilized elegant healthcare systems, combined with artificial intelligence amalgamation, enable healthcare proficients to watch patients' habitual activities in real-time, facilitate on-time reply to patients' healthcare requirements.

Disaster Response Unit

Smart healthcare helps reduce harm by providing fast and safe treatment to patients and doctors in hospitals and remote areas (Fieux, M et al., (2020)).

Remote Healthcare

The aging population and the prevalence of chronic diseases have increased interest in physical fitness equipment in recent years. These development issues have led to the emergence of small devices that have the advantage of lower cost and higher quality. While collecting information to support our data analysis, we created an overall WHD concept (Figure 3) that we like to include various design options that each work for models and products.

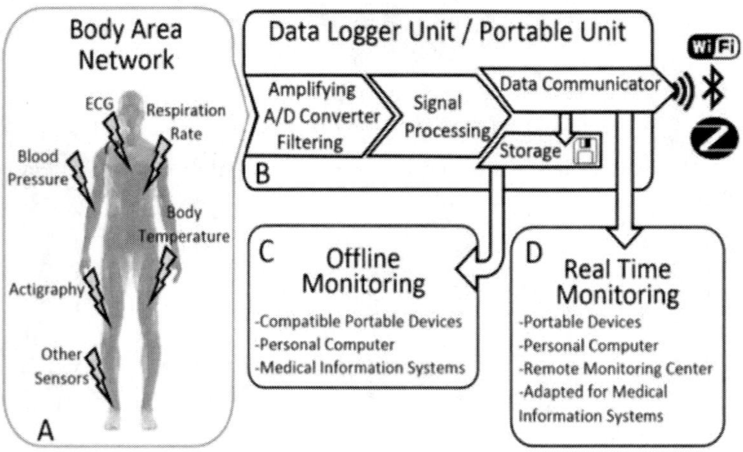

Figure 5. Generic architecture for body area network.

Physical Network (BSN) and Wireless Detector Network (WDN): The terms Physical Network (BSN) and Wireless Detector Network (WDN) are often used interchangeably (WSN). These three words will be utilized for diverse types of clothing based on their design. The term BAN is used in the Figure 5 because it is associated only with sensors around the soma. Connecting these sensors leads to the creation of a sensor network (BAN), which is then sent to the operating room in the same way as a mobile phone. If every part of the grid has or has a counter, we should always call it BSN, not BAN. At the most general level, a WSN may consist of many low-cost, low-power, and small-scale devices, each of which consists of preconfigured systems (sensors, microcontrollers, memories, and radio transceivers, etc.)). This allows All Nodes to perform sensing, computing, storage and communication. The Portable Unit (PU), also known as the Data Collector/ Portable Unit Interface Box or Data Logger Unit, is where all data is stored and contains the WHD outputs and devices. Most of the inputs are the main signals of the sensors. However, it can take delivery of message from additional connected mobile devices. Information transformation among sensors and elements is usually done via cables, making WHD easier and less expensive. The evolution of this communication technology is evident, for example in fine garments with connections (wires) drawn and woven into the material. In WHD this is usually the better way because it doesn't relax on the body, providing greater comfort and freedom of movement. As long as communication is created by biological organisms, as long as the body is used as an energizer, there will be new ways. When the receives analog inputs, these devices amplify and/or filter them before converting them to digital signals. Forwarding is created when data is sent to an item or to different devices. This method eliminates the option of assessing the health of the subject, allowing for detecting abnormalities, predicting disease, or supporting the diagnosis. Raw data collected from the trip will be sent over the air or stored on a memory card in the State of Mount Rushmore. As shown in Figure 3, Element can furthermore collect data from online display gadgets and store up them in memory as well. This two-way message allows different devices to establish a wireless connection with a central device that collects data from various sensors. This method can also be used to mark the time of important events using external devices. Bluetooth, Wi-Fi, ZigBee and recently Lora are the most popular wireless protocols in WHD (Radio Broadband). Bluetooth is a short-range radio-based feature that connects at low power and low cost, is portable and fixed. It is good for WHD and is widely used in products such as smartphones and laptops. The new Bluetooth Low Energy (BLE) technology (version 4.0 and higher) uses less power and is smaller, meaning it fits better in smaller WHDs. For applications where low power consumption is important, a lower layer of the Wi-Fi protocol is used to provide more data. While not inferior to Bluetooth technology, it is still a good technique to use in WHD, especially when longer communication is required. ZigBee is an added less-power, high-speed communication technology that is secured using advanced coding techniques. This enables ZigBee ideal for medical applications because it uses less power than the Bluetooth version. However, the information transfer rates are low. LoRa technology is a wireless system with long service life, low cost and reduced power utilization. In the future, it is predictable to be better than traditional systems such as mobile phones. The disadvantage is that the price is low, and the advantage is that there are many parameters such as transmission power, channel, and the price can be measured and adjusted. This allows to reduce the used space and switch to a special gearbox. Here are a few tips on some of the most popular wireless communication protocols.

Conclusion

The intend of the smart city concept is to advance the excellence of existence of people by offering pioneering and original elucidations in all of the eight pillars of the smart city. One of the needs and most significant, treatment requires the development and improvement of skills that provide simple, weak and real-time feedback with the aim of health care. Designed, tested and demonstrated among accredited customers, the system aims to reduce overall health care costs by preventing unnecessary hospitalizations, increasing the quality of care and ensuring that those in need of treatment are treated as soon as possible. This is a system that measures heart rate and body temperature and, in case of abnormal behavior, transmits them to a monitoring center using GSM, GPS and web technology, taking immediate and potentially future action to save the patient's life. In addition to other important measures, the sensors aim, according to commercially available information, is to provide reliable and effective applications for real-time healthcare. The value of this work depends on two reasons: Foremost, its many exploits and services; By building changes in the software, various diseases such as Alzheimer's, psychosis and sports patients can benefit from this scheme; Further, wireless technology will be employed to eliminate the need for a telephone connection, which may prevent the patient from leaving.

References

Abdelgawad A., Yelamarthi K., Khattab A. (2017) IoT-Based Health Monitoring System for Active and Assisted Living, Lecture Notes of the Institute for Computer Sciences, *Social Informatics and Telecommunications Engineering,* vol 195. Springer, Cham, pp. 11-20.ISBN (Print): 978-3-319-61948-4.

Abdullah, A., Ismael, A., Rashid, A., Abou-ElNour, A., and Tarique, M., Real time wireless health monitoring application using mobile devices‖, *International Journal of Computer Networks and Communications (IJCNC),* Vol. 7, N. 3, May 2015, pp. 13-30.

Anna Gruebler, Kenji Suzuki. "Design of a Wearable Device for reading Positive Expressions from Facial EMG Signals." *IEEE Transactions On Affective Computing*, Vol 5, NO. 3, 2014.

Claudio Pacchierotti, Massimiliano, Vincent Hayward. (2017), Wearable Haptic Systems for the Fingertip and the Hand: Taxonomy, Review, and Perspectives, *IEEE transactions on Haptics*.

Cornelia Stez, Bert Arnrich, Johannes Schumm, Roberto L. A. Marco, GerhandTroster. "Discriminating Stress from Cognitive Load Using Wearable EDA Device." *IEEE Transactions on information Technology in Biomedicine*, Vol. 14, No. 2, 2010.

Coronese, M., Lamperti, F., Keller, K., Chiaromonte, F., and Roventini, A. (2019). Evidence for sharp increase in the economic damages of extreme natural disasters. *Proc. Natl. Acad. Sci. USA.* 116, 21450–21455. doi: 10.1073/pnas.1907826116.

Damian Kelly, Sean McLoone, Terry Dishongh, Mick McGrath, Julie Behan. "Single Access Point Location Tracking for In-Home Health Monitoring." *Proceeding of the 5th Workshop On Positioning, Navigation and Communication 2008.*

El-Nour, A., M. Tarique, D. Foyez, L. Hassan, and A. Rashed. Location Aware Health Monitoring System for Emergency Cases, *International Journal of Scientific & Technology Research* Volume 8, Issue 12, December 2019 ISSN 2277-8616 1247

Emre Oner Tartan, Cebrail Ciflikli. "An Android Application for Geolocation based health monitoring, Consultancy & Alarm System." *2018 IEEE international conference on Software and Applications.* doi: 10.1109/COMPSAC.2018.10254.

Fieux, M., Duret, S., Bawazeer, N., Denoix, L., Zaouche, S., and Tringali, S. (2020). Telemedicine for ENT: effect on quality of care during COVID-19 pandemic. *Eur. Ann. Otorhinolaryngol. Head Neck Dis.* 137, 257–261. doi: 10.1016/j.anorl.2020.06.014.

Hassija, V., Gupta, V., and Guizani, M. (2020). A comprehensive review of the COVID-19 pandemic and the role of IoT, drones, AI, blockchain, and 5G in managing its impact. *IEEE Access* 8, 90225–90265. doi: 10.1109/ACCESS.2020.2992341.

Junaid Ali, Shahid Nasim, Taha Ali, Naveed Ahmed and Syed Riazun Nabi. "Implementation of GSM based Commercial Automobile Tracker Using PIC 18F452 and Development of Google Earth Embedded Monitoring Software." In *Student Conference On Research and Development*, Nov 2009, 16-18.

Marc Bachlin, Meir Plotnik, Daniel Roggen, Inbal Maidan, Jeffrey M. Hausdorff, Nir Giladi. "Wearable Assistant for Parkinson's Disease Patients with the freezing of Gait Symptom." *IEEE Transactions on Information Technology In Biomedicine*, Vol. 14, No. 2, 2010.

Marzencki, M., P. Lin, T. Cho, B. Ngai and B. Kaminsha, "Remote health activity and Asset Monitoring With Wireless Sensor Networks." *2011 IEEE 13th International Conference on e-Health Networking Applications and Services.*

Minh Pham, Dan Yang, Weihua Sheng, Senior. "A Sensor Fusion Approach to Indoor Human Localization Based On Environmental and Wearable Sensors." *IEEE Transactions on Automation Science and Engineering*, 2018.

Muhammad Bilal Janjua, Ahmet Enes Duranay and Hüseyin Arslan, Role of Wireless Communication in Healthcare System to Cater Disaster Situations Under 6G Vision, *Front. Comms. Net.,* 15 December 2020 | https://doi.org/10.3389/frcmn.2020.610879Chamola.

Pratiksha W. Digarse, Sanjaykumar L. Patil "Arduino Uno and GSM based wireless health monitoring system for patients." In *International Conference On Intelligent Computing and Control Systems.* ICICCS 2017.

Pyeong–Gook Jung, Gukchan Lim, Seonghyok Kim, Kyoungchul Kong. "A Wearable Gesture Recognition Device for Detecting Muscular Activities Based On Air-Pressure Sensors." *IEEE Transactions On Industrial Informatics*, Vol. 11, No. 2, 2015.

Samuel B. Schorr, Allison M. Okamura Fellow. "Three dimensional Skin deformation as force substitution: Wearable device design and performance during haptic exploration of virtual environments." *IEEE Transactions on Haptics*, Vol. 14, No. 8, 2015.

Shantanu A. Satpute, Janet Canady, Roberta L. Klatzky, George D. Stetten. "A Vibrotactile Wearable Ring for Assistance with Locating and Reaching Objects in Peripersonal Space." *IEEE Transactions on Haptics*, 2019.

Simone Fani, Simone Ciotti, EdoardoBattaglia, Alessandro MoscatelliMatteo Bianchi. "W-FYD: A Wearable Fabric-based Display for Haptic Multi-Cue Delivery and Tactile Augmented Reality." *IEEE Transaction on Haptics*, 2017.

SudipVhaduri, Christian Poellabauer. "Multi-modal Biometric-based Implicit Authentication of Wearable Device Users." *IEEE Transactions on Information Forensics and Security*, 2018.

Tae-Wook Jo, Yong-Duck, Hoon Choi. "A Bluetooth-UPnP Bridge for the Wearable Computing Environment." *IEEE transactions on consumer Electronics*, Vol 54, No. 3, 2008.

Ya-Li Zheng, Bryan P. Yan, Yuan-Ting Zhang Fellow. "An Armband Wearable Device for overnight and Cuff-Less Blood Pressure Measurement." *IEEE Transactions on Biomedical Engineering,* Vol. 61, No. 7, 2014.

Yoon Sang Kim, Byung Seok Soh, Sang-Goog Lee. "A new Wearable Input Device: SCURRY," *IEEE Transactions on Industrial Electronics*, Vol. 52, No. 6, 2005.

Yunus Ozen, Oguzhan Ozdemir, Necla Bandirmali. "*Android Based Energy Aware Real-Time Location Tracking System.*" 2015.

Chapter 18

Ensemble of Machine Learning and Stream Data Analytics for the Prediction of Weather

Sadhana Tiwari[1], PhD Scholar
and Sonali Agarwal*, PhD
Department of Information Technology, Indian institiute of Information Technology, Allahabad, Prayagraj, India

Abstract

Prediction using time series data is motivating and challenging in weather forecasting. Data mining and stream data analytics both can be applied to achieve good prediction results in weather prediction. Many numerical methods and models have already been developed and applied to generate weather forecasts. Since a lot of weather-related data is being collected each day, understanding how this big data is processed efficiently to predict future weather is pretty interesting. This paper proposes a new approach to predict whether using time series data by applying fusion of spark streaming with machine learning (ML). The more applicable part of this research is that the weather stations could easily process real time data in the form of streams (batch) and process each batch efficiently with low latency. The proposed work can predict and train the data based on several ML models like – Multiclass Logistic Regression, Random Forest and Support Vector Machine to provide a prediction of weather in different classes such as clear, foggy, mostly cloudy, overcast, and partly cloudy. It also compares the various parameters i.e., - Accuracy, F1-score, Precision, Recall and Weighted average of all the models. Overall, Random Forest is the best ML

* Corresponding Author's Email: rsi2018507@iiita.ac.in.
* Corresponding Author's Email: rsi2018507@iiita.ac.in.

In: Blockchain Enabled Secure Big Data Computing …
Editors: L. Ashok Kumar, D. Karthika Renuka, V. Chandrasekar et al.
ISBN: 979-8-89113-243-6

algorithm on static time series data. Furthermore, the socket streaming is also applied on the weather dataset to execute the classification task on streaming mode which makes this work memory efficient.

Keywords: data streaming, big data analytics, prediction, machine learning, classification

Introduction

Stream data analytics and machine learning both play a principal role in analyzing the real time data and generate better results using large amounts of data. These days handling of data is very complex as everybody is generating data and everyone is consuming data. The challenges of stream data analytics are concept drift, sampling, windowing and parameter tuning etc. (Jaynati D. et al., 2017).

To generate accurate predictions after analyzing weather time series data is very essential as many people will be affected due to wrong prediction of weather. In this paper stream data analytics is used for making the batches (streams) of the data to reduce in memory computation of incoming data. Apache Spark is used to generate batches of the weather time series data. Apache Spark includes rich libraries for streaming, event correlation and complex event processing. It can be integrated with java, Scala and python IDE (Bendre, M. R. et al., 2015; Haidar, A., et al., 2018; Sezer, O. B et al., 2016).

Machine learning (ML) models are also very useful in real time data, time series data and IoT data. ML is the branch of Artificial Intelligence (AI), which provides easy implementation with low computation cost, fast training and testing, fast validation and evaluation including low complexity and high performance (Samya, R. et al., 2016; Fathi, M. et al., 2022). The frequent advancement of ML models in the past two decades proves the suitability of ML in weather forecasting using time series data (Samya, R. et al., 2016). Many good and simple prediction models are available in ML such as Linear Regression (Kanavos, A. et al., 2021; Holmstrom, M. wt. al, 2016), Random Forest (RF) (Berk, R. et al., 2007; Choi, S. et al., 2016; Swe, T. T. et al., 2019), Decision Tree (Kanavos, A. et al., 2021; Choi, S. et al., 2016), AdaBoost (Choi, S. et al., 2016), neuro-fuzzy (Haidar, A., et al., 2018), Artificial Neural Network (ANN) (Samya, R. et al., 2016), Support Vector Machine (SVM) (Kanavos, A. et al., 2021), Support Vector Regression (SVR) and convolutional neural network (CNN) (Haidar, A., et al., 2018) for weather prediction. These models can also compute various mathematical parameters like accuracy, precision, recall, F1-score, weighted average by generating the confusion matrix through each model.

Weather Forecasting using various statistical parameters of weather data such as temperature, humidity, wind speed, precipitation level, etc. is very important for hydro-power operation, cyclone detection, thunderstorm, draught, rainfall, and flood identification. As this is the concern of the whole population nowadays many weather stations are connected for sharing their knowledge and data through the internet of things (IoT) devices (Fathi, M. et al., 2022). Stream data analytics clubbed with ML models can produce better results than all the traditional methods. Various machine learning models are used for accurate prediction of weather by training the model consistently and generating the results using test data. In this work, the forecasting of weather data is done by applying various machine learning models likewise multiclass LR, SVM and RF. Streaming of weather dataset is done through Spark and machine learning pipelines. Afterwards the performance of each model will be evaluated and compared. By applying ML techniques and streaming analytics on weather dataset, more accurate weather forecasting can be done (Berk, R. et al., 2007; Choi, S. et al., 2016). Major findings of this work are:

1. Pre-processing of the weather time series data
2. Classification of weather dataset using various machine learning methods
3. (iii) Performance evaluation of various ML classifier and selection of best classification model
4. Divide the weather data in batches of 3 seconds using spark context and ML pipelines
5. Prediction of weather into five specified classes such as Clear, Foggy, Mostly Cloudy, Overcast, Partly Cloudy.

This paper is structured in the following manner: Section 1 explains the introduction of the work. The literature review of the relevant area has been discussed in Section 2. Section 3 provides the details of the material and methods used in this work. Section 5 presents the experimental evaluations and discussions and conclusion followed by future work of the presented work is provided in section 6.

Existing Approaches Used in Weather Prediction

Weather prediction is a complex phenomenon involving dynamic interaction between different parameters and it plays a prominent role in various areas such as agriculture, public planning, industry and stock forecasting. Hadoop Map-Reduce is used to analyze weather data. D. Jayanthi et.al proposed a novel method for analyzing weather data based on spark which results in high temperature, high average precipitation and parameters related to weather across distinct weather stations (Jaynati D. et al., 2017). M. R. Bendre et.al, discusses a scenario for collecting big data using Information and Communication Technology (ICT) in farming environment. Big data analytics in agricultural applications provides new insights into forecasting climate decisions, improving crop productivity and avoiding unnecessary costs associated with harvesting, pesticides and fertilizer use. They develop an algorithm to provide precision in agriculture data by applying big data analytics. They basically focus on how big data is useful to make advance decisions related to weather in the field of agriculture to increase productivity and to reduce cost (Bendre, M. R. et al., 2015).

Ali Haider et.al, proposed an advance forecast system for rainfall prediction on a specific locality of eastern Australia. In this work they used a deep convolutional neural network (CNN) for monthly prediction about rainfall against Australian Community Climate. They also compared their newly designed model with Earth-System Simulator-Seasonal Prediction System (ACCESS). Additionally, the performance comparison between CNN and Multi-layered perceptron (MLP) has been done and better result has been obtained using the proposed model with respect to root mean square error (RMSE), Nash Sutcliff coefficient, mean absolute error, and Pearson correlation (r) (Haidar, A., et al., 2018). Aras Can Onal et.al, designed a new forecast system for weather using clustering and sensor anomaly detection by applying it to a publicly available dataset. In this work, a new learning model is developed through k-means clustering and scikit-learn. The results obtained after applying this framework show that the extraction of significant information from a highly complex dataset is also possible (Sezer, O. B et al., 2016).

A paper published by R. Samya et.al, provides a review on the various forecasting techniques using data mining and ANN. The most commonly used parameters for weather forecasting are temperature, rainfall, wind speed and evaporation (Samya, R. et al., 2016). Another paper says, applying big data analysis to weather forecasting can solve problems associated with existing data management techniques based on technology, technique, and hybrid methods. This paper also compares the previous approaches in terms of accuracy, scalability, runtime, and other quality factors (Fathi, M. et al., 2022). The research work carried out by Kanavos, Andreas, et al., provides weather related information based on precipitation types of winter season. The model proposed in this work receives input data from various weather sensors and forecast the winter weather into classes freezing rain, rain and snow according to the precipitation type using various supervised machine learning classifiers such as Hoeffding Tree, HoeffdingAdaptive Tree, HoeffdingOption Tree, Decision Stump, OzaBag and Naive Bayes. They observed that the OzaBag model gives the best results among all the used classifiers (Kanavos, A. et al., 2021). A research paper published by Sun Choi et.al, applied RF, Decision tree (DT), AdaBoost, and k- Nearest Neighbor's techniques for weather forecasting. They computed delays in airlines services due to weather conditions. For this purpose, weather data between years 2005 to 2015 and U.S. airlines data had been taken and a binary class classification result has been obtained which shows the fight is delayed or on time (Choi, S. et al., 2016).

After reviewing various techniques, it is observed that big data analytics provide good and accurate solutions for predicting weather patterns. Many mathematical algorithms and models have been suggested and used for prediction of weather, but big data technology is very helpful to solve these problems and challenges through distributed computing. It is also observed that spark is a new technology that makes in-memory computing more efficient.

Material and Methods

This section provides the details of the data set used for weather prediction, data pre-processing steps, architecture of the proposed model and step by step processing of the proposed method.

Dataset Description

The weather dataset is freely available on Kaggle online repository. This is a multimodal time series dataset which is suitable for classification and prediction purposes and is having 12 different attributes such as Formatted Date, Summary, Precip Type, Temperature (C), Apparent Temperature (C), Humidity, Wind Speed (km/h), Wind Bearing (degrees), Visibility (km), Loud Cover, Pressure (millibars), and Daily Summary. The data is holding approx. 100000 instances[2].

Architecture of Proposed Model

Figure 1, represents the flow diagram of the proposed model used weather forecasting using static weather data after applying various machine learning models and real time weather data using machine learning pipelines for generating weather data streams.

Weather Sensors

Weather sensors are the hardware sensors which are actually responsible for collection of weather data like humidity, date, time, seasons, precipitation levels etc. These sensors scan data continuously and send it to their nearest weather station. Weather sensors are critical for continuous monitoring of real time weather data and sending it to the correct weather stations (Kanavos, A. et al., 2021).

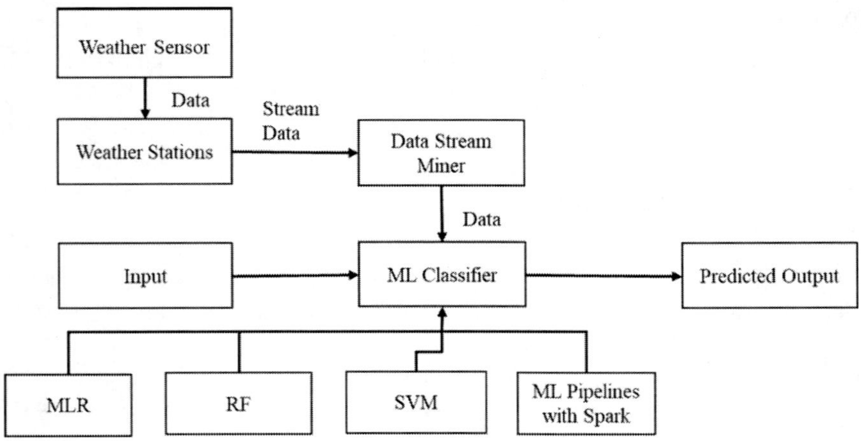

Figure 1. Flow diagram of weather time series data analysis system.

[2] https://www.kaggle.com/muthuj7/weather-dataset, Accessed on March, 2021.

Weather Stations

These are the entities responsible for the accumulation of the scanned data by different sensors, aggregating it, and then sending it to the stream miner for prediction at continuous intervals (Kanavos, A. et al., 2021).

Data Stream Miner

It is the component responsible for handling the stream data appropriately using data streaming technology. For this work PySpark library is used for creating data streams. After continuously streaming data, it sends data in form of batches to the ML classifier (Choi, S. et al., 2016; Swe, T. T. et al., 2019).

ML Classifier/Learner

This is the part which is actually responsible for the prediction and learning from the datasets. It consists of two parts: a classifier which takes a dataset as input and classifies it according to the model which has been trained, and a learner which creates a model based on existing real time datasets. Various algorithms used in this research are as follows.

Random Forest

Random forest mechanism is a kind of ensemble learning method used for classification, regression, predication and other tasks which can be operated by building a number of decision trees at the time of training and displaying the output class that is usually taken as mode of the specified classes or choosing mean prediction of each and every trees. The fundamental concept behind this classifier is a simple but powerful one. Many relatively non-correlated models when operated in conjunction with some other models will outrun in comparison to any individual models (Choi, S. et al., 2016; Pang, B. et al., 2017).

Support Vector Machines

This is a supervised method used for classification, prediction and regression purpose while analyzing any dataset. Identification of a suitable hyperplane in N -dimensional space which clearly classifies all the data objects (where N is the no. attributes) is the primary aim of this technique. To differentiate between both the classes of data objects, multiple probable hyperplanes exist, which can be selected. Our goal is to select the plane having maximum margin that means maximum distance among the data objects of both the classes. Amplifying the margin distance introduces a booster for upcoming data points to be classified correctly with high confidence (Fathi, M. et al., 2022; Kanavos, A. et al., 2021).

Multiclass Logistic Regression

The Multiclass Logistic Regression (MLR) algorithm is basically a supervised learning method which is used for probability prediction of the target variable. The characteristics of the dependent variable or target variable are usually dichotomous. In other words, there are only two possible types. Basically, the dependent variable is basically a binary variable with data encoded as 1 (success/yes) or 0 (indicates a failure/no). Generally, the Logistic Regression model can be divided into two parts- Binomial and Multinomial. This is the simplest method for binary and multivariate classification purposes (Pang, B. et al., 2017).

Machine Learning Pipelines using Spark

Streams of weather dataset will be generated through spark streaming context and machine learning pipelines. In this model the data is divided in the streams of 3ms. Sockets streaming is used to process this data and RF classifier is used to predict the accurate level of the weather[3,4].

Step by Step Processing of the Model

It has been essential and vital to accurately predict the weather because it helps us to be ready for any weather situation, such as be it a normal sunny/rainy day or be it a catastrophe like a typhoon. To make a good and efficient forecast model, ensemble of machine learning, big data analytics and real time streaming have been done and applied on weather time series data. Following are the sequence of steps of the new ensemble model:

- Pre-processing of weather time series data to handle the missing values, redundant values, and null values.
- Some machine learning models such as Multiclass Logistic Regression (MLR), Random Forest (RF), SVM and machine learning pipelines are used in spark streaming environment to calculate the confusion matrix and accuracy, F1-measure, macro average, recall, weighted average and precision etc.

[3] Albert C., September 2016, https://towardsdatascience.com/introduction-to-stream-mining-8b79dd64e460.

[4] https://www.cs.waikato.ac.nz/~abifet/MOA/StreamMining.pdf, Accessed on April, 2021.

- Processing of data in batches i.e., for real time monitoring of weather dataset spark socket streaming is used to divide the data in form of batches of 3 seconds.
- Splitting of weather dataset into train and test set.
- Training of the model according to various well known machine learning classifiers.
- Different machine learning techniques are used for monitoring the weather dataset using both static (offline) and real time (online) methods.
- Comparison of different parameters like accuracy, F1-score, precision, recall, and weighted average has been performed between various models.
- Accurate prediction has been made for any particular day.

Experimental Results and Discussion

In this research, different machine learning models are used to predict the weather dataset in static mode followed by real-time spark streaming context using ML pipelines. Various classifier models used in this work can predict the weather in five different classes which are: Clear, Foggy, Mostly Cloudy, Overcast and Partially Cloudy. Although any classifier can be applied for prediction of weather and tune the parameters for desired levels of accuracy, some models are inherently better for prediction of certain classes of datasets. In this section, the result of three different algorithms RF, SVM and MLR has been produced and compared across various performance measures by tuning parameters of each model. Out of these classifiers RF classifier performs best in terms of accuracy and other classification measures during static weather time series data processing. While comparing the performance of all used ML models in static weather data processing and processing weather time series data in real time environment using spark streaming, it can be observed that the real time processing of the data performs better in terms of accurate prediction and reduced in memory computation.

Models for Offline (Static) Monitoring of Time Series Weather Data

This section presents the results computed for static weather data monitoring using different well-defined ML models like LR, SVM, and RF.

Logistic Regression

Logistic regression can also be extended to classify data which may not have a binary nature, i.e., the output can be more than one, for this purpose the Softmax function is used in this work, instead of using the usual sigmoid function which is generally used in normal logistic regression. The confusion matrix computed weather dataset after applying LR model is given in Figure 2.

Support Vector Machines

Support vector machines are basically used for selecting the most suitable hyperplane which can classify the data objects in N-dimensional object space (where N represents the number of features). The computed result using SVM model is mentioned in Figure 3.

Random Forest

Fundamental concept behind random forest models is simple but powerful one, because it consists of multiple decision trees and these trees will be operated as an ensemble to predict the class. The class having the maximum votes will become the predicted class and the confusion matrix for the same is given n Figure 4.

MultiClass Logistic Regression

	precision	recall	f1-score	support
Clear	0.55	0.07	0.12	3137
Foggy	0.90	0.98	0.94	2173
Mostly Cloudy	0.40	0.39	0.40	8445
Overcast	0.48	0.45	0.46	5082
Partly Cloudy	0.50	0.66	0.57	9502
accuracy			0.50	28339
macro avg	0.57	0.51	0.50	28339
weighted avg	0.50	0.50	0.48	28339

Classification report

```
[[ 206   90  667  413 1761]
 [  17 2124    5   23    4]
 [  43   34 3302 1313 3753]
 [  46   69 1909 2302  756]
 [  62   30 2318  795 6297]]
```

Confusion Matrix

Figure 2. Classification report and confusion matrix using MLR model.

SVM Classification

```
[[ 220   64  836    0 2017]
 [  36  881 1122    5  129]
 [  35   48 3599    0 4763]
 [  43  149 3422    1 1467]
 [  35   54 2453    0 6960]]
```

Confusion Matrix

	precision	recall	f1-score	support
Clear	0.60	0.07	0.13	3137
Foggy	0.74	0.41	0.52	2173
Mostly Cloudy	0.31	0.43	0.36	8445
Overcast	0.17	0.00	0.00	5082
Partly Cloudy	0.45	0.73	0.56	9502
accuracy			0.41	28339
macro avg	0.45	0.33	0.31	28339
weighted avg	0.40	0.41	0.35	28339

Classification Report

Figure 3. Classification report and confusion matrix using SVM model.

Comparison of Various Performance Parameters using Offline Monitoring

The three different classifiers MLR, SVM, and Random Forest are applied on the weather time series dataset in static mode or offline mode and various parameters like accuracy, f1-Score, macro average, precision, recall and weighted average are compared in the following Figure 5.

Random Forest Classification

```
[[ 940   11  488  200 1498]
 [   0 2172    0    0    1]
 [ 213    0 4328  999 2905]
 [  84    1 1455 2975  567]
 [ 433    0 2165  492 6412]]
```

Confusion Matrix

	precision	recall	f1-score	support
Clear	0.56	0.30	0.39	3137
Foggy	0.99	1.00	1.00	2173
Mostly Cloudy	0.51	0.51	0.51	8445
Overcast	0.64	0.59	0.61	5082
Partly Cloudy	0.56	0.67	0.61	9502
accuracy			0.59	28339
macro avg	0.65	0.61	0.63	28339
weighted avg	0.59	0.59	0.59	28339

Classification report

Figure 4. Classification report and confusion matrix using RF model.

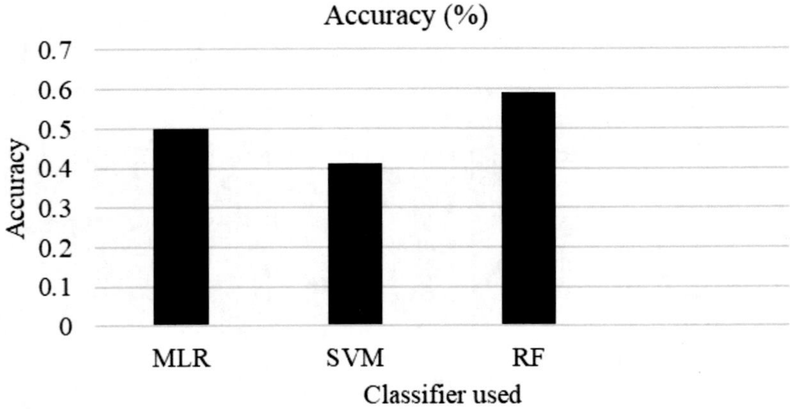

Figure 5. Accuracy of MLR, SVM and RF model.

The graph mentioned above compares accuracy achieved using the MLR, SVM and RF models which shows RF classifier gives best accuracy in static processing weather dataset. The graph given in Figure 6 shows the comparison of f1-score values in case of MLR, SVM and RF models across each five classes clear, foggy, mostly cloudy, overcast and partly cloudy.

The graph given in Figure 7 represents the comparison of precision values of MLR, SVM and RF models across each five classes clear, foggy, mostly cloudy, overcast and partly cloudy.

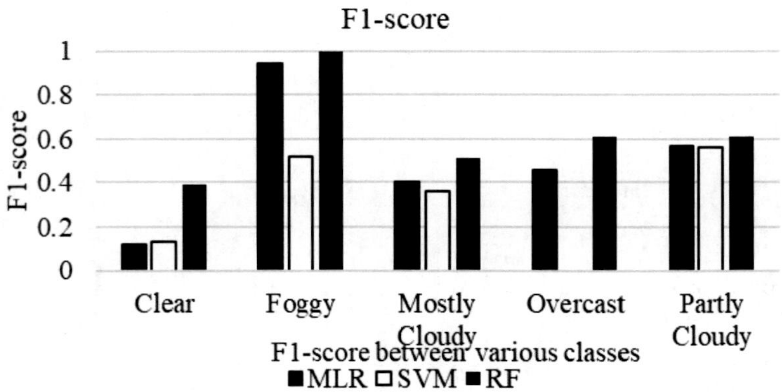

Figure 6. F1-Score across different classes of MLR, SVM and RF models.

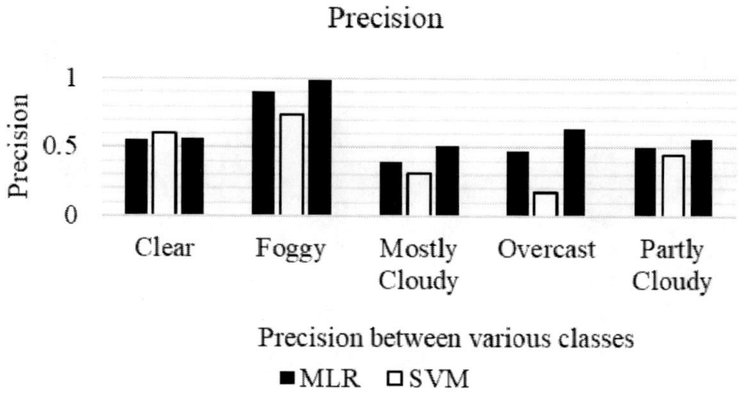

Figure 7. Precision values across different classes of MLR, SVM and RF models.

Figure 8 shows the comparison between recall values obtained through MLR, SVM and RF models across each five classes clear, foggy, mostly cloudy, overcast and partly cloudy.

Figure 9 shows the graph computed for macro average values obtained through MLR, SVM and RF models. It can clearly be observed from the above graphs, that the parameters like recall, Precision, F1-score and accuracy, outperform in case of Random Forest classifier from the other two classifiers and Support Vector Machine performs the worst. Thus, for static weather datasets random forests classifier is the most suitable model.it to the correct weather stations (Kanavos, A. et al., 2021).

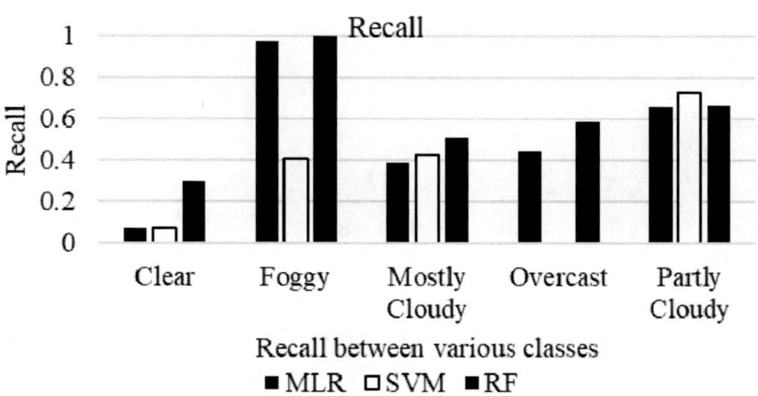

Figure 8. Recall values across different classes of MLR, SVM and RF models.

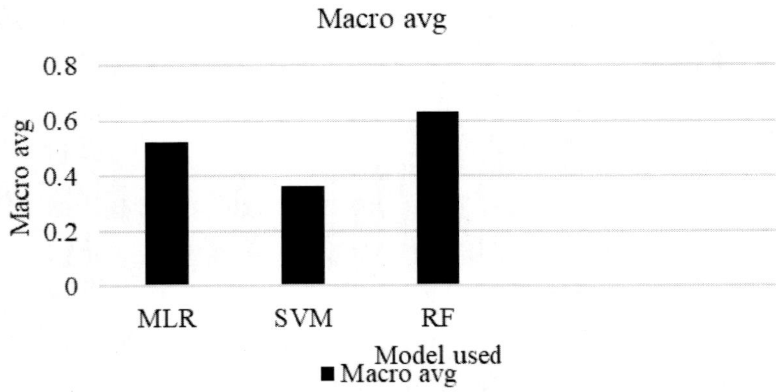

Figure 9. Macro Average of MLR, SVM and RF models.

Model Developed for Real Time Monitoring of Time Series Weather Data

This section discusses the model used for real time monitoring of weather data which is developed through machine learning pipelines.

Machine Learning Pipelines

In this work, batches or streams of weather dataset are generated using socket programming. While for processing stream data in real time using pyspark streaming, a two-stage pipeline is used which is described as.

1st Stage - Vector Assembler

This behaves like a converter which is used to combine a list of columns in the form of a single vector column. In this stage raw features and features generated through various feature transformers will be transformed in the form of a single feature vector, for training the ML models as SVM, logistic regression, random forest and decision trees.

2nd Stage - Application of Model

This is the stage where the machine learning model is applied on the data returned by the vector assembler and the predicted output is analyzed and visualized. Figure 10 shows the screen shot real time weather data processing using machine learning pipelines integrated with the socket pushing in the environment of spark streaming context.

Figure 10. Real time weather data processing using Spark and ML pipelines.

Figure 11. Real time weather label prediction.

Figure 11 displays the real time prediction of weather dataset into five specified classes such as clear, foggy, mostly cloudy, overcast and partly cloudy which is labelled as 1, 2, 3, 4, 5 respectively, using random forest classifier through socket streaming.

Conclusion

In this research, offline and real time monitoring of weather time series dataset has been performed. The model training has been done using various machine learning models such as Multiclass Logistic Regression, Random Forest and SVM algorithm. After training, the proposed model is able to predict the result for test data into five different classes which are clear, foggy, mostly cloudy, overcast and partly cloudy. The performance of the model has been evaluated using many standard metrics of machine learning e.g., accuracy, recall, F1 score, precision and weighted average etc. Spark streaming is used to divide the weather dataset into streams and ML pipelines are used to classify the batches of data into five different classes. Since a huge amount of weather-related data is being collected each day, we proposed a framework to understand how this big data is processed efficiently to predict future weather using spark streaming which is quite interesting. It is clearly observed that online monitoring of time series datasets reduces memory computation in comparison to offline static datasets. Due to limitation of class imbalance problem of the dataset, the work can be improved by handling this.

Disclaimer

None.

References

Bendre, M. R., Thool, R. C., & Thool, V. R. (2015, September). Big data in precision agriculture : Weather forecasting for future farming. In *2015 1st international conference on next generation computing technologies (NGCT)* (pp. 744-750). IEEE.

Berk, R., Deloncle, A., d'Andrea, F., & Ghil, M. (2007). Weather regime prediction using statistical learning. *Journal of the atmospheric sciences,* 64(5), 1619-1635.

Choi, S., Kim, Y. J., Briceno, S., & Mavris, D. (2016, September). Prediction of weather-induced airline delays based on machine learning algorithms. In *2016 IEEE/AIAA 35th Digital Avionics Systems Conference (DASC)* (pp. 1-6). IEEE.

Dogdu, E., Ozbayoglu, M., Onal, A., & Sezer, O. B., (2016, December). An extended iot framework with semantics, big data, and analytics. In *2016 IEEE international conference on big data (big data)* (pp. 1849-1856). IEEE.

Fathi, M., Hashi Kashani, M., Jameii, S. M., & Mahdipour, E. (2022). Big data analytics in weather forecasting: A systematic review. *Archives of Computational Methods in Engineering,* 29(2), 1247-1275.

Haidar, A., & Verma, B. (2018). Monthly rainfall forecasting using one-dimensional deep convolutional neural network. *IEEE Access*, 6, 69053-69063.

Holmstrom, M., Liu, D., & Vo, C. (2016). Machine learning applied to weather forecasting. *Meteorol. Appl,* 10, 1-5.

Jayanthi, D., & Sumathi, G. (2017, April). Weather data analysis using spark - an in-memory computing framework. In *2017 Innovations in Power and Advanced Computing Technologies (i-PACT)* (pp. 1-5). IEEE.

Kanavos, A., Panagiotakopoulos, T., Maragoudakis, M., Kiouvrekis, Y., & Vonitsanos, G., (2021, July). Forecasting Winter Precipitation based on Weather Sensors Data in Apache Spark. In *2021 12th International Conference on Information, Intelligence, Systems & Applications (IISA)* (pp. 1-6). IEEE.

Pang, B., Yue, J., Zhao, G., & Xu, Z. (2017). Statistical downscaling of temperature with the random forest model. *Advances in Meteorology,* 2017.

Rathipriya, R., & Samya, R., (2016). Predictive analysis for weather prediction using data mining with ANN : A study. *International Journal of Computational Intelligence and Informatics,* 6(2), 150-154.

Swe, T. T., Phyu, P., & Thein, S. P. P. (2019). Weather prediction model using random forest algorithm and apache spark. *Weather,* 3(6).

Chapter 19

Privacy Preserving Encryption with Reliable Data Transmission for Cluster Based Vehicular Networks

P. Anandan[1], PhD
N. Krishnaraj[2], PhD
N. Kumaratharan[3] PhD and
N. Anbuselvan[4], PhD

[1]Department of Electronics and Communication Engineering, Saveetha School of Engineering, Saveetha Institute of Medical and Technical Sciences, Chennai, Tamilnadu, India
[2]Department of Networking and Communications, School of Computing, Faculty of Engineering and Technology, SRM Institute of Science and Technology, Kattankulathur, Tamilnadu, India
[3]Department of Electronics and Communication Engineering, Sri Venkateshwara Engineering College, Sriperumbudur, Chennai, Tamilnadu, India
[4]Department of EEE, Saveetha School of Engineering, Saveetha Institute of Medical and Technical Sciences, Chennai, Tamilnadu, India

Abstract

Existing research demonstrates that Cluster-based Medium Access Control (CB-MAC) protocols are good at regulatory and handling Vehicular Ad hoc Networks (VANETs), but they need to be more secure and private. It's for this reason that we came up with a lightweight user authentication structure that works well with WSNs. It allows for mutual authentication and session-key agreement. A person who has a mobile device, like a PDA, can use this scheme to prove who he is before he can use the WSN. Which is based on a secure authentication protocol. In this

In: Blockchain Enabled Secure Big Data Computing …
Editors: L. Ashok Kumar, D. Karthika Renuka, V. Chandrasekar et al.
ISBN: 979-8-89113-243-6

chapter, we show how to use a lightweight information sharing plan (LDSS) for a wide range of distributed computing. It takes on CP-ABE, which is used to keep people from getting into the cloud. They explain in detail how the authentication center's, registration of vehicles, and creation of keys work. In the projected construction, a Global Authentication Center (GAC) stores all of the information about vehicles, and a Local Authentication Center (LAC) keeps a block chain to make it easy to move vehicles between internal clusters. We also want to change the format of control packets in IEEE 802.11 to fix some of the problems with the traditional MAC protocols. Furthermore, cluster formation and selection of cluster-heads, as well as merging and leaving processes, are done while taking safety and non-safety messages into account. This helps improve the performance of the system.

Keywords: VANET, WSN, distributed computing, Medium Access Control, security, privacy

Introduction

Effectual design of a Wireless Sensor Network (WSN) has become a principal area of research in recent years. A Sensor is an instrument that replies and senses some kind of contribution from both the physical and ecological conditions, like pressure, temperature, light, etc. The result is usually an electrical signal that is communicated to a regulator for further processing. Each node in a Vehicular Ad-hoc Network (VANET) has a two-way connection with other nodes (Dansana et al., 2022). These networks are only for cars and trucks. A lot more people are interested in VANET because it has opened a way to make a few new applications, like alerts about travel plans, traffic information, and user-defined apps. There are a lot of unique things about each node in VANET, like a dynamic network structure, a lot of mobility, slow processing speed, and a small amount of memory. Following are some of the features that set VANET apart from similar wireless networks. They add their own special flair to it. A VANET is a very specific subset of the ad hoc network family. VANET is distinct from MANET due to the frequent disconnections in the network's nodes' ability to communicate with one another. Link dependability is affected by factors like network density, vehicle speed, wireless channel congestion, and so on (Schoch et al., 2008). The ability to send and receive messages between vehicles is made possible through vehicular communication. Unlike a centralized system, every node in a VANET may independently create, relay, and get messages. Information messages, safety messages, individual drive control messages, and group drive control messages are the four main categories into which the inter-vehicle communication (IVC) protocols fall. Cloud computing is quickly becoming the dominating technology in the Information and Communications Technology (ICT) industry, promising low-cost data storage and lightning-fast processing for users (Khanagha et al., 2020). It assures its constituents that cutting-edge mobile and Internet of Things (IoT) technology would make their everyday lives easier and more convenient. These new requirements are quickly taking shape as a cornerstone of digital urbanization. Figure 20.1 describes the privacy Privacy-protecting distributed anomaly detection for secure data analysis.

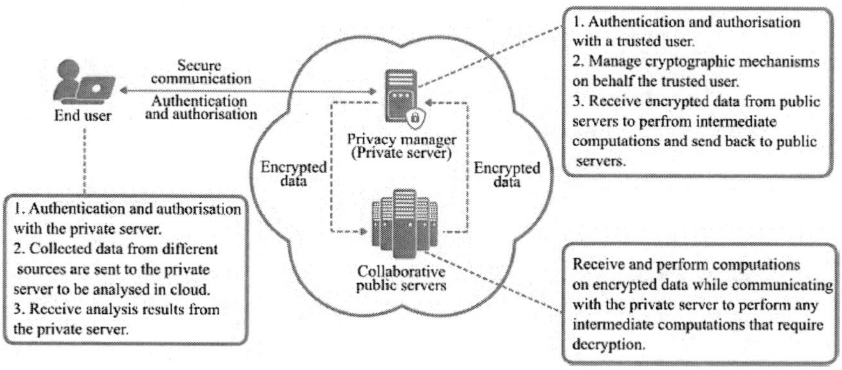

Figure 20.1. Privacy-protecting distributed anomaly detection for secure data analysis.

The illustration provides a visual representation of the fundamental components that comprise our cloud-based approach. Data encryption and decryption are two other major duties of the privacy manager (Alabdulatif et al., 2017), along with ensuring that users from afar are able to use the public cloud infrastructure. Two such instances are the mobile internet and RSS feeds. Services connected to safety (Type 2) include things like emergency message relay, collision warnings, traffic updates, and knowledge of hazards in the road. Due to the sensitive nature of the data handled by Type 2 applications, it is essential that there be minimal delay and minimal propagation losses. Notable cases include air routed tri-copters, groups of vehicles called platoons, and the flight plan of multiple aircraft. Cooperative planning and oversight may be carried out by a group of vehicles or delegated from a central location to others. A tight knit group could coordinate its motion in response to real-time cues from individuals' minds and from one another (Oubbati, 2017).

Figure 20.2. Three basic elements related to security.

The components of the secure cloud architecture are shown in Figure 20.2.

- *Authority:* It is an important part of the system that is in charge of making, distributing, and handling all the secret keys. Other parts of the system trust it.
- *Time Server:* Server for keeping track of the passage of time; it communicates with no other elements of the system. Its job is to provide an accurate estimated timing of release. There are people who may be allowed to verify their identity and get access to the data the data owner has given to someone else. These people are called "data users." It's important to note that only people who have been approved can get their hands on the shared data (Xiong et al., 2014).
- *Data Owner:* The owner of the data must give up data or less that includes some complicated information that they use to share with their friends, but not more than that (data users)
- *Cloud Servers:* Data can be stored and managed on Cloud Servers because there is no limit on storage space. Cloud servers can be used by other organizations with little storage space to store their data.

When all members of a group are not in close contact with one another and top-down orders are not needed, they may form a "virtual head" using distributed consensus rules. The behavior of a VANET may be better understood with the aid of a dependable communication connection between two autonomous cars. Vehicle mobility and traffic density on a route significantly disrupt the dependability among the nodes. The link reliability rating is useful for proving the dependability of wireless connections. From a bird's-eye view, traffic on a road seems to move in one continuous physical stream. Traffic density, traffic flow, and average vehicle velocity are all important parameters to consider (Goodman et al., 2012). It is called td, tf and v and is denoted by x, t. (x,t). Partial Differential Equation (PDE) x signifies a function space and t denotes time. There are benefits and drawbacks to every kind of network, and some are more optimal contexts than others. Administrators have access to a wide range of network topologies, and when making their selection, they must take into account the size of their business, the breadth of its operations, the goals they want to achieve, and their financial resources. Managing configurations, creating visual maps, and keeping an eye on overall performance are all part of what constitutes effective network topology management (Treiber et al., 2013). The key is to understand your business objectives and requirements so that you can design and maintain the network architecture in the most effective manner for your organization.

The Dijsktra's algorithm cannot be directly applied. In order to discover the most trustworthy route, the RG-Dijkstra method is suggested. Finding the safest way is equivalent. The suggested method makes use of a database denoted by the symbol DB, which stores information on all the cars and the most trustworthy value for each. As a group, they are denoted by the symbol c (RD), where 1 represents the origin and represents all other vehicles. The trip begins at the starting vehicle, and further cars are not visited until the end of the trip. From there, we derive the source's dependability rating and choose the car with the highest value to flag as visited. The primary objective of this performance investigation is to examine how well Consistent Routing Protocol (CRP) performs in a highly dynamic setting (Han et al., 2014). A package is developed in NS-2 to display the findings. Five separate iterations are tracked, and their average is shown. Ad-hoc On Demand Vector routing protocol (AODV) and Policy Based Routing (PBR) are two of the most popular routing algorithms currently in use. Data packets sent to a hop-node from other hop-nodes are typically sent to the next hop-node depending on the destination address included in the packets. The packets must be forwarded by the hop-nodes in PBR according to the predetermined networking regulations. A person with administrative rights often sets and enforces rules. Using a cloud environment raises serious privacy and security risks (Obimbo et al., 2014). A third party provides both cloud data storage and cloud-based infrastructure. There is always a chance that a cloud service provider may get private information. As a result, users should always have access to reliable, protected data storage (Fan et al., 2010).

Trust Management in WSN

Sensor networks' rise to prominence may be slowed by their inherent security concerns. Because of this, it's hard to separate this technology from the physical world. Since they display an event, the nodes are also readily available. Anyone can access the wireless channel used for communications. As a result, nodes have a number of drawbacks that include a lack of computing power, memory storage capacity and connectivity. A fixed set of malicious anti-attacks can be launched by anyone, making the network either partially or fully useless. To address WSN security issues, a set of security priorities that can enhance the network's resilience and dependability must be incorporated. Security credentials utilized by cryptographic primitives, such as those required to build secure communication channels, must be disseminated via key management systems. Self-healing and trust management services, for example, should be provided as well. It can assist in safeguarding the aggregation, time synchronization, and routing protocol cores of the network. Distributed computing and safe mobile base station locations must be safeguarded when sensors are part of a sensor network (Zhang et al., 2018).

Literature Survey

This section provides a background study of VANET, related to the mechanism of restraint and security. Initially, it will explore the properties of wireless networks security and privacy approaches in WSN. However, the user does not know where his data is being held because of data outsourcing. New security issues must be addressed because there is no control over data in the cloud.

Secret Key CBT generates a set of secret shares (S) using the Counting Based Secret Sharing Techniques method (SK) (Gutub et al., 2019). There are particular and hidden shares that can be used. To reconstruct the SK, these 'n' valuable shares are allocated to the consumers. This method uses a specific counting-based mechanism to merge 'k' shares. One or two-bit approaches are most used to implement the system. It is impossible to rebuild SK using either approach if one or more shares of "n" are missing (Qi et al., 2022).

In addition to them, Benjamin Since WSNs have their own set of characteristics and limitations, it is impossible to use the various user authentication procedures proposed for traditional unsecured networks. WSNs can't employ traditional methods of user authentication because of their low power consumption and computational limitations. WSNs are also subject to node compromise attacks because they are typically deployed in a dispersed setting. Since this is not the case with conventional user authentication methods, the verifier's (the entity checking the legitimacy of users) job cannot be limited to a single node. To overcome this limitation, the majority of WSN user authentication techniques propose threshold authentication, in which t (the threshold) sensors act as verifiers (Zhou et al., 2010).

A single data owner in a centralized paradigm (Rault et al., 2014) of independent data publication. Publishing data while protecting privacy is a sure way to keep things private. While keeping the actual data in the data set, differential privacy ensures that the record owner has complete privacy. Differential privacy standards are addressed in a framework that ensures maximal data use for classification jobs. However, employing a centralized model has a large computational burden.

(Miyaneh et al., 2016) developed a system for protecting sensitive attributes, degrees, and relationships in directed social network data. A privacy model is a way of describing how private information should be kept private. The ambiguity of social network data (ASN) is dependent on the anatomy of the social network data. To use ASN, groups of people are formed. ASN distributes the specific values of every group member with a group ID among several tables. Hence to modernize the original network, the lossy join of tables relies on the group ID, which introduces uncertainty.

Bhaladhare, PR, and Jinwala, DC developed an algorithm for bacterial foraging optimization based on fractional calculus (FC-BFO). To get the best cluster possible. The BFO algorithm's chemotaxis step made use of the Fractional Calculus (FC) concept. The major objective was to improve the BFO algorithm's optimization capabilities. As a result, the trajectory streams were used to describe the spatial and temporal data. There are two types of privacy breaches that might occur when someone is moving. (Al-Hussaeni et al., 2014) developed an algorithm called Incremental Trajectory Stream Anonymizer (ITSA) for anonymizing sliding windows in a trajectory stream. Individuals can join and leave the sliding windows with an efficient data structure that can hold a huge quantity of information. Using Incremental Trajectory Stream Anonymizer, however, did not increase the dependability.

Hilbert-Huang Transformation and Trust evaluation are used to identify LDoS attacks in Zigbee WSN. The study's three key conclusions are as follows: Scalable LDoS attack detection architecture in Cloud-based IoT network environment; (Chen H et al., 2019). HHT-based LDoS detection in Zigbee WSN. IMF trust rating method using correlation coefficient and KS test for three of the IMF components LDoS attacks in WSN can be detected using a novel LDoS attack detection approach that has been tested. The combined KS test approach for evaluating and selecting high-trust IMF components is extremely efficient.

In a trust architecture, a cross-layer authorization mechanism and SDN (also known as Trust) are integrated. There are two methods for measuring reputation at the node and organizational levels provided by (Chen J et al., 2019). Analysis of proposed reputation evaluation systems demonstrates that they can withstand modification attacks, replay attacks, and message drops as well as achieve a higher detection accuracy for attacked nodes. Results from simulations show that the proposed reputation evaluation methods are capable of being tested.

AES is the cypher algorithm we use. The nonce utilized has a length of 16 bytes. Cipher's average execution time is 237 milliseconds. Although the authentication process takes around 1s to complete, our scheme consumes less than 3 KB RAM, which is less than 30 per cent of available RAM memory. We observe that the verification process is fast because it takes approximately 1s to complete.

Challenges and Limitations of WSN

Obstacles and Problems in WSN

Sensor networks may grow and work together in the real world. The sensors now in place allow for an enormous quantity of data to be gathered. Inaccessible areas might have sensors installed to gather and transmit data. Making the most of sensor networks requires familiarity with the issues they face. The following are examples of network problems that this approach is designed to solve. WSN has more worries and difficulties than ever before.

Energy

The most significant and widespread challenge in the design of a WSN is maximizing power efficiency. Power efficiency enhancement calls for improvements across the board, including sensors, connectivity, and data processing. In most cases, the remaining battery power of the sensor node is the primary determinant of its expected longevity. The restricted energy capacity of individual sensor nodes is a key design consideration for every sensor network. The time required for a procedure change with the kind of programme being utilized.

Privacy Preserving System

One cannot reasonably expect every patient to always carry their whole medical history and report. An online health tracking service that ensures the confidentiality of users' medical records in the cloud. After encrypting all medical records and data, they are kept on a cloud that can be accessed from anywhere on the globe.

Privacy Preservation Access Control Mechanism in Cloud

This method of entrusting the confidentiality of a communication to a certain set of users is known as "Group Key Management." In this case, each member of the group has a symmetric key K, which is referred to as the group key, and any communication they wish to send to the other members of the group must be encrypted using K. All members of the group know K, hence the message can be decrypted. Group keys are generated and distributed when a new user joins or departs the group.

Clustering

Clustering is a subfield of unsupervised learning that involves categorizing data into groups with similar traits. Therefore, a cluster is a grouping of items that share some similarities but also have their own unique characteristics. Using a distance criterion for clustering, for instance, we may group items that are physically near together. This kind of grouping is known as distance-based clustering.

Clustering is a way to find collections of items that have something in common. It may be shown that there is no one "right way" to organize data into clusters, and that the final decision rests with the individual user and the specifics of the job at hand. Criteria for the best possible clustering have been established, however, and they could help users zero in on the best option. A significant difficulty in clustering is determining the optimal number of clusters. The number of clusters may be manually entered or determined by the algorithm, depending on the chosen approach.

The network is broken up into smaller groups, or "clusters," under this architecture, with each cluster performing a distinct function. A CH's responsibilities include maintaining a database of member connections and communicating cluster updates to the BS. As the nodes in a cluster architecture collaborate to transmit data to the BS, they are more efficient and consume less energy than they would if they did so separately. This might perhaps make the gadget last longer. In cluster protocols, nodes may perform two distinct roles: as cluster members or as CHs. The detected data from a member node may be sent either directly to the CH or aggregated by the CH. (Goodman et al., 2012) Because of the larger computational load and greater distances over which messages must travel, CH consumes more energy than member nodes. Algorithms for grouping similar nodes together may save a lot of power and time while also prolonging the life of networks.

Clustering Challenges

The two main procedures in clustering are the construction of a cluster and the selection of nodes. A CH is chosen from the network, and then other nodes that are not CH are chosen to connect to the CH through a cluster node. Below, we'll go into more depth about the challenges associated with clustering.

Challenges of the CH Selection

The effectiveness and lifespan of the network as a whole are negatively impacted by the CH selection process used in the clustering algorithm. Choosing the appropriate CH might have a major impact on your monthly utility bills. The primary considerations while choosing a CH are as follows:

- The BS has the authority to establish a unified process for choosing CHs. It may also spread when nodes take it upon themselves to act as the CH.
- Time taken to choose a channel is referred to as "delay time." Another potential interpretation of this phrase is the time it took for a cluster to develop. This is a crucial setting for a productive clustering process.
- A CH may be selected either at random or with the use of a formula. Therefore, nodes with a higher energy level may have a better chance of being selected as the CH. There is some evidence that the distance parameter has a role in CH selection. Furthermore, cluster size and the number of neighbours may play a role in the CH selection.

If choosing a CH takes too long or is too complicated, the network may have to use more energy than necessary. Because of this expense, choose a CH that is both energy-saving and eco-friendly.

Challenges of the Cluster Formation

The cluster creation phase is complete, and all nodes have indicated their status to the rest of the network. Some nodes assume the part of the CH, while others are merely nodes in the cluster. Clustering algorithms must take into account a slew of issues when executing this stage.

- Most modern probabilistic and randomised algorithms naturally lead to the generation of diverse numbers of clusters. However, in a number of works, the CH set has already been established. Thus, the number of clusters is predetermined.
- Some early clustering systems anticipated that communications between sensors and their designated CH were going to be direct (one-hop). Increasingly, multi-hop intra-cluster communications are required in cases when the number of CH and sensors are restricted, such as when the number of sensors is significant.

Performance Analysis of VSN

Considered the number of vehicles travelling down a multi-lane road to evaluate the effectiveness of safety and non-safety messaging. A comparison of the PDR and delay of sent messages with traditional MAC protocols will be made in this section of the paper. There are no new equations to come up with.20.1. Analyzing Throughput, we used S as the normalized throughput to derive the throughput for safety and non-safety messages, which is represented by the following equation derived in.

$$Sk = \frac{Ps\ Pbusy\ L\ Te = Ps\ Pbusy\ L}{Pi\ Tslot + Pbusy\ Ps\ Ts + Pbusy(1 - Ps)Tc} \tag{20.1}$$

The likelihood of a successful transmission is highest at this location. There is at least one transmission now taking place, and the length of that transmission is L. If a collision occurs, Tc equals Tcspan, which is the temporal span of the slotation systems' capacity to calculate t slot, te, the normalized SMT throughput for the kth cluster can be expressed in terms of this equation:

$$Sksm = \frac{(x-1)P(t-cl)\,(1-P(t-cl)\,)(x-2)\,L}{(1-P(t-cl)\,)(x-1)\,Tslot + (x-1)P(t-cl) + [(1-(1-P(t-cl)\,)(x-1)\,) - ((x-1)P(t-cl)\,(1-P(t-cl)\,)(x-2)\,)]Tc} \tag{20.2}$$

Moreover, for NSMT, it can be obtainable as:

$$Sknsm = \frac{(x-1)P(t-cl)\,(1-P(t-cl)\,)(x-2)\,L}{(1-P(t-cl)\,)\,Tslot + (1-P(x-2)\,Ts-nsm + [(1-(1-P(t-cl)\,)) - (P(t-cl)\,(1-P(t-cl)\,)(x-2)\,)]Tc} \tag{20.3}$$

The normalized system throughput of the system can be calculated using the equations in this section, which is given as:

$$S = j\sum k = 1Sk \tag{20.4}$$

PDR

PDR is the retransmission limit is a factor. As a result, if Mrsm is the maximum retransmission limit for SM, then the PDR of SMT is:

$$PDRsm = (1 - Ps)Mrsm \tag{20.5}$$

If Mrnsm is the maximum retransmission limit, then NSMT's PDR will be equal to the maximum retransmission limit:

$$PDRnsm = (1 - Ps)Mrnsm \tag{20.6}$$

Delay Analysis

The amount of time it takes for a message to be successfully sent is known as a delay. However, the average latency does not take packet drops or collisions into account when assessing unsuccessful transmissions. E[D] can be expressed as the average delay.:

$$E[D] = E[Tinterval] - \frac{Pf\ drop}{1 - Pf\ drop} \times E[Tdrop] \tag{20.7}$$

Here, when two packets are successfully received, we say that Tinterval is the average time between them. A packet drop's possibility is referred to as the "P f drop." Tdrop = time it takes a packet to be lost In this equation, we can express the SM average packet latency as: 2021, 13 400, 18 of 25 years of sustainability

$$E[Dsm] = Te(n - \frac{Pdrop}{1 - Pdrop} \times \frac{2}{+ CW + MrsmCW/2}) \tag{20.8}$$

$$E[Dnsm] = Te(n - \frac{Pdrop}{1 - Pdrop} \times \frac{2}{1 + CW + MrnsmCW/2}) \tag{20.9}$$

Advantages of Cloud Storage

The two primary benefits of cloud storage are

- Low cost and the fact that providers like Amazon provide almost infinite space for storing data.
- Costs associated with storing data in a scalable format are far lower than those associated with storing it digitally.
- No prerequisites are necessary for the installation.
- There is no need to find replacements.

- Because there is no actual contact, it is not necessary to create any kind of controlled environment.
- Each cloud storage service has its own unique mechanism for data backup and restoration.
- There is also no need to contract in outside help.
- Lacking in the need for electronic regulation.
- The use of both virtual and physical data storage capacity is included in the base price.

Record Sharing in Cloud

Organizations of all sizes may benefit from the cloud-based record sharing solution. Staff members who are physically present during therapy may better cater to the patient's individual requirements by quickly and thoroughly reviewing relevant data from any location.

The advantages of sharing medical data are as follows:

i. It improves the quality of care provided, which means patients get their treatments more quickly.
ii. Access to up-to-date information regarding existing medication improves the accuracy of prescriptions and reduces the incidence of avoidable adverse effects.
iii. By allowing physicians to see the results of any X-rays, CT scans, blood tests, etc., remotely, patients may save money. Because of this, no more testing is required.

Easiness of Access

Cloud computing is a simple, straightforward method to improve healthcare administration, despite its flashy moniker. Incorporating cloud computing's speed and convenience into a variety of business models is a natural progression. When compared to the older, server-based method of storing medical records, the cloud-based method is far superior, offering superior speed, accessibility, and availability while experiencing significantly less downtime. Similarly, all medical evidence is accessible via a single system hosted on the cloud, making it possible to retrieve data from wherever there is an internet connection.

Proposed Technique

As a sensor node that serves as a conduit among the user and WSN, the coordinator serves as the user's primary point of contact. Consequently, the coordinator serves as a gateway to the WSN. Because of this, the coordinator is responsible for both the implementation of the access control procedure and the authentication of users. Network security, network management, and responding to user requests are all handled by the coordinator. The suggested WSN-specific user authentication method is presented here. The method by which users are identified. The plan is divided into two parts. We begin the process of registering lawful users in the system so that they can utilize the WSN in the future. When a user requests access to the WSN, the WSN coordinator and the user perform the login and authentication process.

Network Model

A particular sensor node, known as the coordinator, organizes and manages a WSN. The ZigBee/802.15.4 standard can be used as a communication protocol for sensor nodes. As a sensor node that serves as a conduit among the user and the rest of the WSN, the coordinator serves as the user's primary point of contact. Consequently, the coordinator serves as a gateway to the WSN. Because of this, the coordinator is responsible for both the implementation of the access control procedure and the authentication of users. Small sensor networks (Benenson et al., 2004), such as those present in the healthcare system, can be found. However, health-care systems use a system of sensors that are affixed to the patient's body. These sensor nodes are used to keep tabs on the patient's vital signs, such as heart rate and electrocardiogram (ECG). Two types of sensor nodes: coordinators and end devices make up our Wireless Sensor Network (WSN). Network security, network management, and responding to user requests are all handled by the coordinator. Users and questions cannot be handled by end devices, unfortunately. They collect data, transmit it to the coordinator, and carry out orders given to them by the latter.

Intruder Model

Unprotected networks leave users vulnerable to eavesdropping, message forgery, and user deception when wrong data and orders are sent to the WSN. An adversary may compromise a user's security by gaining access to their mobile device and/or credentials (smart card, for instance). The attacker may also have access to a more sophisticated method of interacting with WSN. Since it is reasonable to assume the coordinator's physical security in this scenario, the cooperation attack discussed in is moot. A memorized password is also assumed to be kept secret.

Security Initialization

It's assumed that the WSN administrator loads the relevant secret keys into the WSN and registers new users. A secret key x is chosen by the administrator before the server and coordinator are loaded with this secret key x. This secret key is used to register users on the system server. This secret key is used by the coordinator to confirm the legitimacy of users. At the beginning of the registration process, the system server is contacted by a new user when he or she wishes to register. To avoid the possibility of an administrator from mimicking a user, we recommend that the user login directly to the server and not reveal his login information to the administrator. At this stage, the server administrator should provide access to a genuine user.

Login and Authentication Process

Figure 20.3 depicts the authentication process that must be completed by the user before he or she can connect with the WSN. The following is the procedure for gaining access: 1) The user enters the password he has memorized into the mobile device's login interface. Finally, the user device uses this password and recorded parameters A to compute S, which equals S = PwA. Nu is then generated and sent to the coordinator in the form of M1 = fID; Enc(Nu; S)g, which contains a user identification and an encrypted value of the nonce Nu, as M1 = Enc(Nu; S)g. For example, AES (Advanced Encryption Standard) is an example of a symmetric encryption function, while Dec is its decryption function. Low-rate and low-power networks employ the AES algorithm.

Even lawful users are kept in the dark about the existence of this service. As an added precaution, it is necessary to pick a strong secret key x. Before detailing the security aspects of our method, the following are the characteristics of a one-way hash function: P1: From a given H, it is impossible to determine the value v. (v). To prove that H(v) = H, it is also impossible to discover a different value for v and v0 (v0). The proposed scheme's registration is safe thanks to these features. The secret value S can be known by the user as S = h (IDjjx), but the secret key x cannot be calculated by the user. In addition, the encryption function has the following feature: Encrypted value Enc (v; k) cannot be decrypted without the secret key k. On the login and authentication level, the suggested system is secure. Unauthorized parties are able to listen in on conversations but cannot learn anything about the nonce value or the keys used. We've designed our suggested user authentication system to be resistant to the following type of attacks as well: Impersonation attack: an attacker will be blocked in message M3 if he attempts to impersonate a legitimate user. It is impossible for the attacker to extract Ns from M2 because he does not know the secret S. M3.

Taxonomy of WSN

Sensor Node: Each node in the center of the wireless network is linked. Nodes that act as sensors also act as routers and processors, among other functions.

Clusters: Decisions affecting the whole cluster must be approved by the chosen cluster leader. Data transmission reliability demands a cluster head.

Base Station: The base station is considered the most crucial node in a wireless network with a hierarchical structure. It's a link in the chain that allows data to flow from the sensors to the people who will be using them.

Cluster heads: Their main job is to carry out the responsibilities of the cluster head efficiently. The primary responsibility of the cluster leader is to transport and control the flow of information inside the cluster.

End User: The information gathered by sensor networks has several applications. Sensor networks find applications in many different domains. Getting online is one of the first uses for network data.

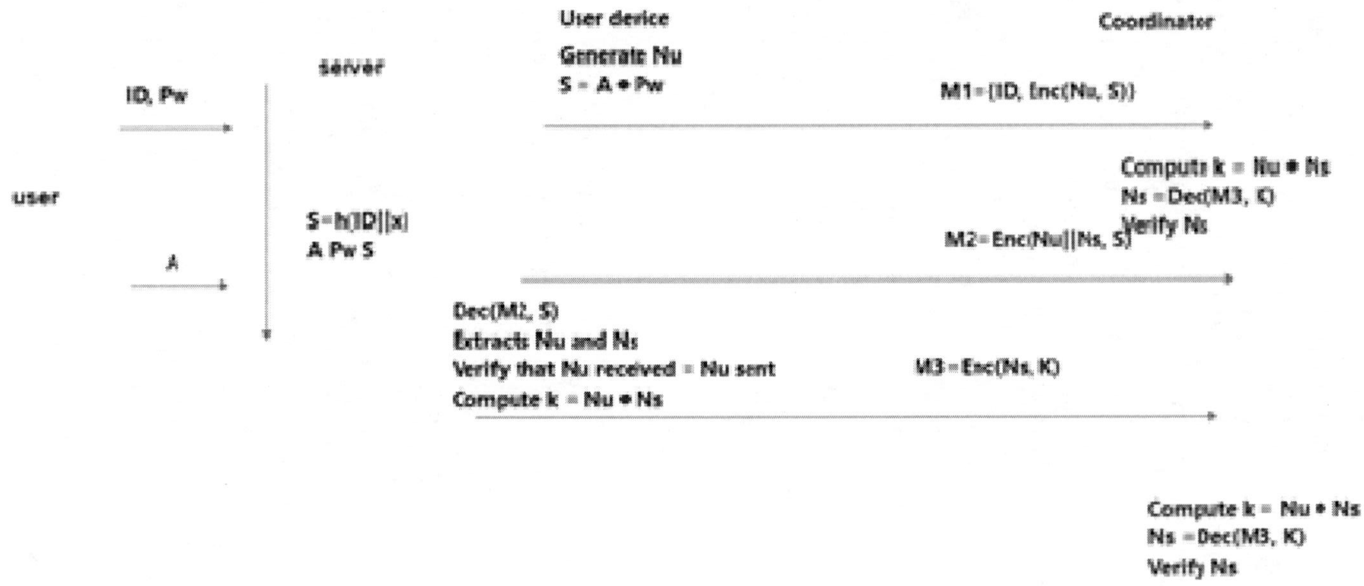

Figure 20.3. Login and authentication process.

Simulations and Experimental Results

In this section the Simulations and Experimental Results of proposed model. In table 20.1 represent that the Proposed model for security Scheme. A network consisting of 100 sensor odes is modelled in MATLAB for evaluation purposes. Uniform and linear positioning of nodes is assumed. The length of the network is 1000m and the base station is located 500m. Nodes are finite, whereas sing has an infinite energy store. Network is considered alive till at least 50% of the nodes are operational

Table 20.1. Proposed scheme security model

	Proposed scheme
Authentication	Mutual
Session key agreement	yes
Data integrity	Maintained
Confidentiality	Maintained
Cryptographic techniques	Encryption and XOR
Infrastructure	No
Scalability	Yes
Target of the query	The coordinator
Vulnerabilities	-
Robustness	Efficiency

Table 20.2 shows that our approach surpasses previous solutions in terms of security features such as confidentiality and integrity. It is also possible to create a session key in our system.

Table 20.2. The encryption and decryption processes at average execution time

Process type	# of data points		
	1200	2400	4700
Decryption	5.04238	11.0920	23.594
Encryption	11.9418	23.7541	46.2950

Anomaly detection and ciphertext cluster analysis at the local clustering level (leaf nodes) with varied data set sizes are summarized in Table 20.3. The graphical representation of average execution time for ciphertext cluster analysis is shown in Figure 20.4.

Table 20.3. Proposed scheme No. clusters with Data points

No. clusters	Data points		
	1200	2400	4700
7	259.17	504.05	1096.57
8	281.97	552.09	1099.93
9	301.30	620.40	1240.22
10	359.10	693.02	1309.23
11	412.81	705.61	1372.39
12	473.15	740.90	1425.19

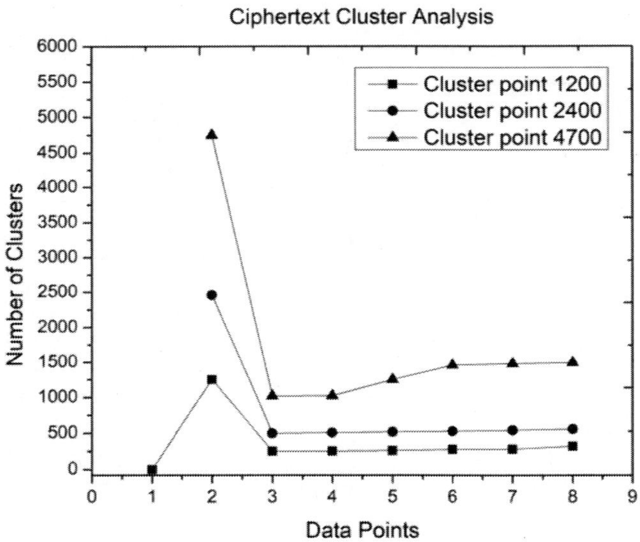

Figure 20.4. Graphical representation of average execution time for ciphertext cluster analysis.

Table 20.4. Experimental of numerous parameters

Parameters	Iteration 1	Iteration 2	Iteration 3	Iteration 4	Iteration 5	Normal
NPCR	99.64	99.70	99.70	99.78	99.63	99.70
MAE	73.0884	74.55	74.4694	73.6108	74.34	74.19
UACI	33.56	33.56	33.46	33.46	33.47	33.46
Mono bit test	10028	9993	10098	10072	10022	10048
Poker-Test	42.1760	16.6656	10.4458	18.2512	22.4587	22.20282
Horizontal - Correlation	0.0093	0.0053	0.0024	0.0044	0.0055	0.0059
Vertical-Correlation	0.0054	0.0054	0.0116	0.0073	0.0041	0.0071
Diag.Correlation	0.0125	0.0765	0.0309	0.0701	0.0194	0.0425

The 20,000-bit stream is used to determine the fastest possible progression of all ones or all zeros in subsequent bits. If the total amount of runs happened inside the corresponding intervals provided in Tables 20.4 then the poker test is passed.

Conclusion

During the cluster joining procedure, one of the cars is designated as the CH (Chief Host) in this chapter to control communication and verify vehicle authenticity. For the benefit of VSN. In this way, CHs are able to save the information about the authenticated vehicles that are stored locally in order to enable speedier member authentication. Adapted to resource constrained WSNs, we developed a user authentication system. The system's safety relies on the user memorizing a password and storing a secret key on his device. There are no infrastructure requirements, and the proposed technique is lightweight and fast to implement. The efficiency of our approach proves that it is lightweight and requires less processing and communication overhead. We hope to use a medical PDA and WSN as a user device in the future to put the suggested system into practice. In addition, as previously said, with a blockchain consensus system, we hope to collect anomalous behavior from neighboring vehicles for behavioral analysis or reputation management in the future. For example, it will eliminate the chance of compromised attacks while also assisting in acting against harmful or anomalous conduct.

References

Alabdulatif A, Kumarage H, Khalil I, Yi X. Privacy-preserving anomaly detection in cloud with lightweight homomorphic encryption. *Journal of Computer and System Sciences.* 2017 Dec 1;90:28-45.

Al-Hussaeni, K, Fung, BCM & Cheung, WK 2014, 'Privacy-preserving trajectory stream publishing', *Data and Knowledge Engineering,* vol. 94, no. PA, pp. 89-109.

Benenson, F. Gartner, and D. Kesdogan, "User authentication in sensor networks, (extended abstract)," In *Informatik* 2004, Workshop on Sensor Networks, 2004.

Chen H, Meng C, Shan Z, Fu Z, Bhargava BK. A novel Low-rate Denial of Service attack detection approach in ZigBee wireless sensor network by combining Hilbert-Huang Transformation and Trust Evaluation. *IEEE Access.* 2019 Mar 8;7:32853-66.

Chen J, Tian Z, Cui X, Yin L, Wang X. Trust architecture and reputation evaluation for internet of things. *Journal of Ambient Intelligence and Humanized Computing.* 2019 Aug;10(8):3099-107.

Dansana D, Sahoo S, Mishra BK. Efficiency and Reliability of IoT in Smart Agriculture. In *Internet of Things and Analytics for Agriculture*, Volume 3 2022 (pp. 301-327). Springer, Singapore.

Fan D, Shi P. Improvement of Dijkstra's algorithm and its application in route planning. In *2010 seventh international conference on fuzzy systems and knowledge discovery* 2010 Aug 10 (Vol. 4, pp. 1901-1904). IEEE.

Goodman KA, Regine WF, Dawson LA, Ben-Josef E, Haustermans K, Bosch WR, Turian J, Abrams RA. Radiation Therapy Oncology Group consensus panel guidelines for the delineation of the clinical target volume in the postoperative treatment of pancreatic head cancer. *International Journal of Radiation Oncology Biology Physics.* 2012 Jul 1;83(3):901-8.

Gutub, A., Al-Juaid, N. and Khan, E., 2019, "Counting-based secret sharing technique for multimedia applications", *Multimedia Tools and Applications.,* vol. 78, no. 5, pp. 591-619.

Han Z, Wu J, Zhang J, Liu L, Tian K. A general self-organized tree-based energy-balance routing protocol for wireless sensor network. *IEEE Transactions on Nuclear Science.* 2014 Apr 2;61(2):732-40.

Khanagha S, Ansari S, Paroutis S, Oviedo L. Mutualism and the dynamics of new platform creation: A study of Cisco and fog computing. *Strategic Management Journal.* 2020 Mar 23.

Miyaneh, E, Haghjoo, MS & Rajaei, M 2015, *Ambiguity in Social Network Data for Presence, Sensitive-Attribute, Degree and Relationship Privacy Protection.*

"*National Institute of Standards and Technology (NIST), Advanced Encryption Standard (AES).*" Federal Information Processing Standards Publications (FIPS PUBS) 197, 2001.

Obimbo, C 2014, 'Privacy Preserving Data Publishing : A Classification Perspective', *(IJACSA) International Journal of Advanced Computer Science and Applications,* vol. 5, no. 9, pp. 29-34.

Oubbati OS, Atiquzzaman M, Lorenz P, Tareque MH, Hossain MS. Routing in flying ad hoc networks: Survey, constraints, and future challenge perspectives. *IEEE Access.* 2019 Jun 19;7:81057-105.

Qi N. Optimization and Simulation of a Dynamic Management System for Building Construction Based on Low-Power Wireless Sensor Networks. *Journal of Sensors.* 2022 Feb 9;2022.

Rault T, Bouabdallah A, Challal Y. Energy efficiency in wireless sensor networks: A top-down survey. *Computer networks.* 2014 Jul 4;67:104-22.

Schoch E, Kargl F, Weber M, Leinmuller T. Communication patterns in VANETs. *IEEE Communications Magazine.* 2008 Nov 25;46(11):119-25.

Treiber M, Kesting A. Traffic flow dynamics. *Traffic Flow Dynamics: Data, Models and Simulation, Springer- Verlag Berlin Heidelberg.* 2013 Nov:983-1000.

Xiong J, Liu X, Yao Z, Ma J, Li Q, Geng K, Chen PS. A secure data self-destructing scheme in cloud computing. *IEEE Transactions on Cloud Computing*. 2014 Nov 20;2(4):448-58.

Zhang DG, Tang YM, Cui YY, Gao JX, Liu XH, Zhang T. Novel reliable routing method for engineering of internet of vehicles based on graph theory. *Engineering Computations*. 2018 Dec 13.

Zhou M, Zhang R, Xie W, Qian W, Zhou A. Security and privacy in cloud computing: A survey. In *2010 Sixth International Conference on Semantics, Knowledge and Grids* 2010 Nov 1 (pp. 105-112). IEEE.

About the Editors

Dr. L. Ashok Kumar was a Postdoctoral Research Fellow from San Diego State University. California. He was selected among seven scientists in India for the BHAVAN Fellowship from the Indo-US Science and Technology Forum and also, he received SYST Fellowship from DST, Govt of India. He has 3 years of industrial experience and 22 years of academic and research experience. He has published 173 technical papers in International and National journals and presented 167 papers in National and International Conferences, He has completed 26 Government of India funded projects worth about 15 Crores and currently 9 projects are in progress worth about 12 Crores. He has developed 27 products and out of that 23 products have been technology transferred to industries and for Government funding agencies, He has created Eight Centres of Excellence at PSG Tech in collaboration with Government agencies and Industries namely, Centre for Audio Visual Speech Recognition, Centre for Alternate Cooling Technologies, Centre for Industrial Cyber Physical Systems Research Centre for Excellence in LV Switchgear, Centre for Renewable Energy Systems, Centre for Excellence in Solar PV Systems. Centre for Excellence in Solar Thermal Systems. His PhD work on wearable electronics earned him a National Award from 1STE and he has received 26 awards in the National and in international level. He has guided 92 graduate and postgraduate projects. He has produced 6 PhD Scholars and 12 candidates are doing PhD under his supervision. He has visited many countries for institute industry collaboration and as a keynote speaker. He has been an invited speaker in 345 programs. Also, he has organized 102 events, including conferences, workshops, and seminars. He completed his graduate program in Electrical and Electronics Engineering from University of Madras and his post-graduate from PSG College of Technology, India, and Masters in Business Administration from IGNOU, New Delhi. After completion of his graduate degree he joined as project engineer for Serval Paper Boards Ltd. Coimbatore (now ITC Unit, Kova Presently he is working as a Professor in the Department of EEE, PSG College of Technology. He is also a Certified

Chartered Engineer and BSI Certified ISO 50001 2008 Lead Auditor. He has authored 19 books in his areas of interest published by Springer, CRC Press, Elsevier, Nova Publishers. Cambridge University Press. Wiley. Lambert Publishing and IGI Global. He has 11 patents, one Design patent and two Copyrights to his credit and also contributed 18 chapters in various books. He is also the Chairman of Indian Association of Energy Management Professionals and Executive Member in institution of Engineers, Coimbatore Executive Council Member in Institute of Smart Structure and Systems. Bangalore, Associate Member in CODISSIA. He is also holding prestigious positions in various national and International forums and he is a Fellow Member in IET (UK), Fellow Member In IETE, Fellow Member IE and Senior Member in IEEE.

Dr. D. Karthika Renuka is Professor in the Department of Information Technology in PSG College of Technology. Her professional career of 18 years has been with PSG College of Technology since 2004. She is an Associate Dean (Students Welfare) and convenor for the Students Welfare Committee in PSG College of Technology. She is a recipient of Indo-U.S. Fellowship for Women in STEMM (WISTEMM)-Women Overseas Fellowship program supported by the Department of Science and Technology (DST), Govt. of India and implemented by the Indo-U.S. Science & Technology Forum (IUSSTF). She was a Postdoctoral Research Fellow from Wright State University, Ohio, USA. Her area of specializations includes Data Mining, Evolutionary Algorithms, Soft Computing, Machine Learning and Deep Learning, Affective Computing, Computer Vision. She has Organized an International Conference on Innovations in Computing Techniques Jan 22-24, 2015 (ICICT2015) and National Conference on "Information Processing and Remote Computing" 27th and 28th February 2014 (NCIPRC 2014). Reviewer for Computers and Electrical Engineering, Elsevier, Wiley Book chapter, Springer Book Chapters on "Knowledge Computing and its Applications". She is currently guiding 8 research scholars for their Ph.D under Anna University. She has published several papers in reputed National and International journals and conferences.

Dr. Sonali Agarwal is working as an Associate Professor in the Information Technology Department of Indian Institute of Information Technology (IIIT), Allahabad, India. She received her Ph. D. Degree at IIIT Allahabad and joined as faculty at IIIT Allahabad, where she is teaching since October 2009. She holds Bachelor of Engineering (B.E.) degree in Electrical Engineering from

Bhilai Institute of Technology, Bhilai, (C.G.) India and Masters of Engineering (M.E.) degree in Computer Science from Motilal Nehru National Institute of Technology (MNNIT), Allahabad, India Her main research interests are in the areas of Big Data, Big Data Mining, Complex Event Processing System, Support Vector Machines, Stream Analytics and Software Engineering. She is having hands on experience on stream computing and complex processing platforms such as Apache Spark, Apache Flink and ESPER. She has focused in the last few years on the research issues in Data Mining application especially in Big Data, Stream Computing and smart cities. She has attended many National and International Conferences/workshops and she has more than 70 research papers in national / international journals and conferences. She has completed her Masters Thesis work at Liverpool John Moores University (LJMU), Liverpool, U.K. during November 1999 to February 2000 under Indo-UK REC Project, a collaboration in between School of Computing & Mathematical Science, LJMU Liverpool UK and Motilal Nehru National Institute of Technology, Allahabad. She has also taken part in Indo Swiss Joint Research Program (ISJRP) and full financial support was awarded to carry out joint research work and to gain knowledge regarding the recent research and experimental facility/work at EPFL, Switzerland, from December 2011 to January 2012. She has also visited Thailand and Sri-Lanka for attending/organizing international level conference/Workshops. She has also a member of IEEE, ACM, CSI and Supervising three Ph.D. Scholars and several graduate and undergraduate students in Big Data Mining and Stream Analytics domain.

Dr. V Chandrasekar (Chandra)**,** Professor of Electrical and Computer Engineering, University Distinguished Professor, Colorado State University has made pioneering contributions in the area of "Polarimetric Radar Observations of the Atmosphere". Dr Chandra has extensive experience in Radar System Design, Radar Network Development, DSP Design as well as RF Communication Systems. He has contributed significantly to the areas weather radar and applications to Atmospheric Sciences. He also conducts research on related topics including Image Processing, Neural Network Applications and Large Scale System Simulation. He has organized and participated in six large multi-agency, national level experiments involving many radars, aircraft and ground instrumentation. He is an avid experimentalist conducting special experiments to collect in-situ observations to verify the new techniques and technologies. Dr Chandra is co-author of two textbooks, Polarimetric and Doppler Weather Radar (by Cambridge

University Press) and Probability and Random Processes (by McGraw Hill). He has been a PI or Co-PI on several national level programs such as the Advanced Communication Technology Satellite (ACTS) program at CSU, DARPA NGI program, the current NASA TRMM mission and the future NASA GPM mission. He is currently a CO-PI of the CSU-CHILL radar facility and plays an important role in maintaining it as one of the most advanced meteorological radar systems in the world available for research, and continues to work actively with the CSU-CHILL radar supporting its research and education mission. He is a Co-PI and the Deputy director of the NSF Engineering Research Center, CASA (Center for Collaborative Adaptive Sensing of the Atmosphere), where he serves as the Research Director and provides leadership for the sensing research thrust. He has also been the Director of the prestigious and long standing Research Experience for Undergraduate Program at Colorado State University. He was elected Fellow of IEEE, The American Meteorological Society and CIRA in recognition of his contributions to "Quantitative Remote Sensing". He is a member of the National Academy of Sciences panel on " Future Radar Systems beyond NEXRAD " as well as the panel to assess "NEXRAD Flash Flood Forecasting Capabilities at Sulphur Mountain, California". He has served as a visiting professor at the National research Council of Italy, as well as Distinguished Visiting Scientist at the NASA Goddard Space Flight Center. He has won numerous awards including, the NASA Technical innovation Award, 2002, Cermack Outstanding Advisor Award, 2001, Abell Outstanding Researcher Award, 2001, Distinguished Minority Service Award, 1999, Deans Council Award for Excellence in Teaching and Research, 1996, Halliburton Foundation Young Faculty Research Excellence Award, 1993, Vaisala Distinguished fellow and Abell Outstanding Research and Graduate Program Award, 2004.

Index